Introduction to Ocean Circulation and Modeling

Introduction to Ocean Circulation and Modeling

Avijit Gangopadhyay

CRC Press
Taylor & Francis Group
Boca Raton London New York

CRC Press is an imprint of the
Taylor & Francis Group, an **informa** business

First edition published 2022
by CRC Press
6000 Broken Sound Parkway NW, Suite 300, Boca Raton, FL 33487-2742

and by CRC Press
2 Park Square, Milton Park, Abingdon, Oxon, OX14 4RN

ISBN: 978-0-367-36597-4 (hbk)
ISBN: 978-0-367-36625-4 (pbk)
ISBN: 978-0-429-34722-1 (ebk)

DOI: 10.1201/9780429347221

Dedication

---

*For my parents and my family
and my teachers and students
and for you, the curious*

Contents

Preface

I would like to begin with a phrase a professor told me when I was graduating from college. "Remember that the four years of undergraduate degree can only teach you where and how to look for information to go deep into a subject. You have not learned the subject; you know where to go to look for answers when you have a deep question."

This book is for undergraduates in college who would like to venture into the workings of the ocean around us – how does it move, what constitutes understanding its basic elements, how do we model such a system? How is it different from modeling the earth or the atmosphere or the climate?

This book is also for those entering graduate students with an expressed interest in any branch of Oceanography (physical, biological, chemical, and geological). What do I need to know about the ocean's physics to do my research better? How do the biological productivity and other chemical parameters depend on or vary with the physical system? How does one model a complex system such as the ocean where all these different aspects and variables of each discipline seem to stand independent of each other? What is interdisciplinary modeling? Where do I start?

In one simple sentence: This is a simple book that introduces any college student with little more than a year of algebra to the seemingly complex field of ocean circulation and modeling. It does not attempt to dive into the complex non-linear dynamics and their very complex and mathematical exposition that have been written by some of the best in the field of Physical Oceanography. Those books will make more sense once you understand the materials presented in this book and have more in-depth questions. We refer the reader to those books at the end of each chapter for further reading and for more exposure to detailed dynamical investigations.

This book is subdivided into two parts. During the first half of the book, we learn some basic physical concepts – the concepts of circulation, namely the wind-driven and the thermohaline components. A brief history of our journey from Franklin to satellites and to numerical models introduces the background of the subject. The concepts of Geostrophy, Sverdrup dynamics, Ekman, Munk, and Stommel problems, two-layer dynamics, and stratification are fundamental to ocean circulation. The ideas of time-dependence and vertical layers are then introduced in detail. Oceanic waves such as Rossby Waves and Kelvin Waves are then discussed with a touch of instability. Many key ideas of Geophysical Fluid Dynamics are explained with ocean applications in this first part in a simple way. In the second half, the reader is exposed to the basics of numerical modeling through the full set of primitive equations and their implementation for different models such as MOM, HOPS, POM, ROMS, MITGCM, and HYCOM. A key chapter is devoted to the ideas of turbulence, eddy parameterization and closure schemes. Then we talk about applications of modeling – on basin scale, on synoptic regional synoptic scale, and on interdisciplinary modeling. Simulation, prediction and data assimilation ideas are discussed with applications focusing on utilizing observations. The final chapters of the book discuss some contemporary topics related to climate change and the future of ocean research.

What is the purpose of this book?

In the digital age, it is easy to lose sight of the basics and get carried away with our sense of curiosity and adventure to do more with what is readily available in front of us. Too often, we forget that what we have today is the result of hard work put together by our previous generations. Too many times have I seen in my teaching career that a student is taking a model as something that is more like a black box that can possibly solve all the unknown problems of the oceans, atmosphere, and climate. A model is as good as our (and your) knowledge and our ignorance, just like a computer — it is a tool to do science. It does not replace and should not limit your (our) scientific curiosity.

This book is an attempt to collect the essential elements of contributions from yesteryears — some elementary concepts of circulation and a few ideas of the modeling framework. We leave more advanced materials for graduate studies and advanced textbooks. We do not claim to be a developer of any of the concepts or the models; we are just trying to create a conduit for those ideas to the young minds to appreciate and develop further.

If this book can entice you to think about those questions in oceanography and climate sciences, then you might want to think about pursuing a career with graduate school or just know where to look for some answers regarding our earth wherever life takes you.

Author's Biography

Avijit Gangopadhyay, PhD, is a Professor at the School for Marine Science and Technology (SMAST) at the University of Massachusetts, Dartmouth, where he teaches ocean circulation and modeling, ocean-atmosphere dynamics and introduction to climate. He previously served as the Interim Dean and Associate Dean of SMAST.

He is known internationally as he has held multiple visiting, distinguished, and honorary professorships at Harvard, the Institute of Oceanography of the University of São Paulo (Brazil), the Indian Institute of Science (Bangalore), Indian Institute of Technology (IIT) at Kharagpur, and IIT Bhubaneswar. He's been involved in the inception of multiple academic programs starting from establishing an international dual-degree PhD program for UMass-USP to the rise of the ocean research centers at IITKGP and IIT Bhubaneswar, as he was one of the founders of the Center for Oceans, Rivers and Land Sciences (CORAL) at IITKGP. His research interests and contributions include the dynamics of western boundary currents, operational ocean modeling and data assimilation, basin-scale climate-related processes, and multi-scale multidisciplinary data-model synthesis studies.

Professor Gangopadhyay holds a BTech in Naval Architecture from IIT Kharagpur, a MTech in Applied Mechanics from IIT Delhi. He earned his PhD in Ocean Engineering from the University of Rhode Island where he first studied the Gulf Stream. Dr. Gangopadhyay was a Research Associate at Harvard University and a Scientist at the Jet Propulsion Laboratory before joining UMass Dartmouth. He is a Fulbright Scholar and has long-standing collaborations with MIT and the Woods Hole Oceanographic Institution. He has numerous publications in many international journals and books during his long career. He resides on the shores of the Atlantic in Dartmouth, Massachusetts.

Acknowledgment

This book is my attempt to present the advanced materials of circulation and modeling to undergraduates. This book would not have been possible without the help and encouragement of many colleagues, students, and friends over the last 30 years. I am thankful to them all.

I wish to thank my students in India and Brazil, who inspired me to write this book. Over the last 20 years, while teaching undergraduates and graduate students there, I always came back to the US with the feeling that there's so much interest to learn, but access is limited because the materials are not in one place. Many of the topics have been chosen because of the discussions that we had during lectures with students and faculty from CORAL at IITKGP, SEOCS at IITBBS in India and a number of institutions in Brazil including IOUSP, FURG, and UFC. Their comments have been most helpful and greatly appreciated.

I'm thankful to my own students who helped me with their questions, research, thesis, and wit to decide on the material of this book. To name a few, Leandro Calado, Rafael Soutelino, Ana-Paula Morais Kreelling, Zhitao Yu, Ayan Chaudhuri, Sudip Jana, Mahmud Monim, Nish Silva, Matt Grossi – thank you.

I owe a big thank you to Adrienne Silver. She has been a source of continuous help with editing and she has drawn many of the figures for this book during her PhD. Thank you, Adrienne. This book would not have been finished without your help.

I wish to thank a number of colleagues from around the world who have always encouraged me to write this book and helped with the history of oceanography. Arnold Taylor from PML/UK, you are a very special colleague for decades. Thank you, Arnold. Thanks to Ilson Silveira and Frederico Brandini from IOUSP/Brazil for detailed comments and edits in many chapters. You and your students have a special impact on the making of this book. Special thanks to Sourav Sil – Thank you for your continuous help and feedback.

A number of people helped me with my journey in the oceans in different places and times in India. I wish to thank them from the bottom of my heart. They are Sishir Dube, Madhusudan Chakraborty, Arjun Malhotra, Puran Dang, Prem Chand Pandey, Subhasish Tripathy, Partha Pratim Chakraborty, Damodar Acharya, UC Mohanty, Raja Kumar, Anil Gupta, Debabrata Sen, Om Prakash Sha, Animesh Das, Prasad Bhaskaran, A. D. Rao and late Tad Murty. Thanks to Arun Chakraborty and Deepak N. Subramani for their helpful comments and feedback on several chapters. Special thanks to Shailesh Nayak (MOES/ISRO),

Mrutyunjay Mohapatra (Director, IMD), and to Sadashib Golder for letting me visit the lab where Sir Gilbert Walker used to work.

My colleagues in the US helped with several chapter reviews, and I am deeply grateful for their suggestions, edits, and words of encouragement. They are Glen Gawarkiewicz, Magdalena Andres, Huijie Xue, Fei Chai, Hassan Moustahfid and Pierre Lermusiaux. I can never thank you enough. Ruth Musgrave from Canada was a tremendous help for the Turbulence and Modeling Chapters. Thank you all.

I am grateful to a number of people at UMass Dartmouth who have helped building thoughts and ideas for this book's material. Thanks to my SMAST colleagues – Steven Lohrenz, Mark Altabet, Brian Howes, Miles Sundermeyer, Lou Goodman, Wendell Brown, and Cindy Pilskan for keeping me away on a sabbatical so I could focus on writing. Thanks are also due to Mike Marino, Andre Schmidt, Mike Deignan, Eric Lyonnais, Gail Lyonnais, Cindy Costa, Arlene Wilkinson, Susan Silva, and Chris Fox for helping out on administration and computing so that I could write. Thanks to all others in the SMAST fisheries group for enticing me to think about including a fisheries modeling unit.

I consider myself very lucky to have some very special mentors throughout my academic life. The challenges of the Gulf Stream separation problem were introduced to me by Peter Cornillon, Randy Watts, and Tom Rossby. While at Harvard, Allan Robinson, his group, and his collaborators helped me appreciate how numerical models can enlighten our understanding of circulation and processes. I learned time-series analysis from Lester LeBlanc and Leland Jackson. Thank you. As a faculty at UMass Dartmouth, I was fortunate to discuss interesting ideas with Brian Rothschild, Ken Brink, and John Farrington, and many others. I will never forget those discussions. In my early days at UMass, I received tremendous opportunities to teach various undergraduate and graduate courses in Physics and in Marine Science when Chancellor Peter Cressey and Chancellor Jean MacCormick were at the helm. For that, I am grateful.

I have benefited from multiple research and teaching topics discussions over the last two decades with many colleagues, which have helped me in writing this book. With the risk of missing some of them, I would like to mention a few of those wonderful and supportive colleagues whose thoughts and ideas have shaped some of the well-described parts of this book. They are in no specific order: Debasish Sengupta, Ravi Chandran, Ramaswamy Venkateshan, VSN Murty, Abhijit Mitra, Karim Hilmi, Alex Warn-Varnas, Art Miller, Glenn Flierl, Mike Spall, Dale Haidvogel, Enrique Curchister, Scott Glenn, Yi Chao, Lee Fu, Tim Liu, Tommy Dicky, David Pierce, Belmiro Castro, Paulo Polito, Olga Sato, Joao Lorenzetti, Mark Wimbush, Lew Rothstein, Karen Tracey, Hyun-Sook Kim and Kathy Donahue. Thank you.

Many research grants over the years from multiple agencies such as NASA, NOAA, ONR, NSF, IAI ESSO-INCOIS, CNPq-Brazil, and Massachusetts State initiatives have supported my research and teaching activities in the US and abroad. I am grateful for their support in my research and student funding. A number of different grants have supported visits and lectures, and interactions with students and colleagues abroad. The support of MOES Chair Professorship at IITKGP during 2012–2015, the Distinguished Visiting Professorship at IOUSP during 2014, Visiting Professorship at IITBBS during 2015–2020, and the Fulbright US Scholar Fellowship in 2013–2014 are gratefully acknowledged.

This manuscript evolved from my lecture notes from multiple introductory graduate and undergraduate courses. Frank Smith helped me create the very first draft. Thank you, Frank. This book was finally prepared and completed in LaTeX on Overleaf. The continuous and prompt attention of the Overleaf support stuff was critical for the timely completion of this book and is gratefully acknowledged. The support staff of PLS and Copyright.com and IPCC were very helpful in seeking and approving many figures and tables that are republished in this book. I wish to thank them and the authors and publishers to allow us to use those for the benefit of the students. A special thanks to Ameya Pathare, who shared his nature photography album for a few beautiful images in the book.

The editorial staff at Taylor and Francis was very helpful in providing early and useful information regarding seeking permission for all the figures that were republished from our own work and from other works. Their constant help and support and understanding of the delays associated with the pandemic were truly exceptional. Thank you.

Finally, I wish to thank my family for bearing with me over all these years of constant travel and multiple research and teaching assignments away from home. I am forever grateful to my wife, Mita for always supporting me and taking care of our immediate and greater family, she keeps us all together. My mother-in-law, Jaya, has always had an impact on our family, as her strength inspires us all. Lastly, a special thanks to my daughters, Amritaa and Onkita for providing inputs, edits, and constant encouragement to get out of those writing blocks! Thank you all.

February 2021
Dartmouth, Massachusetts, USA

Sunrise at Wailua, Kaua'i

1 Our Home – The Earth

> This most beautiful system of the
> sun, planets and comets, could
> only proceed from the counsel and
> dominion of an intelligent and
> powerful being.
>
> Isaac Newton (1643–1727)

OVERVIEW

In this first chapter, we introduce the ocean as a physical subsystem of the larger Earth system. We will learn about the physical characteristics or properties that define and help quantify our ocean environment. Oceanic and atmospheric phenomena happen on multiple scales of space and time. The role of the ocean as a regulator of Earth's temperature is discussed. This is achieved by the redistribution of excess heat from the equatorial regions to the polar regions by tropical storms and oceanic currents. Finally, a brief history of oceanography from the time of Benjamin Franklin to the present-day numerical models is presented.

1.1 THE BASICS

We all live in a place called Earth. This is our home and let us get to know our home a little better. I think we all know that our Earth is like an inflated soccer ball, it has two poles, North Pole and South Pole, and it inflates in the middle region along the equator. Our home, this earth rotates around the sun in an elliptic orbit, and receives its energy from the sun through light and heat. As we orbit around the sun in one year, our earth also rotates around its own axis in 24 hours and that makes our day and night cycles.

Now, think about our home without us for a minute – it still exists without us. It rotates, gets energy from the sun, and maintains a balance of heat around the house for us to live in it, if we choose to. So, this home of ours can be scientifically called a "system" or part of a larger "climate system." Just like our residential house can be divided into multiple subsystems, and then put together in a complex integrated system, the climate system can be thought of as an integrated system of four individual systems:

DOI: 10.1201/9780429347221-1

- Geosphere (the solid earth system)
- Hydrosphere (the oceans)
- Atmosphere (the air and winds) and
- Biosphere (humans, animals, plants, and the living ocean)

We have come to understand that there are three basic physical subsystems: the earth, the atmosphere and the oceans. These three components together make up what we call the physical climate or physical weather systems. And, these physical systems together affect us, the humans, the animals living on this planet, and the fish-to-biota in the oceans – or the biosphere. In a convoluted sense, our behavior in the biosphere also affects the atmosphere and the hydrosphere, and those changes, in turn, affect our biosphere, or our habitat. However, this book focuses on the ocean subsystem. Thus, while we study the basics of ocean circulation and modeling, we need to appreciate the integrative nature of the climate subsystems and their dependencies on each other and beyond.

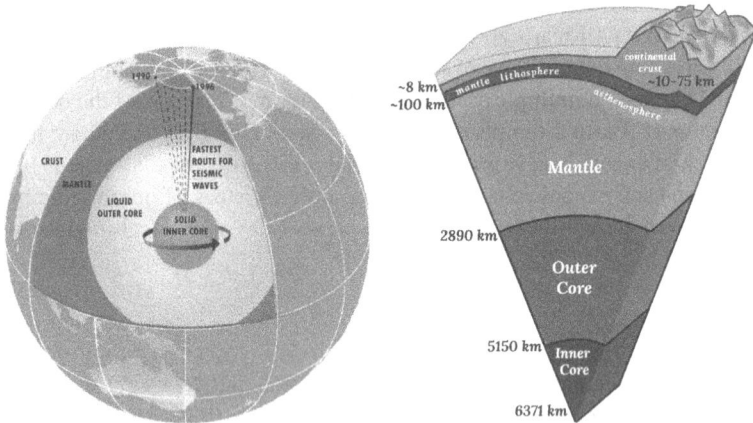

Figure 1.1: Left: A cut-away illustration of Earth's interior with inner core and outer core. The change in the magnetic field is also plotted from 1900 to 1996. (NASA, Dixon Rohr, https://www.nasa.gov/sites/default/files/images/608134main_world-orig_full.jpg.) Right: Cross-section of Earth's Interior. This file is licensed under the Creative Commons Attribution-Share Alike 4.0 International license, author: Volcan26.

Let us introduce our home to ourselves in a curious way! You must have seen a globe, if not, this is the time to see it. And then, think about that globe for a minute with me. What do you think is the radius of Earth? Yes, the radius of the Earth – it's about 4000 miles or about 6400 kilometers. That's how big of a sphere our planet is. And it is solid earth, its inner core (about 1000 km) is solid iron, its outer core (another 2300 km) is liquid iron, topped by about

3000 km of mantle over which there is 40 km of crust. See Figure 1.1. This is our geosphere.

Now, as you know and have heard many times, our planet's surface is full of water. Actually, 73% of the earth's surface is covered by water or oceans. What is the average or typical depth of the oceans? It is about 2.3 miles or 4 kilometers or 4000 meters. That is our hydrosphere. A thin ribbon of fluid water of 4 km laced around the big geosphere of a radius of 6400 km.

What is on top of the water on the surface? Yes, it is the atmosphere. And, what is the average thickness of the atmosphere that matters most to us? Remember the pilot in your last flight announcing that "we are now cruising at a safe altitude of 33,000 feet." That's it. The average thickness of the atmosphere that matters or, within which most of the wind actions like storms and hurricanes and typhoons and cyclones and tornadoes happen, is about 10 kilometers or about 33,000 feet.

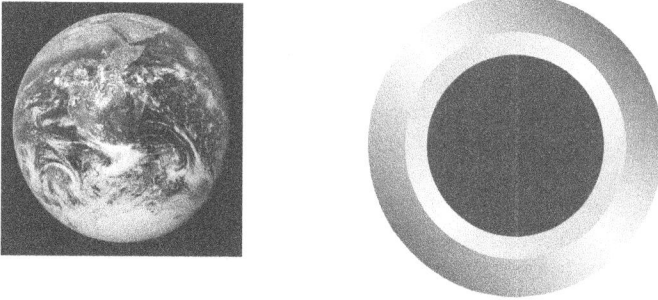

Figure 1.2: Left: Apollo 17 hand-held Hasselblad picture of the full Earth. This picture was taken on 7 December 1972, as the spacecraft traveled to the moon as the last of the Apollo missions. A remarkably cloud-free Africa is at the upper left, stretching down to the center of the image. Saudi Arabia is visible at the top of the disk and Antarctica and the south pole are at the bottom. Asia is on the horizon is at the upper right. The Earth is 12,740 km in diameter. (Apollo 17, AS17-148-22727). NASA NSSDC image. Right: A section through our spherical physical climate system on the right. The solid earth (on the left) is shown by its red core of 6400 km radius in the right panel. It is surrounded by water as a thin ribbon of 4 km of thickness. The other blue fluid is our atmosphere with an average thickness of 10–30 km. Not to Scale.

There you have it. The three concentric spheres of our physical home, the physical climate system, one on top of the other – three different medium, solid earth, liquid water, and gaseous air; with relative densities of 1000 to 1 to $1/1000$ kg m^{-3} — all in sync in one system, rotating around its own axis once a day, and going around the sun once a year. See Figure 1.2. This is our climate system. And, the ocean is in the middle of it all. It is forced by the winds in the atmosphere, supported and guided by the bathymetry of the ocean bottom (the geosphere), and gets its energy from the sun, stores and distributes it as needed to balance and maintain the temperature for the habitats of the biosphere.

1.2   THE OCEANS – AN INTRODUCTION

So, long story short, we are here to start thinking about the workings of the oceans. Imagine that you are taking a morning walk on a seashore enjoying the sunrise, or strolling on the beach enjoying the sea breeze, or watching the sunset; there is this big, endless ocean in front of you. And, you start wondering how it all works!

So, you want to touch the ocean. You walk near the waves breaking in towards you. First things first, you pick it up, Or you touch it. What do you feel? Temperature. So the ocean water is characterized by its thermal signature or temperature.

That's one thing. And then there is the other thing about seawater. The ocean has that smell – it's the smell of salt. Now we're getting to understand that it's the nature of the ocean water to be salty. This salt comes from years, millions of years of this whole system rotating around and diffusing minerals from the solid earth. This system gets fed by precipitation which lowers the salinity, and by river runoff, which also freshens it up a little. But there is a global balance of salt. And so, this is what we call the two properties of the seawater—temperature and salinity. Typical vertical profiles of temperature and salinity in the North Atlantic near the Gulf Stream region (35°N, 70.5°W) are shown in Figures 1.3a and 1.3b. Note how the temperature decreases with depth and salinity or salt content increases with depth.

Like any other matter or substance, water is generally characterized by its density. So, the temperature and salt dictate the density, which also depends upon pressure. Pressure is synonymous to depth. In a stable water column, the deeper the water, the denser it is. That is, lighter water overlays above denser water. See Figure 1.3c. So, eventually you define the density as rho (ρ), which is a function of salt (S), temperature (T), and pressure (p). Mathematically, this is represented by what is called the Equation of State:

$$\rho = \rho\,(S,T,p) \tag{1.1}$$

While temperature is generally given in Celsius (C), the salinity is expressed in something called practical salinity units (psu) or parts per thousands (ppt). For example, the average salinity of seawater is about 35 ppt or 35 grams per kilogram (g/kg) of seawater.

There is no straightforward relationship between density and its three independent variables: salinity, temperature, and pressure. There is an empirical relationship (UNESCO, 1980), which is described in discussed in various papers such as (UNESCO & SCOR, 1981), (Millero & Poisson, 1981), and (Millero,

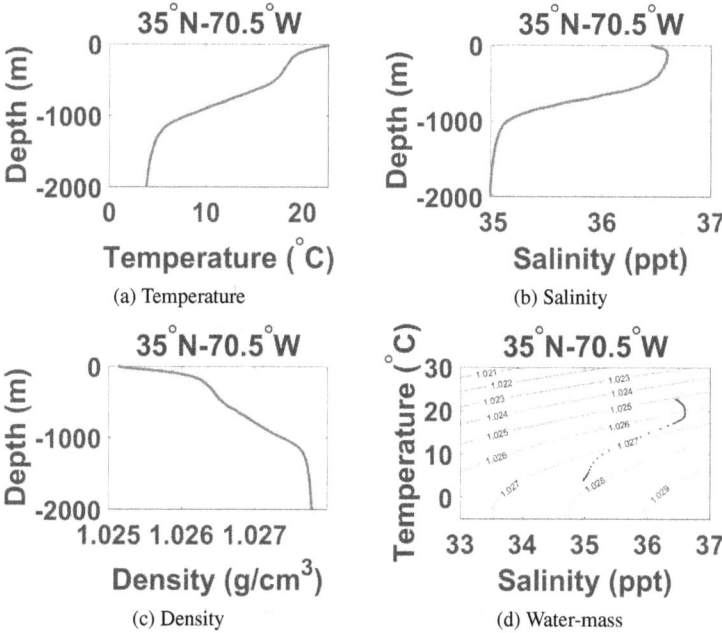

Figure 1.3: Typical vertical profiles for temperature (T), salinity (S), and density (ρ). All profiles are shown for a region in Northwest Atlantic near the Gulf Stream. The bottom-right panel shows the water-mass diagram in the t-s space, with density contours superposed. Figure credit: Adrienne Silver.

2010), that is generally accepted as a given diagnostic equation that connects the density to the underlying temperature, salinity, and pressure of a particular water mass. See Figure 1.3d. Note that temperature and salinity are independent variables that contribute to the density of the water. Generally, density increases with increasing salt and cooler temperatures going down the water column.

1.3 SPATIAL AND TEMPORAL SCALES

As you stand near the ocean and watch the waves come towards you, you want to understand how these waves might work. How are they generated? How do they propagate? Is there a pattern of propagation? Why do some of these waves break? To answer such questions, you need to first characterize the waves. A typical way to characterize the waves, for that matter, any process or a natural phenomenon is by characterizing their preferred or observed length scales and time scales. A length scale is a representative length by which you can describe the process or phenomenon. A time-scale is the representative duration of time over which the phenomenon's major changes can be described. Now, think about the spatial scales of these waves and how frequently you see the crests coming one after the other. Their typical spatial scales are meters and temporal scales are around seconds and minutes. And since these waves are generated and supported by gravity, they are called surface gravity waves.

Let's talk about one more thing — the atmosphere. On a typical fall day, you hear about these depressions, hurricanes, cyclones. What is the typical radius of a cyclone? Few hundred kilometers, two hundred kilometers? Yes. It is on the order of hundreds of kilometers. So suddenly, you now have a phenomenon that is big. Hundreds of kilometers – but it's one physical feature of atmosphere. What is actually a cyclone in scientific terms? Anything that rotates like this is called a vortex. So it has a rotational speed and the radius. How long does it take for the cyclone to come and make landfall? A few days. Two days, two-three days. So now we are talking about a time scale of about two days from the formation to the propagation and making landfall on an area of hundreds of kilometers with its influence. That's a vortex moving through the atmosphere. And it's not generated in some haphazard manner. There's a system that generates it. It's called an instability of a front that gets out and distributes energy (heat and moisture) through the atmosphere. Instabilities are generated by small disturbances within a front, which can then grow in time to large amplitudes that the original front cannot sustain anymore, leading to detachment of such features in vortices out of the frontal system. We will learn about them later in Chapter 7.

For the time scales, there are days and for the space scale there are hundreds of kilometers, correct? We now went from shallow water gravity waves of seconds and meters to these huge cyclones, which are days and hundreds of kilometers.

Let us take a closer look at Figure 1.4. So, this is what we call a space/time scale of a selected set of processes that are important in the atmosphere. The x-axis is time. We are discussing space (or length-scale) in y-axis and time in x-axis. The time goes from microseconds, to one second … then to a minute, then an hour – a day – then about a week or ten days – next a year, then ten years or decadal scale … then a hundred years … that's a century … and then the millennium. See, it goes from microseconds to millennium in the temporal axis.

In length-scale-wise (y-axis), there is a centimeter, a meter, a kilometer – it's in a log scale, and then thousands of kilometers. And now let's see how we can characterize processes with their spatial and temporal scales, so … about a second to minute and on the spatial scale of centimeters to meter, we have microturbulence. For those of you who are familiar with Newton's second law, all of these processes are governed by "$F = ma$" framework, each set up within its own specific space-time scales. This is needed because different physical things happen at different space and time scales.

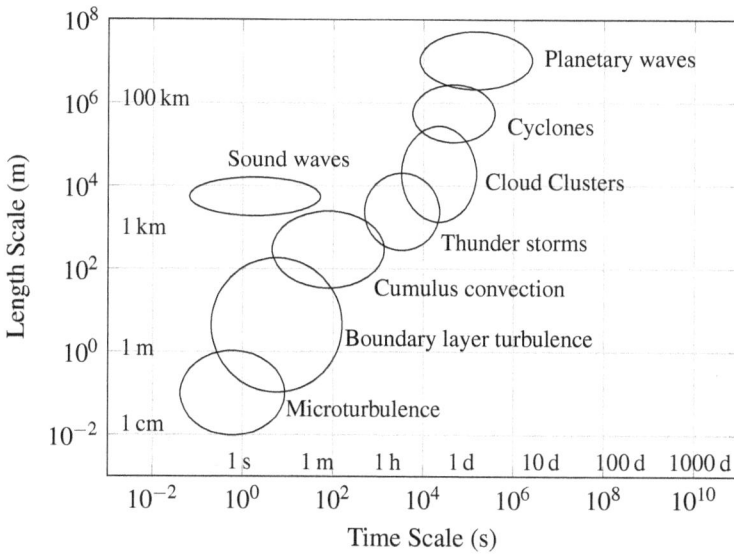

Figure 1.4: Representative temporal and spatial scales for a selected set of atmospheric processes. After (The COMET program).

And then there is a boundary layer, and cumulus convection, now clouds come in. So, microturbulence, boundary layer turbulence ... that's still about meters to 500 meters and a minute scale. Turbulence is occurring at that scale. But then you have cumulus convection – this is an hourly and kilometer scale. Then you have thunderstorms. These are all atmospheric phenomena ... the clouds clusters ... cyclones are right there. Now, look at the scale for cyclones. You're already in thousands of kilometers, and you're on the day-to-week time scale. So, if you want to understand cyclones as well as turbulence, you need to have a model that will need to resolve the turbulence, and it will have to let you analyze cyclones.

Now the problem is: if you want to understand turbulence, you need a model that will resolve every second. So, that means your time-step will have to be even smaller than that, probably microseconds or milliseconds. And you need to be resolving centimeters or less to get your microturbulence correct. To set that up in a kilometer-by-kilometer spatial domain, now you divide micro-centimeter, kilometer by micro-centimeter, you already run into millions of grid points! And similarly, on the time scale. So, you have to have your time-stepping very small so that you can resolve these small scales. But, then the computers do not have the power to solve for such a massive gridded system (at least not yet!). Then, you cannot resolve a thousand-kilometer scale. You understand the problem here because your computing power is not there to resolve this as well as that because you need then tens of thousands of kilometers divided by micro-centimeter scales to have a whole grid, which will be millions and billions of grid points. So, you need to decide what kind of problem you want to do, and it is that kind of a model you have to implement to study your particular process that you are interested in.

So similarly, you can have a process diagram broken up in their spatial and temporal scales for the oceans. See Figure 1.5. So, if you look at the oceans ... again, the same thing ... x-axis has the time scales ... so this is milliseconds to minutes, hours, days, year, and century, and length scales of centimeters all the way to thousands of kilometers. So, micro-turbulence is right there in the seconds-to-minutes time-scale and centimeter-to-meter spatial scale. Then there is the boundary-layer turbulence ... these are ocean-atmosphere interaction kind of issues. Now you have wind sea breeze kind of things, which are about one meter to hundred meter range. Swells are in the hundred meter range, with minutes, fifteen minutes, that kind of frequencies. But then you have internal (gravity) waves and inertial waves. These are just less than a day. We have tides here – these are six-hourly to diurnal (twice-daily) periods that you see, and they are thousands of kilometers long, depending upon where you are looking at it. Then you have eddies, which are somewhere in the few kilometers to hundreds of kilometers range, depending on where you are, and their time scales vary from days to months. These are the eddies that you see in the open ocean.

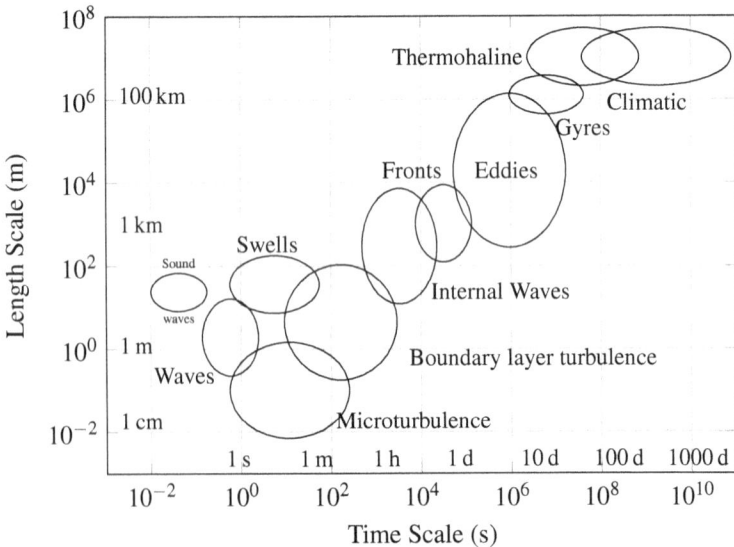

Figure 1.5: Representative temporal and spatial scales for a selected set of Oceanic Processes. After Dickey and Bidigare, 2005.

You look at a satellite picture, you'll see these eddies. See Figure 1.6 for the region of Western North Atlantic. After a month or two or more, these eddies will be moving and maybe even gone, and new eddies will appear. It takes that amount of time for the eddies to demise. Then, the longest scale for time and space is for the thermohaline circulation. It's the density-driven part of the overall oceanic circulation, which has very long time scales. There are sinking regions of water in Labrador Sea. Dense water sinks and then goes all around the world, and it takes about thirty to forty years to come back to the surface. These waters can have many distinctive chemical properties, and you can trace them – and that tells you how they move around. So, in short, this is the description of features and processes in a space-time manner. Again, the same thing: what is it that you want to understand? You have to choose something where you can contribute, and then if you want to use a model or if you want to use the data. If you use a model, then you have to make sure that the model can resolve those scales. If you're using data, you need to make sure that your data has those frequencies and spatial resolution so that you can understand those processes. You cannot just take any data and start addressing any of these processes. Your observational data set needs to capture those processes in a space/time sense so that you can address or explain their behavior.

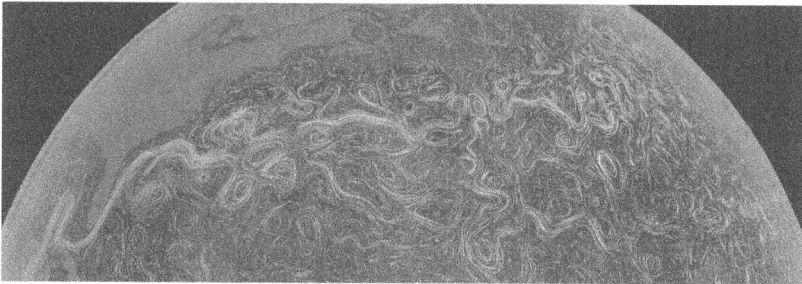

Figure 1.6: Visualization of the Gulf Stream stretching from the Gulf of Mexico to Western Europe. (NASA Scientific Visualization Studio: Gulf Stream Sea Surface Currents and Temperatures; Greg Shirah, lead animator.)

1.4   OCEANS – THE REGULATOR OF TEMPERATURE

So, how does the climate system work? Where does the energy come from? Well, that is easy. The sun. But then what? How do the ocean and atmosphere react? How do they transfer and redistribute the energy coming from the sun around and across the planet to keep the earth habitable at a comfortable temperature of $18 - 19°C$?

Let us first find out some details about the global heat budget. What's the temperature of the sun? It is 5750° Kelvin. Kelvin is a special scale shifted by 273°C; it is the absolute temperature scale. In this scale, the absolute zero temperature is at −273°C, which is the state of the lowest possible energy for any molecule. The kinetic energy or the movement of molecules is negligible or zero at this temperature. The sun radiates heat as a black body (perfect radiation) and following the Stephan-Boltzman law of radiation, which is given as follows:

$$Q = \sigma T^4 \tag{1.2}$$

where Q is heat (Wm^{-2}), T is the temperature in °Kelvin and

$$\sigma = 5.67 \times 10^{-8} Wm^{-2} K^{-4}$$

Using this simple formula in equation 1.2, we get a radiated heat from the sun as $6.2 \times 10^7 \, Wm^{-2}$.

Now, some of that heat energy from the sun is coming to us – that is the total energy received by our home, the Earth to feed on. Given the Earth-Sun distance, the size of the sun, and the earth's area exposed to the sun, the earth receives about $344 \, Wm^{-2}$, which needs to be redistributed around the planet so that we can inhibit this planet at a comfortable $18 - 19°C$? How does this work?

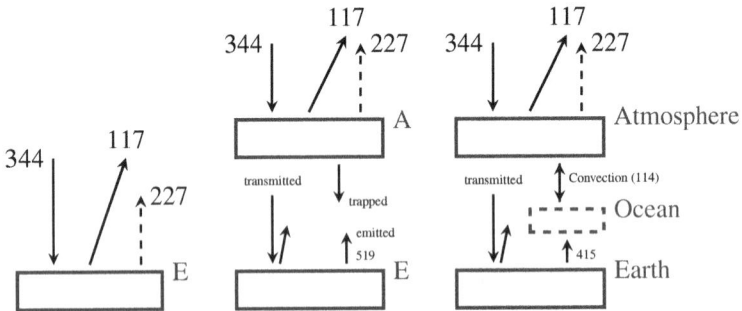

Figure 1.7: Three different models of the climate system. Left: Only Earth $(T = -21°C)$; middle: Earth + atmosphere $(T = +36°C)$; right: Earth + ocean + Atmosphere $(T = 19°C)$. In all cases, the Earth (E) absorbs and emits equal amount of heat which keeps our equilibrium temperature.

We can think about two models. See Figure 1.7 – left two panels for these models. The first model (on the left) has only the Earth – without any atmosphere or ocean. The second one in the middle of Figure 1.7 has an atmosphere that absorbs, reflects, transmits, and reabsorbs the incoming radiation from the sun and outgoing radiation from the Earth. In the first model (only Earth), part of the incoming/incident flux will be reflected by the earth, which is about $117 \, Wm^{-2}$, and that leads to an effective temperature on the earth to be equal to a temperature that can be obtained using equation 1.2 again, with $F = (344 - 117)Wm^{-2}$ or $227 \, Wm^{-2} = \sigma T_1{}^4$. This results in an equilibrium temperature for the earth as $T = 251°K = -21°C$. That is too cold for our habitation! And way cooler than the present-day annual average of $19°C$.

Now, let us introduce an atmosphere around the Earth with its present-day characteristics of absorption, transmission, emission, and reflection (middle panel of Figure 1.7). After going through some algebra (see Cushman-Roisin and Beckers, 2011 for details, if interested), this model yields an equilibrium temperature of $36°C$ from a total emitted radiation of $519 \, Wm^{-2}$. Now that is too hot for us to live.

So, what just happened? In our first model, when we did not have an atmosphere, it was cold at $-21°C$. We had $-21°C$ in the first model, without atmosphere, without ocean. We just added the atmosphere in the second model, and we got to $+36°C$, which is too high! The fact that we put in an atmosphere and got higher temperature than the present time means that the atmosphere works as the greenhouse. It just traps the heat. This is the so-called "greenhouse effect" of the atmosphere. What it does is it traps the heat and then sends it back to the earth. So, these two simple models just showed you the effect of atmosphere on our planet in terms of warming the earth. Realize that, if we did not have any atmosphere, we would not have lived in a warmer world. Realize that carbon dioxide, water vapor and methane are three of the major constituents of the atmosphere, which when increased in their composition will raise the porosity of the atmosphere and will trap more heat than necessary and warm our planet even more. This is in short what happens in the greenhouse effect and these gases which help the atmosphere become more susceptible to trap heat like the greenhouse are called the greenhouse gases.

Now let us think about yet another simple model of the Ocean-Atmosphere-Earth system. See the right panel of Figure 1.7. Since the solid earth model was too cold and the atmosphere-earth model was too hot; it appears that the role of the ocean is more like a regulator for the planet. A third model can now be conceived in which the ocean enters between the earth and the atmosphere just like on the right panel in Figure 1.7. Remember that the atmosphere in the second model is only radiating. So we got incoming short wave radiation through it and reflected outgoing long wave radiation. The oceanic heat transfer is a different process than radiation or emission. It is convective heat transfer which depends on latent heat of water – changing state from liquid to vapor. The heat out of the ocean goes back into the atmosphere, which returns the heat back to the ocean in terms of precipitation (rain) and thus the atmosphere gets to cool down. The convective heat exchange between the atmosphere and the ocean is estimated to be about $114 \, Wm^{-2}$ and the emission from the Earth's surface now becomes about $415 \, Wm^{-2}$. After adding the ocean in this model, the temperature of the earth is found to be around 19°C corresponding to the $415\,Wm^{-2}$ pf emission and using Equation 1.2. Well, there it is! Now then, the impact of the greenhouse effect (warming due to atmosphere) is balanced by the hydrological cycle (through convection). This is why the ocean is so important for the climate change scenario. It regulates the overall heat balance of the earth-atmosphere-ocean system and its circulation is critical to understand how the heat balance is maintained and heat is distributed throughout the different oceans and their feature around the world. In summary, we just learned the ocean's role in our climate system. The ocean acts as a regulator of the heat transfer between the earth and the atmosphere and cools it down. Had there been no ocean, our earth system would be too warm. This is where the greenhouse effect comes in.

So, if your oceans are warmer than before, then there will be less heat transfer, and the atmosphere will be warmer than what it is now, and that in essence will warm up the climate. In a sense, this heat radiation has to be properly kept under control. Otherwise, we will run into an atmosphere which is much warmer than it is because the oceans cannot function as a regulator like it has been doing for years. So this, in effect, is the science of climate in a nutshell. There are complicated processes that describe the details, but from the heat budget point of view, this was a simple explanation of how the ocean works as being a heat sink, if you will, for the whole climate system.

1.5   CARRIERS OF HEAT – CYCLONES AND CURRENTS

Because of the spherical surface of the Earth, naturally, we have an excess of heat near the equator. And we have a deficit of heat near the poles. But the earth is in a thermal equilibrium. Clearly, over an annual period, there is not much change of temperature of habitation. So, there has to be a transfer of energy between the equator and the poles during the year and on a continuous basis. So the excess heat needs to be transferred from the equatorial region to the Polar regions. See Figure 1.8. How does this happen in reality? It is the response of the atmosphere and oceans that organize themselves to carry this excess heat through their own circulation features. So both the ocean and the atmosphere takes the heat from the equatorial region out to the poles. That's how the whole ocean-atmosphere-earth system has to work to be in equilibrium. It's the sense of energy transfer and redistribution on the earth's surface.

So the heat that comes in from the sun is disproportionately warming the equatorial region – it's getting more heat than other latitudes; and that this excess heat needs to be shunted out to the poles, and the way that is done is through the circulation of atmosphere and oceans. They carry different amounts of heat at different latitudes depending on the overall circulation patterns which are briefly discussed here.

Let's talk about atmosphere a little because that is what forces the ocean. What happens in the air on the equator where you get most of the heat? It will

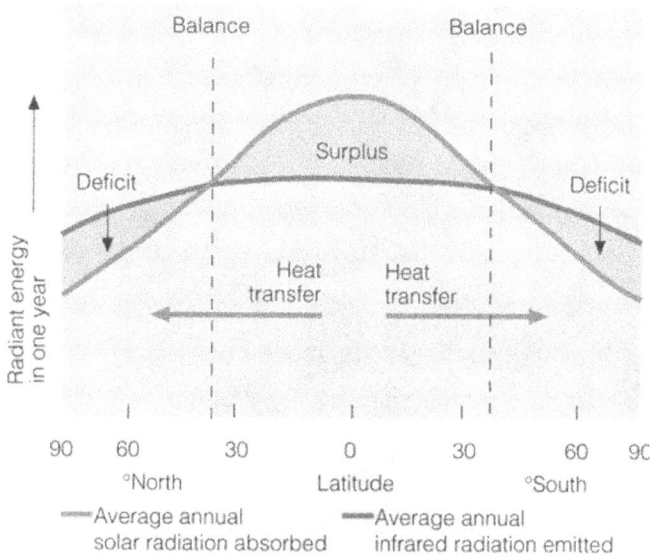

Figure 1.8: Heat excess on the equator. Source: NASA.

rise – warm air rises up. What will happen to the air at the poles? It will sink, right? So, you have the air rise on the equator, then go northward, and then sink near the poles and come back along the surface of the Earth. See Figure 1.9. These are called direct cells or Hadley cells. It's a straight thermal convection – a closed convective circulation system.

This was OK until we realized that, well, it really does not happen that way because the Earth is rotating. So, because of the rotation of the Earth, these cells get short-circuited. And the way it gets short-circuited is that the cell closes much before the upper-level air can reach 90° north or 90° south. Consider that the equatorial rising and polar sinking have to happen and then one can imagine two direct cells near the equator and the pole in each hemisphere – banded by a mid-latitude region between 30 and 60 degrees of latitudes. So, if the equatorial rising is fine, but then the air has to dip down at around 30°N, and then rise up again around 60°N and then it will again sink at the North Pole. So this is the short-circuit – these are between 30°N and 60°N, and similarly on the southern hemisphere. So, in the subtropics between 30 and 60 degrees, we now have an opposite cell, correct? These cells in the subtropics are called indirect cells and also called Ferrel cells. However, the atmosphere is obviously more complicated, but this is a simple idea of what happened. And effectively, you can see that these regions of indirect cells — 30° to 60° in both

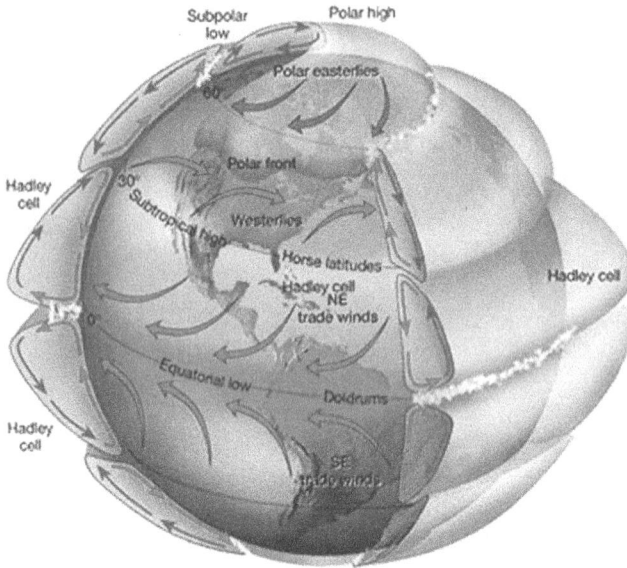

Figure 1.9: A depiction of atmospheric cells around the earth. The direct or Hadley cells are on the equator and near the poles. The indirect or Ferrell cells are on the subtropics. Image from NASA's Remote Sensing Tutorial: The Water Planet – Meteorological, Oceanographic and Hydrologic Applications of Remote Sensing. Permanent URL: https://serc.carleton.edu/download/images/10044/3d_hadley_md.v3.jpg.

hemispheres — are the regions of the subtropics where there is much instability growth in the atmosphere. The tropical depressions and instability propagates poleward and grows in these regions. When these vortices propagate over the warm oceans, they gather their energy from the heat content in the upper layers of the ocean and grow from a small depression to a fully developed storm. Heat content is the integrated heat storage in the upper mixed-layer of the ocean. It is the instability that drives eddies in the atmosphere – these eddies are the vortices, also known as cyclones (in the Indian Ocean), Typhoons (in the Pacific), and Hurricanes (in the Atlantic).

So, the subtropical region is the region of instability, eddies, vortices, fronts, movement. The fronts are primarily within the latitudes at 60 north and at 30 north. And for the equatorial circulation, if you now think about the surface winds, then you have the trades, which are predominantly from east to west in the tropical region. These are southeast trades. And then in the subtropics, it goes from west to east or westerlies. So the westerlies are in the subtropics, and the northeast trades and the southeast trades are mostly in the tropics. And then you have polar easterlies. These are mean flows, But there'll be lots of instabilities because of the rotation of the earth and the movements of the flow field. And those are your cyclones and hurricanes and typhoons. Now, this is a simple way of thinking about how the wind system works in three different bands: tropics, subtropics, and the polar regions. In between these, there are fronts. Like the polar front, the ITCZ (Inter Tropical Convergence Zone), etc. There is the southern hemisphere ITCZ, there is a northern hemisphere ITCZ.

How about the ocean? How does the ocean carry the heat? Currents. There are well-organized and fast western boundary currents which carry the heat from the equator to the poles. Name some of the currents. Gulf Stream. Kuroshio. These are called the warm water currents. The Gulf Stream flows along the coast of eastern United States and then separates from the coast and meanders northeastward towards the European continent. The Kuroshio flows along the east coast of Japan and similarly meanders in the northwest Pacific. They carry a lot of heat – all the way from the equatorial regions to the subpolar and polar regions. The Gulf Stream carries about 100–150 times the water in the Amazon River. Think about that! In the South Atlantic, there is the Brazil Current. What about the South Pacific? The East Australian Current, remember Nemo? It's a swift current that hugs the coast, and it transfers a lot of heat. Similarly, there is the Mozambique Current/Somali Current in Indian Ocean/Arabian Sea. Similarly, on the east coast of India, you have the Bay of Bengal Western Boundary Current (WBC), and that carries heat from equator out in the springtime. And its counterpart, the East India Coastal Current (EICC) in winter brings back freshwater from the head bay all the way into the Arabian Sea in autumn time. So this process of circulation is the way the energy gets transferred from one region of imbalance to the other where it is needed. It's Nature's own way of circulating energy and water masses of different heat content. So this is what is the dynamics of oceans and atmosphere that we are going to be studying.

An idea to keep in mind is that warmer currents flowing (western boundary currents) from equator to the Poles would carry warmth and colder currents flowing from polar regions to the equatorial regions (eastern boundary currents) also redistribute heat in terms of carrying cooler water out of the Polar regions and putting them in the equatorial regions to cool these regions.

Ocean circulation gyres are formed primarily under the influence of atmospheric winds and the earth's rotation. While being bounded by continents on either sides, the ocean circulation is generally manifested in terms of well-known boundary currents, which help transport excess heat from the equator to the poles via organized, narrow, swift, and deep Western Boundary Currents (WBCs) and bring colder waters from polar regions to the equatorial regions via unorganized, broad, slow and shallow Eastern Boundary Currents (EBCs) to maintain a global heat balance (Figure 1.10).

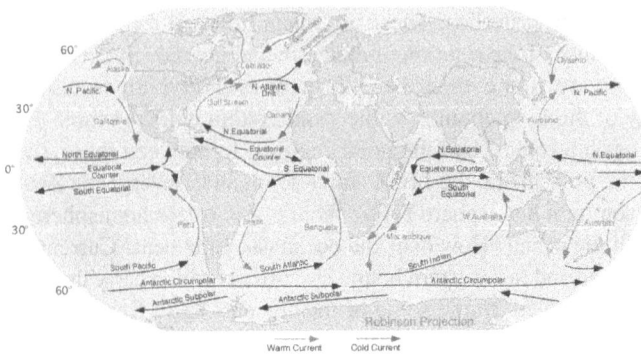

Figure 1.10: Major ocean currents around the world. U.S. government publication. References for the currents: Curtis Ebbesmeyer map at Wired; credit: Dr. Michael Pidwirny www.physicalgeography.net.

Do you see these organized currents on one side of the ocean? These are the western boundary currents. There is an asymmetry in ocean circulation. The ocean does not circulate in a symmetric manner. There is this asymmetry that the western boundary of all the oceans ... there are these strong organized poleward currents along the western boundary of all the major ocean basins. In the North Atlantic, it is the Gulf Stream. In the South Atlantic, it is the Brazil Current. In the North Pacific, it is the Kuroshio. In the South Indian Ocean, it is the East Australian Current (EAC). In the Bay of Bengal, there is the springtime Western Boundary Current (WBC) and autumn-time East India Coastal Current (EICC). In the Arabian Sea, it is the Somali Current. Then there is the Agulhas Current in the South Pacific.

Typically, the western boundary currents are formed in response to large-scale wind forcing. Basin-wide wind-stress curl (the rotational effect of the zonal and meridional winds combined) integrated over a period of time creates the organized WBCs along the east coast of a continent. In contrast, the shallow and broad EBCs are generally extremely eddy-rich and known for their prevalent upwelling regions around the capes and promenades along the coast.

1.6   A BRIEF HISTORY FROM FRANKLIN TO SATELLITES

Let us now take a brief walk through the history and development of modern oceanography. Let us start with the story of Benjamin Franklin and the Gulf Stream. The first well-known chart of the path of the Gulf Stream called the "Franklin Folgers chart," was crafted in the 1760s by Benjamin Franklin and his nephew, Timothy Folgers, a sea captain from Nantucket Island in Massachusetts, United States. The story is simple and elegant, worthy of repeating to any newcomer to oceanography.

Benjamin Franklin was a scientist, a politician, and did many other things. Let us go back to the 1760s, about 260 years from now. He was the Postmaster General of the first post office in Philadelphia, and at that time, the territory of what is now the United States of America was a British colony. Franklin was actually posted in Philadelphia, looking after all the mails coming from all over the British colonies to London. So he got these complaints from his compatriots there, the Britishers who were settled in United States, especially in New York and Boston, that when they sent letters to London, the people in London received them in two weeks. But when the people in London, their relatives, wrote letters, they often took three or four weeks to come. There was a need for someone to figure it out.

Benjamin Franklin, being the Postmaster General, sitting there in his office and reading through all these letters, said "This is a real problem because everyone is complaining about this." The problem was in his mind for some time, two/three years maybe, and until he was having a drink one day with his nephew, Timothy Folger, who was a ship captain in Nantucket (in the state of Massachusetts). Franklin explained the one-way delay of the mail transportation to Tim. He listened and then said, "Uncle Ben, I actually might know what is happening, I think. There is this thing in the ocean – kind of like a warm river that kind of helps us when we go from New York to London, and then when we come back, we follow the same route, and this thing actually slows us down like anything." Then, Benjamin Franklin, being the scientist and mind that he was, started studying the log books of ships to find out how much time it took. And he actually calculated the speeds of these ships and how much power they were consuming and all that.

You know what he did next? He told Timothy Folger that "OK, I will come with you, and I'll sample the ocean on your way to and from Britain." He actually took a bucket and a thermometer, lowered them down, and went across the current – and did this over a year, measured the temperature and calculated the speed of the current in very rudimentary ways, and then created a map for the ships. It's called the Franklin-Folger chart of the Gulf Stream, and shows that there is indeed a warm river that flows from the Atlantic coast of the United States to England on the European side all the way. See Figure 1.11. And we tried to model it in 1996 (Chao et al., 1996) after two hundred and some years, and we were pretty close to what he did by hand in 1760. And we were happy. This was done with all our computers and everything else! It is curious to note that the Franklin-Folgers chart was missing for about two hundred years, until Phillip Richardson of the Woods Hole Oceanographic Institution found two copies of the chart in the Bibliothéque Nationale in Paris (P. L. Richardson, 1980). For more interesting facts about this chart including a temperature plot from Franklin's data, please see the excellent book by Arnold Taylor (A. H. Taylor, 2011).

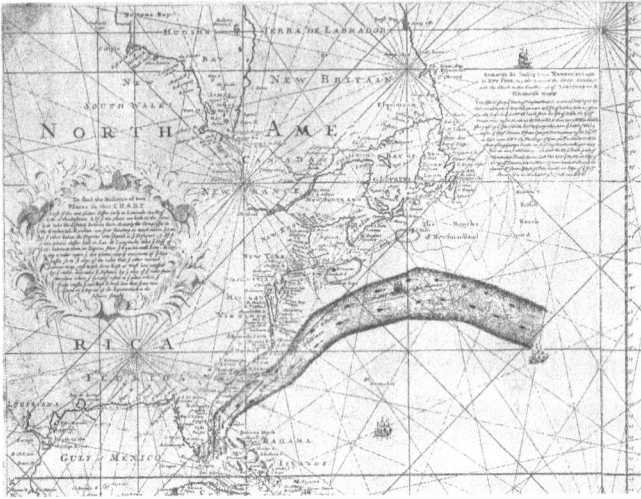

Figure 1.11: This is the upper left portion of the first Franklin-Folger chart of the Gulf Stream printed in London in 1769 and held by the Library of Congress (LOC). The commons has this chart as File:Franklin-Folger chart of the Gulf Stream LOC 88696412jpg 30 May 2018, publication authorized by Benjamin Franklin and Timothy Folger. Gulf Stream.

The next century (1800–1899) was the century of navigators and observers, and it was pioneered by Fridtjof Nansen – the only Oceanographer to be awarded a Nobel Prize! Do you know why he was awarded the Nobel? Before we talk about that, let us talk oceanography. In the late nineteenth century, American, British, and other Europeans carried out coordinated expeditions to explore ocean currents, ocean life, and the ocean bottom. The first systematic scientific expedition to explore the world's ocean and seafloor was the Challenger expedition from 1872 to 1876, onboard the British 3-masted warship "HMS Challenger." Nansen also spearheaded a cross-Greenland expedition between 1881 and 1896 to understand the polar currents and Greenland ice coverage.

Nansen had a keen insight into oceanography and how things work. Nansen and his fellow explorer, Hjalmar Johanesen were on a trek in their vessel *Fram* in September of 1895, when the *Fram* was frozen into the Arctic ice drifting slowly in an eastward direction as the ice was being pushed by the surface currents. Nansen noticed that the direction of the drift was not following the wind; instead was about 20–40° to the right of wind! Nansen perceived that this was due to the rotation of the earth. He also realized that the angle of the drift of the current was reducing as they go deeper in the water. When Nansen came back to Norway, he discussed this idea with a Swedish physicist, Vagn

Walfrid Ekman. Ekman developed the first theory of the wind-driven currents and described the veering of the currents from surface to layers below in the so-called "Ekman Spiral" – that oceanographers are so familiar with today. We will discuss the Ekman Spiral in Chapter 4. For more interesting stories about Nansen, Ekman, and Franklin, please see A. H. Taylor, 2011 and other books mentioned at the end of this chapter.

Prior to this, James Rennell (1742–1830), who has been called the Father of Oceanography, started as a Shipmate in an East India Company ship at the age of 21. He surveyed the Indian subcontinent to map the countries and waters in the Indian subcontinent (Rennell, 1788). He was the first Surveyor General of Bengal. He started his oceanographic scientific research when he was traveling by a sailing ship with his family from India to Britain after his retirement in 1777. During the extraordinary prolonged voyage around the Cape of Good Hope he mapped the banks and currents at the Lagullas and published in 1778 the work on what is today called the Agulhas Current (Rennell et al., 2014). This was one of the first contributions to the science of Oceanography. He was the first to explain the causes of the occasional northern current found to the south of the Isles of Scilly, which has since been called as Rennell's Current. In 1830, before his death, Rennell was one of the founders of the Royal Geographical Society in London. His work on the Atlantic was published later by his daughter (Rennell, 1832).

Some also consider Matthew F. Maury, as the founder of modern oceanography, who also stressed the importance of the ocean circulation for the climate by influencing the winds, the air temperature, and the hydrological cycle in his book on the physical geography of the sea published in 1856 (Maury, 1856). Maury first started a systematic survey of ocean currents. Coming back to Nansen and the Nobel – well, he was a great administrator and with a great heart for humanity. After the first World War, Nansen was asked to find a way to preserve the identity of people who got displaced from their own country, because of war, famine, or other human tragedies. He designed the "Refugee identification Card" – a card that was issued to people in such situations. It was called the "Nansen passport". Nansen passport was given to stateless refugees to enable them to cross national borders after WWI. Nansen received his Nobel in 1922. For a detailed look at his remarkable life, please read Huntford, 2012.

Roald Amundsen, a Norwegian explorer, reached the South Pole in 1911 and flew over in an airship to the North Pole in 1926. There was a first South Atlantic Expedition by the German Ship Meteor in 1925–27. During the first half of the twentieth century, a number of theoretical developments advanced our understanding of the oceans and their workings. Notably, Ekman, 1905, Rossby, 1936, Sverdrup et al., 1942, and H. Stommel, 1948 laid the foundations of modern oceanography linking winds and the rotation of the earth to the ocean currents. Wind-driving of the currents, the importance of earth's rotation in terms of Coriolis force led to the establishment of a physical understanding of simple phenomena such as Ekman transport to upwelling and biological processes. Such understanding led to the appreciation that the study of the seas needs to be done in an integrative and interdisciplinary framework. The first comprehensive work on our understanding of the oceans was compiled by H. U. Sverdrup with Fleming and Johnson in 1942 (Sverdrup et al., 1942).

After the Second World War, between 1948 and 1970, quick theoretical development with Henry Stommel (WHOI), Walter Munk (Scripps), Allan Robinson (Harvard), Jules Charney and Ed Lorenz (MIT), and many others paved the way for modern oceanography to enter into the new era of the computational domain. It was Sverdrup's wind-driven gyre idea, Stommel's westward intensification, Munk's Boundary layer theory, Charney's instability theory, and Robinson's operational model development for the U.S. Navy – to name a few contributions – which created the foundations of modern oceanography, as we know it. The first numerical model for the world ocean was developed by Micheal Cox and Kirk Bryan in 1969–1971 at the Geophysical Fluid Dynamics Laboratory (GFDL), Princeton, USA. The development of climate models (Manabe & Bryan, 1969; Manabe et al., 1975) was progressing in parallel to the ocean modeling efforts. From 1950 to 1972, there was a rapid growth of observational oceanography worldwide. The International Geophysical Year (IGY) was launched during 1957–1958 and has shown the value of the combined and collaborative efforts in our understanding and researching the oceans. The International Indian Ocean Expedition (IIOE) became an official entity with its activities between 1 September, 1959 and 31 December, 1965. The first multi-ship multi-nation survey of the Indian Ocean was done during 1960–1963, was the observational period of the first IIOE. Upon recommendation from the first IIOE, the National Institute of Oceanography in Goa, India, and the National Institute of Oceanography in Karachi, Pakistan were established in the early sixties. Following up after 50 years, there has been a recent initiative for the second IIOE led by the United States, India, and other countries during 2015–2025. International efforts and science plans related to this initiative are outlined by R. R. Hood et al., 2016 and R. Hood et al., 2018.

The University of Sao Paulo approved Brazil's first post-graduate program at their Institute of Oceanography in 1973 and admitted the first batch of students in 1974. China's First Institute of Oceanography in QingDao was established in 1958. Argentina became part of the modern oceanographic campaign with its programs in Buenos Aeros. Multiple Polar Research Institutes and Polar Oceanography programs have started around the world from the 1970s with satellite observations revealing amazing synopticity and unprecedented views of changes occurring in the polar regions.

In the mid-seventies (1970–1980), the Mid Ocean Dynamics Experiment (MODE) and PolyMODE established the fact that the oceans are full of eddies – synoptic eddies at mesoscale resolution (a spatial scale where advection is balanced by the earth rotational effects) – the dominant variability in the ocean is governed by the eddies. Around the same time, the first ocean viewing satellite (SeaSAT) was successful to show the first synoptic view of a large area of the ocean surface through infra-red imagery (Advanced Very High Resolution Radar) at 1.1 km resolution! It showed a clear synopticity and a rich dynamical character of the mesoscale eddies and their interaction with the currents and fronts, which was never possible in the early days of oceanography. A satellite picture of the western North Atlantic is shown in Figure 1.12, which has a resolution of 1.1 km. You can see the Gulf Stream, its eddies (called Rings) on both sides, and other features of circulation at multiple scales.

During the 1970–1990s, rapid advances also happened with our understanding of coastal processes and modeling of the coastal ocean and its ecosystem. Theoretical understanding of tides (Newton and Kelvin) and storm surges were already well developed due to their immediate societal impact for a long time. Advances in satellite and in situ observations and easier access to computing infrastructure in the 1980–1990 period led to a new understanding of a coastal phenomenon in terms of theoretical and interdisciplinary modeling at a higher resolution than ever before. Examples of such coastal phenomenon include: (i) coastally trapped waves (K. Brink, 1982), (ii) coastal upwelling indices and their applications for fisheries (Bakun, 1973); (iii) river plume dispersion dynamics (Garvine & Monk, 1974); and (iv) cross-shelf exchanges between the deep and coastal oceans across shelfbreak fronts (Linder & Gawarkiewicz, 1998). Physical and interdisciplinary processes and methods of multiple coastal phenomenon around the global coastal oceans have been well documented in the three-volume series of The Sea – Vol 10 (The Global Coastal Ocean; Brink and Robinson), Vol 11 (Processes and Methods; Robinson and Brink) and Vol 12 (Physical-Biological interactions; Robinson, McCarthy, and Rothschild).

Natural and accidental disasters such as the 2004 Tsunami (Indonesia), Deepwater Horizon Oil Spill in 2010 (Gulf of Mexico) and Fukushima Tsunami and the subsequent nuclear plant accident in 2011 (Japan), have resulted in countries becoming more aware of the importance of protecting our coastal resources and thus led to more investment in science in the twenty-first century. Add to this the clear signatures of warming on these already-stressed coastal regions. Thus, multiple new technologies for observing and monitoring the coastal ocean have come to fruition over the last 30 years. Among these are the High-Frequency Radars, Gliders, and LIDAR. Coastal ocean observation and modeling are now complementing each other for societal applications in an operational setting through multiple "coastal ocean observation system" (COOS) network around the coasts of the United States (called the Integrated Ocean Observation System – IOOS), Canada, Europe, and almost all maritime nations in the world. Multiple global efforts are underway and we will discuss some of these efforts including the Pioneer Array in the northeast coast of the United States (Gawarkiewicz & Plueddemann, 2020) in Chapter 11 later.

Three important developments during 1970–1990, – the two in situ experiments – MODE and PolyMODE (multi-national effort between US-USSR-Europe) and the first satellite mission for Oceans SeaSAT, helped us understand the dominant spatial patterns in the ocean. These were the days during the cold war and before PCs, Laptops, Internet, and Cell phones. However, high-powered computers were becoming available to researchers, which led to the development of "Ocean Models" at multiple scales and for different purposes and for solving/understanding different oceanic processes. The idea of forecasting or prediction came first from the Navy's need for "operational forecasting" following the footsteps of the meteorological "weather forecasting" efforts.

Seasat (1978) was a landmark development and success in satellite oceanography. It was a proof of concept mission. It was launched on June 26, 1978 and managed by NASA/Jet Propulsion Laboratory. Seasat was built by JPL to test a variety of oceanographic sensors including imaging radar, altimeters, radiometers, and scatterometers. It had five instruments – Radar Altimeter (for sea surface height), SAR (Surface Aperture Radar), SMMR (Scanning Multichannel Microwave Radiometer), Microwave Scatterometer (for Winds), and VIRR (Visible and Infrared Radiometer). Many later dedicated satellite missions are built from the experience of Seasat.

During 1980–1990, there have been satellite missions to sense parameters like sea surface height (through Altimeter) and sea surface chlorophyll (through reflectance) in addition to sensing temperature. Missions such as NOAA-AVHRR (SST), GEOSAT (SSH), and Coastal Zone Color Scanner (CZCS) came into operations. Utilizing such synoptic multiparameter datasets in numerical models led to new mathematical tools such as data assimilation and nested modeling system. New and more efficient modeling techniques using new computing architecture (power of parallel processors) led to different modeling framework – Modular Ocean Model (MOM), Parallel Ocean Program (POP), Harvard Ocean Prediction System (HOPS), Princeton Ocean Model (POM), Regional Ocean Modeling System (ROMS), Navy Layered Ocean Model (NLOM), Navy Coastal Ocean Model (NCOM), Finite Volume Community Ocean Model (FVCOM), NEMO, MITGCM, HYCOM, MSEAS, MOM6, and others.

The first operational ocean modeling system for Navy used "synoptic feature models" for fronts and eddies detected from satellites and projected through the water column with analytical-empirical formulation (A. Robinson et al., 1988,1989).

The first Indian satellite for ocean research was named OceanSAT I and II. These were launched as experimental remote sensing satellites in 1979 and 1981, respectively, for Oceanography and Hydrology research. See Ulaby et al., 1986 for more details. Later on, the Indian Research Satellite (IRS-P4) was developed for very high resolution (250 m) sensing of temperature and chlorophyll, and was launched in 1999 by the Indian Space Research Organization. India has since sent multiple satellites for observing chlorophyll, winds, SSH (Saral-Altika), and parameters related to clouds and Monsoon. India has also established multiple institutions in the late twentieth century to explore and understand the oceans around the Indian Ocean including Antarctic

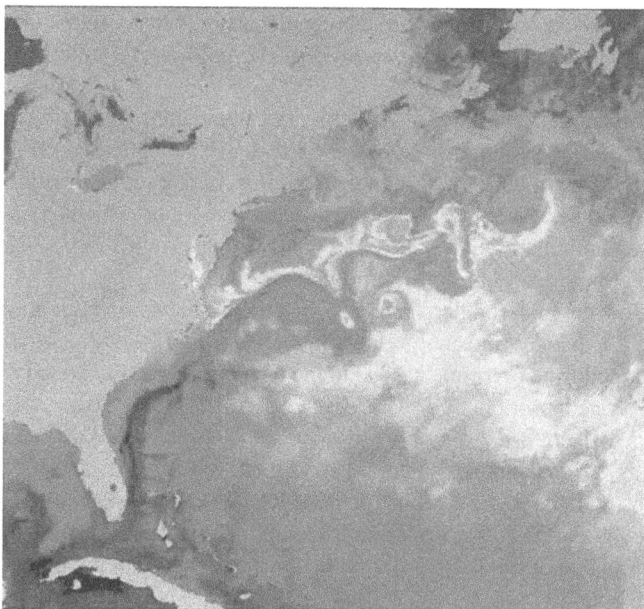

Figure 1.12: A satellite view of the Gulf Stream (1980s). https://commons. wikimedia.org/wiki/File:Golfstrom.jpg

and the Arctic Oceans. To name a few of these, the National Center for Polar and Ocean Research (NCPOR), the INCOIS (Indian National Center for Ocean and Information Services), NIOT (National Institute of Ocean Technology) are at the forefront of ocean research. The Southern Ocean research focusing on Antarctica was started in 2004 by NCPOR using RV Sagar Kanya; and since then, this institute has become a major partner for the Indian Ministry of Earth Sciences (MOES). Details of the history of activities of NCPOR can be found in the article by Pandey et al., 2006. The INCOIS efforts were very successful in establishing the first Tsunami Warning System (TWS) in the Indian Ocean region. The first Tsunami Atlas for the Indian Ocean was prepared by an academic institute, the Indian Institute of Technology at Kharagpur, India (Bhaskaran et al., 2005) after the 2004 Indonesian Tsunami.

In addition to these and many other smaller educational centers, there is the well-established Indian Meteorological Department (IMD), which has a long history with Sir Gilbert Walker from the early 1900s. More about Sir Walker and his contributions to the Walker Cell and Southern Oscillation, North Atlantic Oscillation, and Auto-regressive Modeling are discussed in Chapter 6. All that discovery happened while Sir Gilbert Walker was the Director General of IMD in India. A nice dynamical framework and explanation of many interconnected ocean-atmospheric processes are captured in the book by Arnold Taylor of PMEL, UK (A. H. Taylor, 2011).

The first Brazilian satellite for collecting oceanic data is SATCAT (in Portuguese, Satellite de Coleta de Dados) or SCD-1 was launched in 1993. The South American nations (Brazil, Chile, Argentina) are actively working together to develop new satellites for the next decades. A Chinese-French mission CFOSAT was launched in October 2018 from China to study ocean surface winds and waves at a very high resolution.

With the advent of computers, internet, numerical models, and our ability to collect, process, display, and analyze data in real-time and assimilate such data in numerical models in real-time has led to the new efforts of "Integrated Ocean Observation Systems" or "IOOS." An Ocean Observation system comprises of a network of people engaged in observations and developing tools for monitoring and predicting the ocean environment for the benefit of the society. In 1998, the U.S. Congress called for the establishment of IOOS, which is one system with seven societal goals. The IOOS mission is to routinely provide data and model-based forecasts and outlook for different ocean and ecosystem parameters for use by the people.

During the last thirty years (1990–2020), there have been extensive progresses made in interdisciplinary and climate modeling. Many of these modeling efforts have their seed in the Numerical Weather Prediction (NWP) efforts conceived in 1922 by L. F. Richardson. More history on these efforts is briefly described in Chapters 13 and 14 later in this book.

After the United States established its own IOOS with 11 regional components (from Hawaii and Alaska to around Pacific and to the northeast of the United States via the Caribbean), other countries have established their own IOOS-equivalent. And then there is an effort to coordinate all of these around the world of one ocean – called the Global Ocean Observation System (GOOS), under the auspices of the United Nations International Oceanographic Commission (IOC). At the same time, a strategic plan for the U.S. Integrated Earth Observation System (IEOS) was developed to monitor the whole earth with a climate system perspective. Naturally, both of these later efforts (GOOS and IEOS) have synthesized into a Global effort called the "Global Earth Observing System of Systems" or GEOSS. This global effort plans to align the regional and country-wide efforts on a global scale and monitor the earth. The future of the Ocean and Earth system understanding depends on careful and thoughtful utilization of such observations and on our ever-evolving modeling capabilities.

CONCLUSION

In this chapter, we have learned that our home, the ocean, and atmosphere are physical subsystems of the larger Earth system. We learned about the physical characteristics or properties that define and help quantify our ocean environment. We realize that different phenomena in the ocean and the atmosphere happen on multiple scales of space and time and it is important to understand them individually as well as interacting across these scales. The role of the ocean as a regulator of Earth's temperature was established by using three simple radiation models. It is important to realize that such regulation is achieved by redistribution of excess heat from the equatorial regions to the polar regions by instabilities such as cyclones, hurricanes, and typhoons in the atmosphere and by oceanic currents in a systematic way. We realize that the rotation of the Earth plays a major role in governing the circulation and naturally occurring phenomena in the ocean and atmosphere. Finally, a brief history of oceanography from the time of Benjamin Franklin to the present day with established numerical models is presented. This journey of oceanographic development from the permanency of asymmetric oceanic currents to the realization of the importance of Earth's rotation, to the abundance of eddies, to satellite depiction of a synoptic view of the ocean led us to develop data-assimilative dynamical numerical models, which are useful for the society for interdisciplinary and climate-related applications.

FURTHER READING

Ekman, V. W. (1905). *On the influence of the earth's rotation on ocean-currents.* Almqvist & Wiksells Boktryckeri, A.-B.

Huntford, R. (2012). *Nansen: The explorer as hero.* Abacus.

Millero, F. J. (2010). History of the equation of state of seawater. *Oceanography, 23*(3), 18–33.

Richardson, P. L. (1980). Benjamin Franklin and Timothy Folger's first printed chart of the Gulf Stream. *Science, 207*(4431), 643–645.

Robinson, A. R. (2012). *Eddies in marine science.* Springer Science & Business Media.

Stommel, H. (1948). The westward intensification of wind-driven ocean currents. *Eos, Transactions American Geophysical Union, 29*(2), 202–206.

Stommel, H. M. (1965). *The Gulf Stream: A physical and dynamical description.* University of California Press.

Sverdrup, H. U., Johnson, M. W., Fleming, R. H., et al. (1942). *The oceans: Their physics, chemistry, and general biology* (Vol. 7). Prentice-Hall, New York.

Taylor, A. H. (2011). *The dance of air and sea: How oceans, weather, and life link together.* Oxford University Press.

Ulaby, F. T., Moore, R. K., & Fung, A. K. (1986). *Microwave remote sensing: Active and passive. Volume 3. From theory to applications.* Addison-Wesley Publishing Company, Advanced Book Program/World Science Division.

Walker, S. G. T., & Bliss, E. (1928). *World weather, III.* Edward Stanford.

Breaking Waves at Pahoa, Haw[...]

2 Responses and Forces

> Although to penetrate into the intimate mysteries of nature and thence to learn the true causes of phenomena is not allowed to us, nevertheless it can happen that a certain fictive hypothesis may suffice for explaining many phenomena.
>
> Leonhard Euler (1707–1783)

OVERVIEW

In the first chapter, we introduced our the earth as a four-component system of systems of the ocean, atmosphere, geosphere, and biosphere. The ocean plays the role of regulator of temperature in the multiply-connected system of ocean, atmosphere, and land surface within which the biosphere thrives. The regulatory radiation balance is maintained by heat transferred from the equatorial regions to the poles by currents and eddies in the ocean and by hurricanes (in the Atlantic), typhoons (in the Pacific), and cyclones (in the Indian Ocean) in the atmosphere. The history of oceanographic development followed a path of discovery and better understanding of natural phenomena. The recognition of the importance of earth's rotation for oceanic and atmospheric motions, realization of combined impact of wind and earth's rotation on oceanic flow, existence of abyssal flow, and their dynamical explanation from a density-difference perspective led us to theorize and develop mathematical models of ocean circulation. Advent of satellites allowed us to view the synoptic scale of the ocean at one instant of time. Advances in computing methods and technologies then encouraged us to think about possibilities of developing predictive numerical models in operational settings for societal applications. The rest of the book will take you through some of these ideas in detail.

In this chapter, we focus on the basic laws of nature that govern the physical world, specifically the ocean. We will do this from a perspective of representation of appropriate force and response variables that are required to characterize and quantify the oceanic motion.

DOI: 10.1201/9780429347221-2

This is achieved by first introducing laws of motion presented by Newton, which connect the response variables with the forcing functions. We then consider a fluid element and identify the major forces acting on it. The forces arising from pressure gradient, earth's rotation (Coriolis force), and friction are discussed. The overarching principle of mass conservation is presented, which closes the system of equations with equal number of response variables to solve. There are finally seven governing equations for a fluid element, which connect seven primitive variables.

2.1   INTRODUCTION

A good way to think about the question "how does something work" is to try to understand the way that "something" responds to the external and internal forces acting on it. That "something" could be a country's economy, a small community, a particular human being, an animal, a disease, or a physical system such as the earth, the oceans, the atmosphere, or the climate system. On a smaller scale, that "something" could be bacteria, a virus, or a habitat, a group of fish, a species, or a fluid particle.

So, to understand our oceans, let us consider a small regional piece of the ocean as that "something" or our system. First, we need to define this system. We need to represent this system in terms of a set of variables so that we can understand the "response behavior" of these variables in space and time as it is subjected to a variety of "forces." The response might be characterized by a series of equations connecting the response variables with the forcing functions. The connection needs to describe the variables' evolution in space and time under the influence of a combination of forces that the fluid element is subjected to. The physical concepts that describe the zero-order or fundamental processes that connect the response variables to the forcing functions in a quantitative mathematical framework are, in essence, the story of this book.

In summary, we need to represent two aspects of the system under study – the response variables and the forcing functions. Let us discuss and define these for the oceanic system in this chapter.

2.2   THE RESPONSE VARIABLES

As mentioned in Chapter 1, it is natural to think about the ocean in terms of its two touch-and-feel properties – its temperature (T) and salinity (S). At different depths or (pressure, p) in the water column, the temperature and salinity combination lead to the density (ρ) of the fluid, which determines the layering or stability of the local water with respect to its surrounding water masses. So, these four properties (T, S, p, ρ) can be considered as a set of response variables. Note that these are also connected by the Equation of State 1.1 defined in Section 1.2 on page 4.

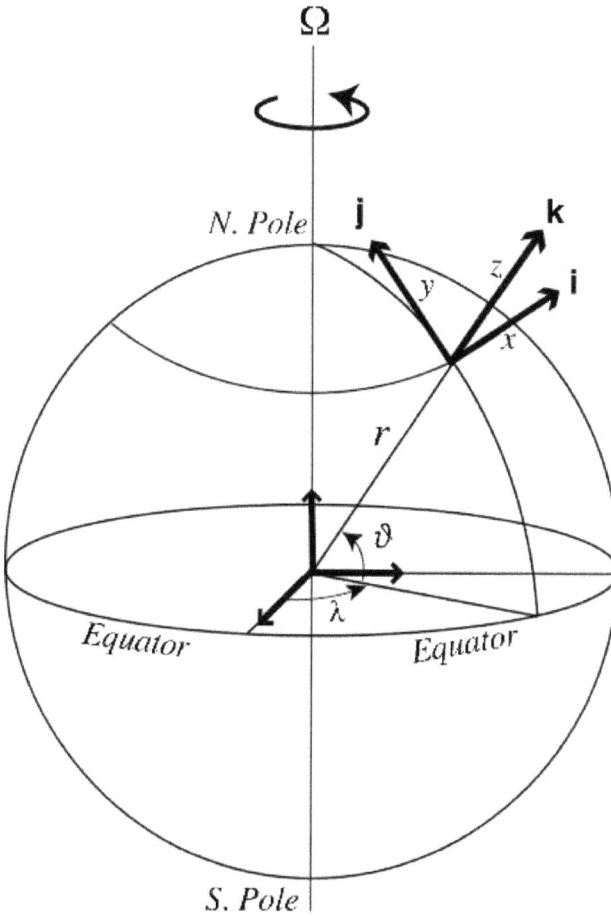

Figure 2.1: Spherical Coordinate system and the rotating frame of reference.

Now, we need to consider a space-time coordinate system that is used to define the field of these response variables (and the forcing functions) and their variation and evolution. As the earth is spherical, it is natural to define a spherical coordinate system such as that shown in Figure 2.1. Assuming the origin at the center of the earth, we can choose the latitude (θ), longitude (ϕ), and radius (r) as the three independent coordinates that can define any position and property and their variation and evolution.

However, we live on the surface of this earth and our ocean and atmosphere is very thin (about 4 km and 10 km, respectively) compared to the radius of the earth (about 6400 km) as discussed in Chapter 1. So, it is convenient for us to think about describing the oceanic and atmospheric motions on flat surfaces with small thickness along the vertical, which is directed towards the center of the earth. To put it simply, these three directions would be East/West, North/South and vertically up/down. Let us choose the Cartesian coordinate system (x, y, z) and time (t) – for a small oceanic region. This is shown in Figure 2.1 on the surface of the spherical earth.

The ocean basins are large in horizontal extent and the earth has the sphericity that makes the horizontal distance between any two fixed longitudes reduce from about 110 km near the equator to zero at the poles. Realizing this, we, however, would like to connect the Cartesian coordinate system with the spherical coordinate system keeping the sphericity (or conversion from latitude/longitude to distance) in mind. So, for a large tropical and mid-latitudinal ocean and atmosphere, the Cartesian coordinate representation on the surface of the earth works just fine. The three directions would then be the zonal (across-longitude, along-latitude) x-direction, meridional (across-latitude, along-longitude) y-direction, and the local vertical (toward the center of the Earth) is the z-direction. Time (t) is the fourth dimension.

So, we can now define the functional forms of the *response* variables in the Cartesian coordinate as follows:

$$T(x, y, z, t); \ S(x, y, z, t), \ p(x, y, z, t) \text{ and } \rho(x, y, z, t).$$

See Figure 2.2 for an example of a typical vertical profile of temperature and it's seasonal evolution.

Clearly, the temperature (T) and salinity (S) of the water define its characteristics or water type and thus called water-mass properties. Together with pressure (p), they (T and S) uniquely define the density (ρ) in a regional ocean and thus, such water-mass characterization helps us identify and distinguish waters from one region to another, as well as its formation region or even the process by which such water-masses have formed. The range of temperature varies from $-4°C$ in the polar regions to about $30°C$ in the equatorial regions. The salinity varies from near zero (very fresh waters) to almost 40 psu (practical salinity unit) in some abyssal regions of the oceans. There are thus different ways of looking at these water-masses – one is by using the T-S diagram, and the other is by looking at their depth profiles. Typical temperature, salinity, and density profile for the Western North Atlantic region are shown in Figure 1.3(a,b,c). Associated T-S diagram for the North Atlantic ocean is shown in Figure 1.3(d). Look at the variation of different properties with depth in the two representations. We will discuss these figures in other parts of this book. Sometimes it is convenient to get the density σ_0 as function of (T, S) only at the surface $(p = 0)$. This σ_0 is generally obtained by using the International Thermodynamic Equation of Sea Water (TEOS-10)(McDougall and Barker, 2011).

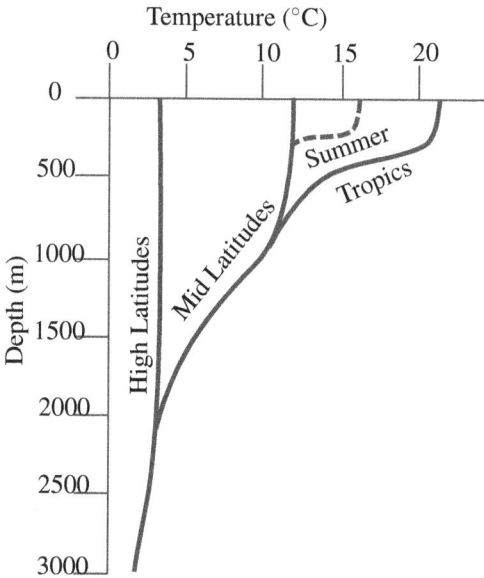

Figure 2.2: Typical temperature profiles in the open ocean. Below a relatively shallow surface layer, the ocean is uniformly cold. After Knauss and Garfield, 2016.

Now that we defined the variables related to the water-masses in the ocean, let us think about its motion – the currents, the eddies, the fronts, the upwelling regions of the coastal ocean or along the equator, the freshwater plumes running into the ocean – how would one characterize their movement? Well, you need to define the motion in three dimensions and in time, just like we did it for the water-mass properties. So, the fluid motion is typically represented by its three-dimensional velocity components at every location of the ocean: u (zonal or x-component), v (meridional or y-component), and w (vertical or z-component). And, mathematically, each one of these response variables is a four-dimensional function of $(x, y, z, \text{and } t)$, i.e., they vary (change and evolve) in the four-dimensional space-time framework and are represented by

$$u(x,y,z,t), v(x,y,z,t), \text{and} w(x,y,z,t).$$

A typical zonal velocity (u) structure of the Gulf Stream is shown in Figure 2.3.

See Figure 2.1. So, together with the water-mass variables, we have now defined a total of seven response variables (T, S, p, ρ, u, v, w) that represents

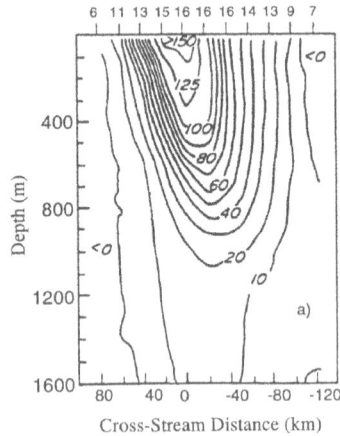

Figure 2.3: A typical velocity structure across a current. Gulf Stream velocity section from Pegasus average synoptic profile. Republished from "Gangopadhyay, A., Robinson A. R., and Arango H. G., 1997: Circulation and Dynamics of the Western North Atlantic, I: Multiscale Feature Models. *Journal of Atmospheric and Oceanic Technology*, 14(6), 1314–1332." American Meteorological Society. Used with permission.

the oceanic system. These are called the "primitive variables," which represent the responses of oceanic circulation.

2.3   THE FORCING FUNCTION

Let us now think about the various forces acting on the oceanic fluid. First, there is the internal pressure of the fluid which acts on any volume (and mass) of fluid element, because the pressure varies with space and there is thus a gradient (difference) of pressure that exerts this force on the fluid itself internally. Then there is frictional forces between fluid surfaces in motion. There is also friction at the bottom of the water column, wind friction on the surface, and lateral friction on the sides. Then, there is gravity which is acting on the fluid.

Then there are thermohaline (temperature and salinity) forcing, or sources and sinks of heat and salt, which will affect the temperature and salinity distributions of the ocean. Think about the seasonal heat input from the sun and loss due to radiation from the ocean surface, penetration of heat in the upper layers, and then slow diffusion into the interior. Think about evaporation, precipitation, and river runoff as major sources and sinks of the salinity distribution.

One of the most important elements of understanding ocean circulation is the recognition that the ocean is part of the Earth (3/4 of it actually on the surface) which is rotating on its own axis once a day. This rotation of the earth translates into an "apparent force" called the Coriolis force acting on the fluid, if the fluid were to be moving in a non-rotating or fixed frame of reference. The non-rotating frame is a typical way of understanding a system's motion which is not affected by the system's rotation. This Coriolis force represents the effect of rotation on an otherwise non-rotating frame formulation. We will discuss this later in Section 2.5.

2.4 NEWTON'S LAWS OF MOTION (CONNECTION BETWEEN FORCE AND RESPONSE)

Now that the forcing and response variables are identified, we need to link them together in a mathematical way so that we can develop an understanding of the evolution of the system (in terms of its response) subjected to such forcing. These linkages between the forcing and the velocity response variables are called the "Equations of Motion" or "Momentum Equations." They describe the "changes" in the system variables in response to the applied forcing. How does the system's velocity change when you apply a force to it?

This is exactly the question that led Newton to develop the "Laws of Motion." It is probably a very common knowledge that Newton's 2nd Law of Motion defines this link between the force and the matter's response. The response is the acceleration of the matter (mass) or the rate of change of velocity with time. Mathematically,

$$F = ma \tag{2.1}$$

where F is the total force acting on the mass "m" and the resulting response is the acceleration "a."

Let us go a bit deep into this equation. Look at the equation. Why is it "ma" on the right-hand side (RHS)? Why is it not $F = ma^2$? Or $F = m^2 a^3$? Or something else? Why is it not $F = mv$, where v is the velocity?

The answer lies in the First Law of Motion! And that is why there is the First Law of Motion. So, what is the First Law of Motion? An object remains at rest if initially at rest; or remains in constant motion (constant velocity) if initially moving at a velocity, unless subjected to an external force. And this last phrase is critical – so, the first law states that an object (system) would move at a constant velocity unless an external force is applied to it. This begs the question: "What happens when the force is applied? Is there a change in velocity? How does the velocity change?" And that leads to the Second Law of Motion.

"*The time-rate of change of velocity is proportional to the applied force.*" And the constant of proportionality is the mass of the body or the system, or how much stuff is in it. The rate of change of velocity with respect to time is called the acceleration, and naturally then $F = ma$ implies that the same amount of force would accelerate a smaller mass much more than it would to a heavier mass. And, for a fixed amount of mass, a larger (weaker) force would result in faster (slower) acceleration.

2.5 RATE OF CHANGE

So, what is this rate of change? How do you quantify it?

Let us consider the functional form of any of the response variables. Let's choose the temperature "T." It is a function of space (x, y, z) and time (t). So, its changes can be represented by its partial changes in the three dimensions and with respect to time (locally). These are $\frac{\partial T}{\partial x}, \frac{\partial T}{\partial y}, \frac{\partial T}{\partial z}$, and $\frac{\partial T}{\partial t}$. These are the individual partial changes along a dimension when the changes in the other dimensions are neglected.

However, the total change $\frac{dT}{dt}$ is a composite for of the local (temporal) derivative and partial spatial changes advected by the three velocity components (u, v, w). This is simply given by,

$$\frac{dT}{dt} = \frac{\partial T}{\partial t} + u\frac{\partial T}{\partial x} + v\frac{\partial T}{\partial y} + w\frac{\partial T}{\partial z} \qquad (2.2)$$

It is important to understand how this comes about. Let us look into this now.

In order to understand the oceanic processes, we need to understand and quantify the changes that occur in these processes. Mathematically, then we need to quantify the changes in density, velocity components, and thermohaline variables. Thus it becomes necessary to derive a relationship between the rate of change of a field (response) variable following the motion and the rate of change at a fixed point. The former is called the substantial or total or material derivative ($\frac{dT}{dt}$). The second one is called the local derivative ($\frac{\partial T}{\partial t}$).

If a water parcel move from a value of T_0 at a location (x_0, y_0, z_0) and at a time t_0 to a new value $T_0 + \delta T$ at a new location $(x_0 + \delta x, y_0 + \delta y, z_0 + \delta z)$ at a new time $(t_0 + \delta t)$, then the change of the property T, or δT can be represented by

$$\delta T = \left(\frac{\partial T}{\partial t}\right)\delta t + \left(\frac{\partial T}{\partial x}\right)\delta x + \left(\frac{\partial T}{\partial y}\right)\delta y + \left(\frac{\partial T}{\partial z}\right)\delta z \qquad (2.3)$$

+ higher order terms (which are neglected). This is known as the Taylor series expansion.

Dividing both sides by δt, we get

$$\frac{\delta T}{\delta t} = \frac{\partial T}{\partial t} + \left(\frac{\partial T}{\partial x}\right)\frac{\delta x}{\delta t} + \left(\frac{\partial T}{\partial y}\right)\frac{\delta y}{\delta t} + \left(\frac{\partial T}{\partial z}\right)\frac{\delta z}{\delta t} \qquad (2.4)$$

Taking the limit for $\delta t \to 0$ and noting that at this limit, $\frac{\delta x}{\delta t} = u$; $\frac{\delta y}{\delta t} = v$; and $\frac{\delta z}{\delta t} = w$.

We get the equation for the total derivative $\frac{dT}{dt}$ as the sum of the local (temporal) derivative and the three advective terms:

$$\frac{dT}{dt} = \frac{\partial T}{\partial t} + u\frac{\partial T}{\partial x} + v\frac{\partial T}{\partial y} + w\frac{\partial T}{\partial z} \tag{2.5}$$

Similarly, we can now write the total change of the three velocity components (accelerations) as

$$\frac{du}{dt} = \frac{\partial u}{\partial t} + u\frac{\partial u}{\partial x} + v\frac{\partial u}{\partial y} + w\frac{\partial u}{\partial z} \tag{2.6}$$

$$\frac{dv}{dt} = \frac{\partial v}{\partial t} + u\frac{\partial v}{\partial x} + v\frac{\partial v}{\partial y} + w\frac{\partial v}{\partial z} \tag{2.7}$$

$$\frac{dw}{dt} = \frac{\partial w}{\partial t} + u\frac{\partial w}{\partial x} + v\frac{\partial w}{\partial y} + w\frac{\partial w}{\partial z} \tag{2.8}$$

The above three are the Total Acceleration of fluid velocity components decomposed in their respective local (temporal) and advective accelerations (last three terms in each equation's RHS).

And for salinity, its total change can be given by,

$$\frac{dS}{dt} = \frac{\partial S}{\partial t} + u\frac{\partial S}{\partial x} + v\frac{\partial S}{\partial y} + w\frac{\partial S}{\partial z} \tag{2.9}$$

and similarly for density,

$$\frac{d\rho}{dt} = \frac{\partial \rho}{\partial t} + u\frac{\partial \rho}{\partial x} + v\frac{\partial \rho}{\partial y} + w\frac{\partial \rho}{\partial z}. \tag{2.10}$$

2.6 FORCES ON A FLUID ELEMENT

Now that we understand and can express the acceleration term or the rate of change of any property such as temperature, salinity, density, or pressure, let us investigate the quantification of the different forces acting on a fluid element. As mentioned in Section 2.2, these include (but not exhaustively): internal pressure-driven force, friction/viscous forces, gravity and the very special 'apparent force' called the Coriolis Force due to the rotation of the earth.

Before we dive into the forces, let us look at the equation $F = ma$ from a slightly different perspective. We can also write this as "$a = F/m$" – or the acceleration = Force per unit mass. So, if we think of a fluid element of unit mass (density times unit volume), then its acceleration (rate of change of velocity) can be expressed simply as the force applied to it!

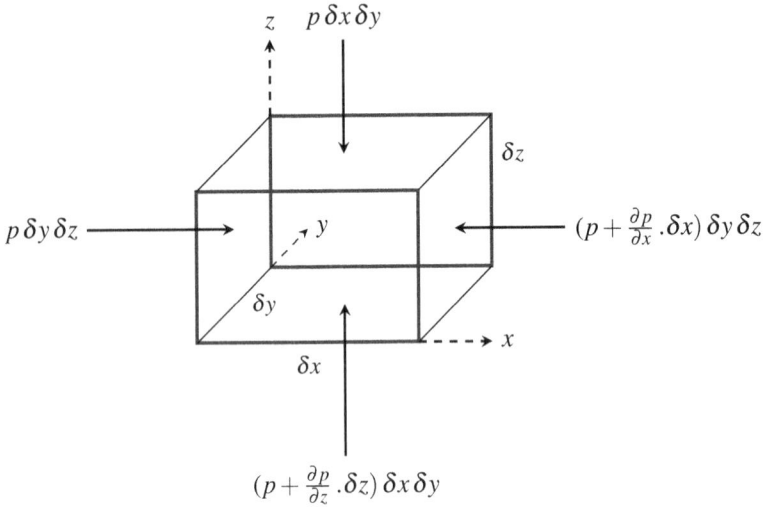

Figure 2.4: Pressure gradient forces on a fluid element.

2.6.1 THE PRESSURE-GRADIENT FORCES AND HYDROSTATIC BALANCE

Now let us consider a fluid element as in Figure 2.4. Its widths are given by $\delta_x, \delta_y,$ and δ_z, along the three different axes. So, we are thinking about the internal pressure, p, which is a function of (x,y,z). How do you express its change in space? Let's see. What would be the change with respect to x? How would you describe it mathematically? Well the rate of change in the x-direction is given by the partial derivative of $p(x,y,z)$ with respect to x. Yes. Its $\frac{\partial p}{\partial x}$. So, the change along x-direction would be simply the x-width or the distance between the two sides of the element along the x-axis times the rate of change in that direction – or $\frac{\partial p}{\partial x}\delta_x$.

So, if the pressure on the near side is $p(x,y,z)$ then the pressure on the far side along the x-axis is simply $p + \frac{\partial p}{\partial x}\delta_x$. Since pressure is always acting inward the forces are acting opposite to one another. The force on the near side is $p\,\delta y\,\delta z$, and the force on the far side is $(-(p + \frac{\partial p}{\partial x}\delta x)\,\delta y\,\delta z)$. The difference between these two forces is the net force in the x-direction due to the pressure gradient (the difference in pressure between the two faces) and now can be written as:

$$F_x = -(p + \frac{\partial p}{\partial x}\,\delta x)\,\delta y\,\delta z + p\,\delta y\,\delta z \qquad (2.11a)$$

$$= -\frac{\partial p}{\partial x}\,\delta x\,\delta y\,\delta z \qquad (2.11b)$$

Similarly,

$$F_y = -\frac{\partial p}{\partial y}\,\delta x\,\delta y\,\delta z \qquad (2.12)$$

$$F_z = -\frac{\partial p}{\partial z}\,\delta x\,\delta y\,\delta z \qquad (2.13)$$

Note that the total volume of the element is $\delta x\,\delta y\,\delta z$, and thus, dividing the pressure-gradient forces by the total mass ($\rho\,\delta x\,\delta y\,\delta z$), we get the net pressure-gradient force per unit mass as: simply $\frac{1}{\rho}\frac{\partial p}{\partial x}, \frac{1}{\rho}\frac{\partial p}{\partial y}$, and $\frac{1}{\rho}\frac{\partial p}{\partial z}$ in the x, y, z directions. These can then be thought to be contributing towards producing accelerations in different directions for the fluid element.

We will discuss how the horizontal pressure-gradient forces are balanced by other external and internal forces later. It is worth thinking about the vertical pressure-gradient ($\frac{1}{\rho}\frac{\partial p}{\partial z}$) here. If there are no other forces in the vertical and the fluid is in static equilibrium, these pressure-gradient forces must be balanced by the gravitational acceleration acting on the fluid element downwards. Thus,

$$\frac{1}{\rho}\frac{\partial p}{\partial z} + g = 0 \qquad (2.14a)$$

$$\frac{1}{\rho}\frac{\partial p}{\partial z} = -g \qquad (2.14b)$$

This is known as the Hydrostatic balance equation.

2.6.2   THE VISCOUS FORCE (SHEAR STRESS)

In a very similar manner, we can think about the viscous forces or shear stresses on the fluid element. So, what is the internal friction within the fluid? It acts at the molecular level, caused by the friction between molecules as they move around continuously within the fluid, and depends on the fluid's internal property called the 'molecular viscosity.' Let us think about how it works.

Consider two imaginary parallel plates within the fluid at a distance l in a flow field which is given by $u(z = 0) = 0$; $u(z = l) = u_0$.

The setup is shown in Figure 2.5. For simplicity, let us assume $u(z)$ varies linearly and increases from 0 to u_0.

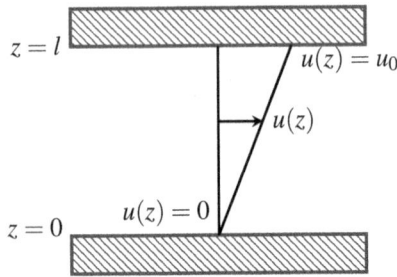

Figure 2.5: Newton's law of friction.

Now think about the forces required to keep the plate moving at a velocity u_0 at $z = l$. This force must be proportional to the velocity (u_0) and the area of the plate (A) and inversely proportional to the distance (l) of the plate from the bottom stationary plate at $z = 0$. Mathematically, this frictional force is then

$$F = \mu A \frac{u_0}{l} \tag{2.15}$$

– This is Newton's law of friction. The constant of proportionality μ is a property of the fluid's internal friction, or called the molecular viscosity, or the coefficient of friction.

In an infinitesimal sense, the force per unit area (F/A), or the shear stress (note the shearing nature of friction) is given by $\tau_{zx} = \mu \frac{\partial u}{\partial z}$. Note the replacement of $\frac{u_0}{l}$ term with its continuous representation of $\frac{\partial u}{\partial z}$.

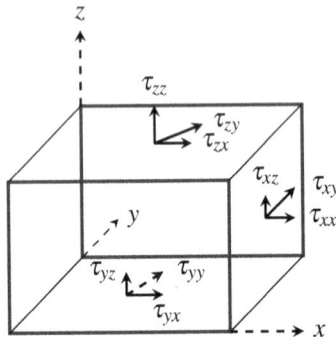

Figure 2.6: Shear stresses on a fluid element.

Now, one can draw similarity to the pressure-gradient force derivation and think about all the faces of a fluid element and construct Figure 2.6.

Note that the shear stresses are denoted by two subscripts of the τ field. It is important to recognize the placement of the directional axes in the subscript.

There are two important considerations to think about how the shear stresses are working. The area of the face on which the shear stress is being applied (represented by the area vector direction normal to the area) and the direction in which the stress is acting. In fact, the first subscript represents the direction to which the area is perpendicular. The second subscript represents the direction of the force itself. Thus, it is easy to see that $(\tau_{zx}, \tau_{zy}, \tau_{zz})$ are the three shear stress components acting on the surface A (which is perpendicular to z-direction) and directed toward x, y, and z directions, respectively. Similar logic is applied to all other faces and these are indicated in Figure 2.6.

Another way of representing the shear stress is what is called "tensors." We can represent all the nine components of the shear stress in a matrix form. This form is easy to remember and application of tensor becomes very useful when working with turbulence and other non-linear dynamics of the fluid flow. Tensor form of shear stress is given by:

$$\begin{bmatrix} \tau_{xx} & \tau_{xy} & \tau_{xz} \\ \tau_{yx} & \tau_{yy} & \tau_{yz} \\ \tau_{zx} & \tau_{zy} & \tau_{zz} \end{bmatrix}$$

The rows represent the surfaces, on which the three components of shear stresses in the three columns (denoted by the second index) are acting.

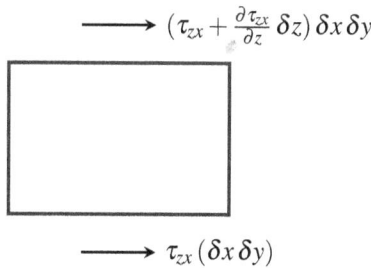

$$\longrightarrow \left(\tau_{zx} + \frac{\partial \tau_{zx}}{\partial z}\, \delta z\right) \delta x\, \delta y$$

$$\longrightarrow \tau_{zx}\, (\delta x\, \delta y)$$

Figure 2.7: Shear stress component on one side of a fluid element.

Next, we need to derive the 'Force per unit mass' for these different shear stresses. Similarity with the pressure-gradient derivation is obvious. Let us do this quickly for the x-component of the shear-stress, which is acting on the (horizontal) area perpendicular to the (vertical) z-direction, or τ_{zx}. The simple gradient of the shear stress (which varies in x, y, z like pressure) is shown in Figure 2.7.

The force per unit mass is thus given by

$$\text{Force/unit mass} = \frac{\frac{\partial \tau_{zx}}{\partial z} \delta z \, \delta x \, \delta y}{\rho \, \delta x \, \delta y \, \delta z} = \frac{1}{\rho} \frac{\partial \tau_{zx}}{\partial z} \qquad (2.16)$$

If we use $\tau_{zx} = \mu \frac{\partial u}{\partial z}$ from Newton's law of friction before, then force/unit mass $= \frac{\mu}{\rho} \frac{\partial^2 u}{\partial z^2} = \nu \frac{\partial^2 u}{\partial z^2}$. The ratio of molecular viscosity to density (ν) is also a fluid property and called the kinematic viscosity. The units of molecular and kinematic viscosity are $kg \, m^{-1} s^{-1}$ and $m^2 s^{-1}$, respectively.

Finally, collecting the terms in the $x, y,$ and z directions for all of the shear-stress driven forces, we get the following three forces:

$$F_x = \nu \left[\frac{\partial^2 u}{\partial x^2} + \frac{\partial^2 u}{\partial y^2} + \frac{\partial^2 u}{\partial z^2} \right] \qquad (2.17)$$

$$F_y = \nu \left[\frac{\partial^2 v}{\partial x^2} + \frac{\partial^2 v}{\partial y^2} + \frac{\partial^2 v}{\partial z^2} \right] \qquad (2.18)$$

$$F_z = \nu \left[\frac{\partial^2 w}{\partial x^2} + \frac{\partial^2 w}{\partial y^2} + \frac{\partial^2 w}{\partial z^2} \right] \qquad (2.19)$$

Note that the ocean is highly *turbulent* and the molecular viscosity numbers do not allow the forces to be compatible with pressure gradient and other forces acting on the oceanic fluid. Thus, an idea of "eddy viscosity" originated, in which the coefficients like kinematic viscosity are replaced by eddy viscosity coefficients. The concepts of turbulence and eddy viscosity play very important roles in understanding and modeling ocean circulation. We will explore some of these challenges later in Chapter 9. The values of these viscosity coefficients are generally found from sea-going, lab, or numerical experiments and are an area of active research. Not one number fits all the oceans or their simulations or different coastal regions and deep oceans.

2.6.3   THE CORIOLIS FORCE

The rotation of the earth is a driving force for all fluid things on the earth's surface like the oceanic water-masses and the atmospheric air-masses. The time-scale of this effect has to be comparable to the rotational period of 1 day (called the inertial period). So, any phenomenon that spans or evolve, over a period of 24 hours or more would significantly feel the rotation of the earth. In fact, any phenomenon from 3 hours to 24 hours would also feel the effect of the earth's rotation. Tides are great examples of such super-inertial motions.

Well, that is somewhat easy to understand. But, how does the earth's rotational effect transition into the Force-response framework? Newton's second law is applicable for a fixed reference system, not for a rotating reference system! So, the observed fluid flow on the rotating earth has to account for the earth's rotational effect in some way. This is where the idea of an "apparent force" due to earth's rotation was put forward by French mathematician, engineer, and scientist, Gaspard-Gustave Coriolis (1792–1843). Let us follow his thinking in a heuristic way. We spare the rigorous mathematical derivations of linking a rotating frame to a non-rotating frame to other advanced books in this field.

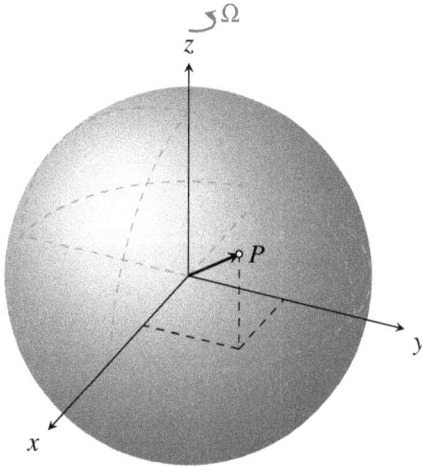

Figure 2.8: The rotating coordinate system and the Coriolis force.

There are two things to consider here for the rotational effect of the earth on a moving oceanic fluid element. The first is what happens when the fluid element is not moving or is stationary on a point on the rotating frame. It will then feel only the effect of rotation on it as a centrifugal force (much like what a bicyclist feels on a circus rim or when you are on a merry-go-round or a joyride wheel). The second one is a bit more complicated. The rotation of the earth affects any moving object due to the object's motion, or in other words deflects the moving object from its original path. Simply put, if the earth was not rotating, you would go from point A to point B (due north) in a straight line with a velocity v. As the earth is rotating as you are moving, you will be deflected away from point B and reach point B'! This deflection can then be thought of as being caused by an "apparent force" acting on the moving object! If we can quantify this deflection of a moving object in terms of a "rate of change of velocity" – we can then include that expression in our Equations as a "force per unit mass" or the Coriolis acceleration (Coriolis force per unit mass).

Let us look at Figure 2.8. The earth's rotation is given by its angular frequency Ω. Assume that the particle is at a latitude ϕ. The original of earth is at O and the particle is at A. The local x-y plane or the regional ocean is perpendicular to the local vertical (along OA). The earth's angular velocity Ω

can now be decomposed in two orthogonal components on the fluid element: one along the horizontal plane ($\Omega\cos\phi$) and the other along the local vertical ($\Omega\sin\phi$). This latter component is the one that provides the rotational effect on the fluid on the horizontal plane moving with velocity (u,v), which are the zonal and meridional velocity components.

Any particle on the earth's surface at a latitude ϕ would be affected by the earth's rotation differently, which is given by $\Omega\sin\phi$. At the equator, when $\phi = 0$, the particle feels no rotational effect ($\sin\phi = 0$); at the poles, the effect of rotation is the highest ($\sin\phi = 1$) and is equal to the earth's rotational rate of 1 day^{-1}.

Figure 2.9: Effect of Rotation on a moving particle on Earth.

Now, let us look at Figure 2.9 for the moving particle on the earth's surface going from A to B with a speed u toward East. The distance between A and B is given by ut where t is the time taken to reach B. Now while the particle is moving from A to B in time t, the earth has rotated by an angle ($\Omega\sin\phi \times t$) as shown in the figure. Naturally, the particle will reach point B$'$ and the arc displacement BB$'$ is simply given by ($ut\ \Omega\sin\phi\ t$). Mathematically, the displacement s can be expressed as $ut^2\Omega\sin\phi$. And the acceleration is given by the second derivative of s with respect to time t, or d^2s/dt^2 or ($2\Omega\sin\phi\ u$).

Clearly, the displacement or the acceleration is perpendicular to the original direction of the particle's movement. In the northern hemisphere, ϕ is positive, and thus the displacement is to the right of the original direction of the particle. Similarly in the southern hemisphere, the force acts to the left of the motion, because ϕ is negative.

So, the rotation has an impact on the velocity field to create an acceleration in a direction normal to its motion (velocity). And that effect of the earth's rotation depends on the latitude at which you are. So, it's not a simple function of just the rotational frequency of the earth. Depending on where you are on the earth's surface (latitude-wise), you will feel the rotational effect differently. On the equator, you won't feel it; at the Poles, you will feel it the most. This local rotation is called the Coriolis Frequency or f, which is equal to $2\Omega\sin\phi$. The value of f varies from -2Ω at the South pole to zero on the equator to $+2\Omega$ at the North Pole.

Since the rotation of the earth, Ω is along the earth's axis and the local vertical on the surface of the earth (where our Cartesian plane is located) is at the latitude ϕ, the Coriolis force has three different components in the three directions (x,y,z). Let us consider this on a plane tangential to the earth's surface (at latitude (ϕ)), perpendicular to the local vertical (Z). See Figure 2.10. The earth's rotation Ω is always along the axis of the earth and can be projected along the local vertical (Z) as $\Omega \sin \phi$ and along the meridional direction (y) as $\Omega \cos \phi$. Realize that the angle between the local vertical and the vector Ω at latitude ϕ is simply $(\pi/2 - \phi)$. So, the local rotational vector due to earth's rotation at any point P on the earth surface is given by $(0, \Omega \cos \phi, \Omega \sin \phi)$. Note the effect of earth's rotation along the local vertical $(\Omega \sin \phi)$ reduces to zero on the equator and is maximum $(= \pm \Omega)$ at the poles.

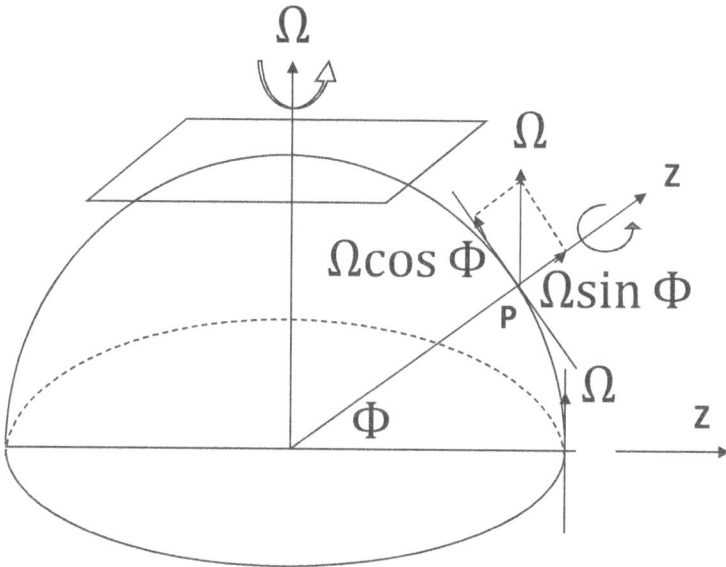

Figure 2.10: Variation of Coriolis deflection with latitude. The tangential plane at any point P on the earth's surface experiences different magnitudes of rotation which depends on its latitude.

In vector notations, the Coriolis acceleration can then be easily represented by the cross-product as follows:

$$2\vec{\Omega} \times \vec{R} = 2 \begin{bmatrix} \mathbf{i} & \mathbf{j} & \mathbf{k} \\ 0 & \Omega \cos \phi & \Omega \sin \phi \\ u & v & w \end{bmatrix} \qquad (2.20)$$

which yields the three components as follows:
$(2w\Omega \cos \phi - 2v\Omega \sin \phi)$ along the x-direction;
$(2u\Omega \sin \phi - 0)$ along the y-direction;
$(0 - 2u\Omega \cos \phi)$ along the z-direction;

Using the Coriolis frequency, $f = 2\Omega \sin\phi$, the Coriolis force per unit mass in the x and y-direction are $-fv$ and $+fu$, respectively. Note that since the vertical velocity w is small compared to the horizontal velocities, we have neglected the first term in the x-component above. The Coriolis force per unit mass in the vertical direction is generally absorbed within the acceleration due to gravity acting on the parcel.

Note that in the Northern hemisphere (where $f > 0$), the eastward velocity component (u) results in a southward Coriolis acceleration of ($-fu$), and the northward velocity component (v) results in an eastward ($+fv$) Coriolis acceleration. The opposite is true for the Southern hemisphere (where $f < 0$). For more details on the Coriolis force and its derivation, please see the books by (J. Holton, 2016; Knauss & Garfield, 2016; Pond & Pickard, 1983).

2.7 THE CONSERVATION EQUATIONS

Now that we have discussed the force-response framework, it is time to quantify the system itself. We generally think of the system as a conservative system that preserves its mass, heat content, and salt content. For a control volume of flow field, this means that whatever mass comes into a volume, should exit the volume; similarly, the heat (or salt) input should balance the output in a way to redistribute the temperature and salinity of the volume. Let us look at the mass, heat, and salt conservation for a fluid element.

2.7.1 THE CONSERVATION OF MASS

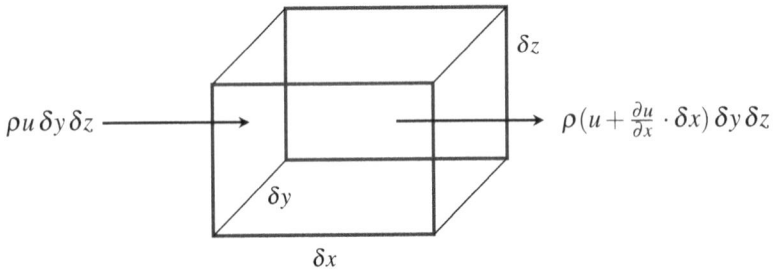

Figure 2.11: Conservation of mass, heat, or salt.

Consider the control volume as in Figure 2.11. The x-directional velocity changes from u on one side (A) to $u + \frac{\partial u}{\partial x}\delta x$ on the other side (B). So, the inward flow rate (mass per unit time) is given by $\rho u\,\delta y\,\delta z$, where $\delta y\,\delta z$ is the area of the face A or B. Similarly, the outward flow rate through side B is given by $\rho(u + \frac{\partial u}{\partial x}\delta x)\,\delta y\,\delta z$. Thus, the net flow out of this volume in the x-direction is given by $\rho \frac{\partial u}{\partial x} \delta x \delta y \delta z$. Similarly, the net flow out of this volume in the y and z directions are times $\rho \frac{\partial v}{\partial y} \delta x \delta y \delta z$, and $\rho \frac{\partial w}{\partial z} \delta x \delta y \delta z$, respectively.

Thus if the system conserves mass, the net flow out of the system is given by,

$$(\rho)(\frac{\partial u}{\partial x} + \frac{\partial v}{\partial y} + \frac{\partial w}{\partial z}) = 0 \tag{2.21}$$

or,

$$\nabla \cdot \vec{V} = 0 \tag{2.22}$$

where $\vec{V} = iu + jv + kw$, in vector notation with (i, j, k) as the unit vectors in the three (x, y, z) directions.

This is also called the mass conservation equation or equation of continuity or the divergence equation. Why divergence? It is easy to see that by rewriting the equation of continuity as follows.

$$\frac{\partial u}{\partial x} + \frac{\partial v}{\partial y} = -\frac{\partial w}{\partial z} \tag{2.23}$$

So, if the left-hand side (LHS) is greater than zero (positive horizontal divergence) then $\frac{\partial w}{\partial z} < 0$ – or w increases upward – meaning upwelling – thus the flow at the surface will diverge from each other. In the opposite scenario, if $\frac{\partial u}{\partial x} + \frac{\partial v}{\partial y} < 0$ (negative horizontal divergence or convergence) then $\frac{\partial w}{\partial z} > 0$, or w increases downward – meaning downwelling – thus the flow at the surface will converge together to result in a downwelling situation.

Note that if the density is considered to be varying within the control volume, then

$$\nabla(\rho u) = 0 \tag{2.24}$$

would be the continuity equation, which leads to a density conservation and a divergence equation separately.

2.7.2 CONSERVATION OF HEAT AND SALT

Simply put, the total rate of change of temperature and salt should be zero to conserve heat and salt. However, similar to Newton's second law, one would assume that the total change of temperature would balance the heat input per unit mass and the total change in salt would balance the salt input per unit mass.

Mathematically, $\frac{dT}{dt} = 0$ and $\frac{dS}{dt} = 0$ in the absence of any heat or salt input. $\frac{dT}{dt}$ and $\frac{dS}{dt}$ are the total derivatives.

If the heat input is Q, then

$$\frac{dT}{dt} = Q_{source} + Q_{sink} \tag{2.25}$$

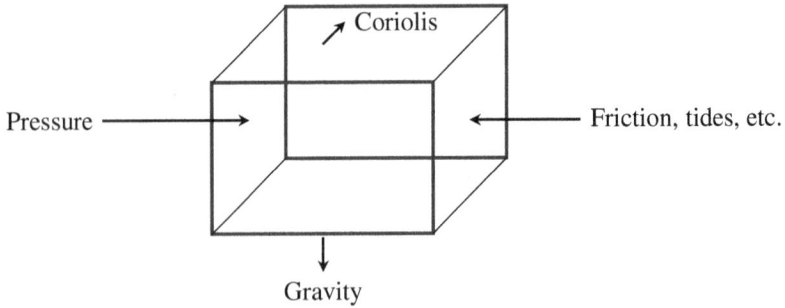

Figure 2.12: Forces on a fluid element.

And If there is salt input/output, in terms of evaporation (E), precipitation (P), and river runoff (R), then

$$\frac{dS}{dt} = E - P + R \qquad (2.26)$$

2.8 THE GOVERNING EQUATIONS FOR THE OCEANIC SYSTEM

Now that the response variables and the forcing functions are described on the fluid element, we can use Newton's second law of motion ($F = ma$) to connect them. This is usually done by rewriting the second law as

$$a = F/m \qquad (2.27)$$

for each component of the velocity (u, v, w).

The acceleration a is the total change of velocity component as described in Section 2.5. The different forces (pressure, gravity, frictional forces, Coriolis) acting on the fluid element were derived in Section 2.6 and is summarized in Figure 2.12. Note that there are other frictional forces such as lateral friction, bottom friction, winds forcing, internal diffusion, tides, and waves, which can be derived and added to the RHS of the $a = F/m$ construction.

Putting it all together, let us collect equations 2.11 through 2.26 appropriately within the various acceleration terms on the right side of equation 2.27.

The three momentum/acceleration equations:

$$\frac{du}{dt} = \frac{\partial u}{\partial t} + u\frac{\partial u}{\partial x} + v\frac{\partial u}{\partial y} + w\frac{\partial u}{\partial z} \qquad = Pr + Cor + Fric + Others \qquad (2.28)$$

$$\frac{dv}{dt} = \frac{\partial v}{\partial t} + u\frac{\partial v}{\partial x} + v\frac{\partial v}{\partial y} + w\frac{\partial v}{\partial z} \qquad = Pr + Cor + Fric + Others \qquad (2.29)$$

$$\frac{dw}{dt} = \frac{\partial w}{\partial t} + u\frac{\partial w}{\partial x} + v\frac{\partial w}{\partial y} + w\frac{\partial w}{\partial z} \qquad = Pr + Cor + Fric + Gravity + Others$$

$$(2.30)$$

The *Others* include forces such as astronomical tides, earthquakes, waves, and similar forces unaccounted by the pressure, Coriolis, and friction. For motions where rotation, pressure, and friction are important (for longer-than periods of earth's rotation), the contribution for other forces are generally small and are not considered for such motions.

Using the derivations in section 2.7, the two tracer equations for heat and salt become

$$\frac{dT}{dt} = \frac{\partial T}{\partial t} + u\frac{\partial T}{\partial x} + v\frac{\partial T}{\partial y} + w\frac{\partial T}{\partial z} \qquad = Q_{source} + Q_{sink} \qquad (2.31)$$

$$\frac{dS}{dt} = \frac{\partial S}{\partial t} + u\frac{\partial S}{\partial x} + v\frac{\partial S}{\partial y} + w\frac{\partial S}{\partial z} \qquad = -E + P + R \qquad (2.32)$$

and the continuity equation

$$\nabla \cdot (\rho \vec{V}) = 0 \qquad (2.33)$$

and the equation of state relating the density to the temperature and salt

$$\rho = \rho(S, T, p) \qquad (2.34)$$

comprise the system connecting the seven primitive variables u, v, w, T, S, ρ, and p.

These equations represent the force-response system as well as the thermohaline and mass conservation of the basic oceanic element. We will look at some of these balances in the next few chapters before getting into modeling the ocean on the basis of these seven primitive equations.

CONCLUSION

In this chapter, we learned about the basic laws that govern oceanic motion starting from Newton's laws, now on a rotating frame. A total of seven response variables were identified – the three velocity components (u, v, w), the pressure (p) and density (ρ), and the two scalar variables temperature (T) and salinity (S). Newton's law connects the force per unit mass with the acceleration or the rate of change of velocity. We discussed this idea of rate of change in detail before diving into the details of how to think about the forces acting on the fluid element. We then considered a fluid element and identified the major forces acting on it. The forces arising from pressure gradient, viscous forces (internal friction) are similar. The Coriolis force is an apparent force due to the earth's rotation and acts on a moving particle on the rotating frame. Gravity is responsible for the hydrostatic balance of a stable water column in the vertical. Adding them up in three different directions led to the three equations of motion. The overarching principle of mass conservation is then presented along with the conservation of temperature and salt. Together with the non-linear equation of state these provide four additional equations to close the force-response framework. There are finally seven governing equations (three momentum equations, conservation of mass, temperature and salt, and the equation of state) for a fluid element, which connect seven primitive variables (u, v, w, p, ρ, T, S).

FURTHER READING

Cushman-Roisin, B., & Beckers, J. (2011). *Introduction to geophysical fluid dynamics: Physical and numerical aspects*. Elsevier.

Gill, A. E. (1982). *Atmosphere-ocean dynamics (international geophysics series)*. Academic Press.

Holton, J. R. (1973). An introduction to dynamic meteorology. *American Journal of Physics*, *41*(5), 752–754.

Knauss, J. A., & Garfield, N. (2016). *Introduction to physical oceanography*. Waveland Press.

Pond, S., & Pickard, G. L. (1983). *Introductory dynamical oceanography*. Gulf Professional Publishing.

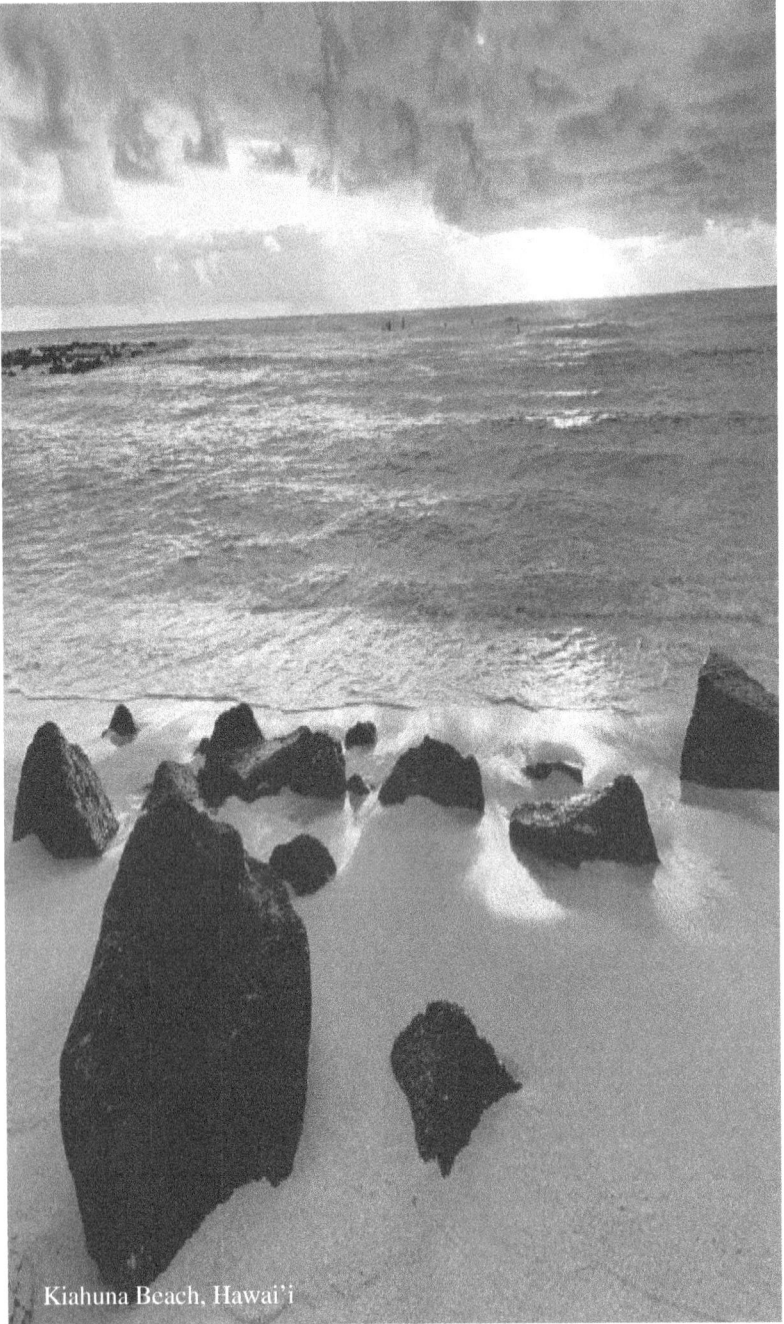
Kiahuna Beach, Hawai'i

3 Geostrophic Equilibrium

A secret turning in us
makes the universe turn.
Head unaware of feet,
and feet head. Neither cares.
They keep turning.

Rumi (1207–1273) (The Secret
Turning) Compiled by Coleman
Barks

OVERVIEW

In the last chapter, we learned how Newton's laws of motion could be applied to connect the forces on a fluid element to the response variables. We discussed how the governing equations are derived, with acceleration being on the left-side of the equation and the various forces as force per unit mass on the right-side. Then we showed how the different types of forces (pressure gradient, Coriolis, gravity, viscous, friction, shear) are acting along and across the six surfaces of the fluid element. Summing them up in three different directions led to the three equations of motion. Additional four equations (two tracer conservation equations, mass conservation, and density equations) resulted in a system of seven equations for seven variables (u, v, w, T, S, p, ρ).

In this chapter, we will focus on two of these forces, pressure-gradient, and the Coriolis force, and assume steady-state (no acceleration or time-dependence on the left-side of the governing equations). This balance between the pressure gradient and Coriolis due to Earth's rotation leads to what is called the *geostrophic equilibrium*. This idea of geostrophy is very powerful in oceanographic and atmospheric circulation. It allows one, to a very good approximation, to infer surface velocities along a front or around an eddy and also estimate a velocity section from temperature and salinity profiles. We will see some large-scale application of the geostrophic balance for some well-known natural phenomenon near the end of this chapter.

We will begin with the geostrophic equations and develop understanding of a key variable called the specific volume anomaly. This helps us link the density and velocity of a water column using the hydrostatic balance. Then we will discuss the application of the geostrophic balance along two vertical profiles of temperature and salinity to derive a velocity profile of flow across the

DOI: 10.1201/9780429347221-3

section, given a level of no (or known) motion. This would lead to a first-order understanding of large-scale balances of pressure gradient and rotation for the oceanic gyres and of boundary currents. We will also look into two types of simple balances known as the inertial motion and thermal wind balance.

3.1   INTRODUCTION

One of the most important issues in understanding the earth's balance is to realize that many aspects of this system are in a so-called rotational balance of the ocean and atmosphere. In other words, the ocean and the atmosphere are kind of in an equilibrium on the rotational frame of the earth, which is rotating around its own axis with the rotational rate (angular frequency) of once a day. Mathematically, it is $\omega = 1$ cycle/day $= 2\pi\,\mathrm{rad/day} = 1/864,000\ Sec^{-1}$.

This rotational equilibrium has a scientific name: "geostrophy." What is the meaning of geostrophy? The word came from two Greek words: "Ge," meaning the earth; and "strophe" meaning "turned." So, the word "geostrophy" effectively means "earth turned" or "Rotating Earth."

In the equation $F = ma$, when $F = 0$ (no external force), then $a = 0$. This tells us that the velocity would be constant if and when all forces acting on the body sums up to zero (a balanced state). This velocity is either zero (state of rest) or a non-zero constant (a state of constant motion). This was Newton's first law, remember? But, that was for a fixed reference frame.

For a rotating frame, the rotation (or the angular velocity) makes any parcel on the rotating frame with non-zero velocity feel the Coriolis acceleration in the normal direction of its velocity, as was shown in Section 2.4. This Coriolis acceleration is often grouped with other terms on the right-hand side of the equation ($a = F/m$). The left-hand side can then represent the acceleration of the parcel on the rotating frame itself.

Remember that in the Northern hemisphere, the eastward velocity component (u) results in a southward Coriolis acceleration of ($-fu$), and the northward velocity component (v) results in an eastward ($+fv$) Coriolis acceleration.

Thus the equation of motion with the pressure gradient force (neglecting friction) becomes

$$\frac{du}{dt} = -\alpha\frac{\partial p}{\partial x} + fv + \cancel{Friction} + \cancel{} \tag{3.1}$$

$$\frac{dv}{dt} = -\alpha\frac{\partial p}{\partial y} - fu + \cancel{Friction} + \cancel{} \tag{3.2}$$

Strictly speaking, there are other components in the three-dimensional velocity fields coming from the z-component of the velocity and other components from u and v with the quotient $\cos\phi$. Many of these terms are an order of magnitude smaller and thereby neglected in this balance.

Furthermore, the balance equation in the z-direction is given by:

$$\frac{dw}{dt} = -\alpha \frac{\partial p}{\partial z} + 2\omega \cos \phi u - g + \cancel{Friction} \tag{3.3}$$

setting $\frac{dw}{dt} = 0$ and neglecting the small Coriolis acceleration. yields the so-called Hydrostatic balance:

$$0 = -\alpha \frac{\partial p}{\partial z} - g \tag{3.4}$$

on integration from a depth H to the surface,

$$dp = \int -\rho g dz \tag{3.5}$$

This is the vertical balance. This basically means that buoyancy is "holding" gravity – the deeper you go, the denser it is and the pressure increases with depth, which then balances the dense fluid to stay below. The hydrostatic balance assumption fails when there is vertical acceleration ($\frac{\partial w}{\partial t} \neq 0$) or convection.

3.2 THE GEOSTROPHIC BALANCE

Let us now consider the case when the Coriolis force is balanced by the horizontal pressure gradient force in a steady state (no change with respect to time). The fluid feels an apparent force due to its motion in a rotating frame perpendicular to its direction of motion. In the absence of any other force and in steady state, the pressure gradient force has to balance this apparent force. In this situation, the balance is called the geostrophic balance and is simply given by the following two horizontal equations of motion.

$$0 = -\alpha \frac{\partial p}{\partial x} + fv \tag{3.6}$$

$$0 = -\alpha \frac{\partial p}{\partial y} - fu \tag{3.7}$$

Now consider a sea surface slope to see how the balance between the pressure gradient and the Coriolis force works. See Figure 3.1. The pressure gradient force ($\alpha \frac{\partial p}{\partial n}$) acts normal to the surface, as shown for the left node in Figure 3.1. The gravity acts vertically down at that point. If we decompose the pressure gradient in its two orthogonal components in the vertical and horizontal directions as shown on the right node, gravity clearly balances the upward component of the pressure gradient force. This is simply given by the equation:

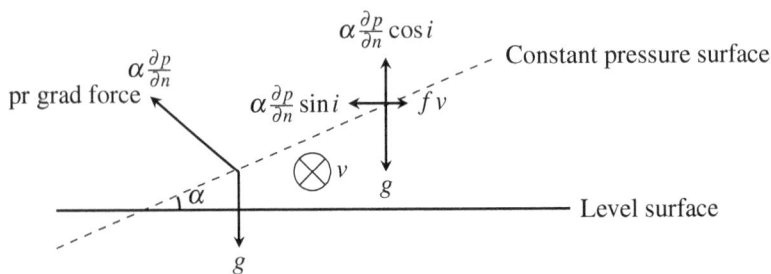

Figure 3.1: Balance of pressure gradient, Coriolis and gravity on a surface across a flow.

$$\alpha \frac{\partial p}{\partial n} \cos i = g \tag{3.8}$$

where i is the sea surface slope. Note that $\frac{\partial p}{\partial x} = \frac{\partial p}{\partial n} \sin i$ and $\frac{\partial p}{\partial y} = \frac{\partial p}{\partial n} \cos i$.

But this horizontal term $\alpha \frac{\partial p}{\partial n} \sin i$ needs to be balanced by something. How about the Coriolis? Otherwise, you would not have a slope of the sea surface anymore and move toward a state where the surface is flat. It turns out that the Coriolis forces acting up or down the slope balance the pressure gradient along the slope. The velocity of the parcel is through the plane of the slope or normal to the slope in the non-vertical direction.

The Coriolis force is to the right of the flow. The flow is directed into the paper in the above picture. See the right node in the above figure for the balance in the horizontal direction. This is the balance between the pressure gradient and the Coriolis force. The pressure gradient component in the x-direction $\alpha \frac{\partial p}{\partial n} \sin i$ is balanced by Coriolis force (fv) resulting from the y-directional flow (v).

For the horizontal balance at the node, we can now write the following equation:

$$fv = \alpha \frac{\partial p}{\partial n} \sin i \tag{3.9}$$

$$= \alpha \frac{\partial p}{\partial n} \cos i \tan i \tag{3.10}$$

$$= g \tan i \tag{3.11}$$

We have used the vertical balance with gravity to arrive at the final equation. Similarly, the pressure gradient in the y-direction $\frac{\partial p}{\partial y}$ is balanced by Coriolis force ($-fu$) resulting from the x-directional flow (u).

Let us look at a few examples of how simple this equation is to use for realistic flows such as the Gulf Stream, the Kuroshio, the Brazil Current, and the springtime WBC in the Bay of Bengal. Satellite altimetric observations across the Gulf Stream (width = 100 km) indicate a sea surface height difference of about 160 ± 35 cm. So, at 45°N latitude, the above equation becomes: $f = 10^{-4}s^{-1}$. $fv = g \tan i$ yields, $v = 1$ $ms-1$. The slope $\tan i$ is nothing but the ratio of the sea height difference between the two sides of the current and its width.

For the Brazil Current, in the southern hemisphere, at 25°S, $f = -0.615 \times 10^{-4}s^{-1}$, with a height difference of 50 cm over a 100 km width, $v = -81 \, cm \, s^{-1}$.

For the springtime WBC in the Bay of Bengal at 18°N, $f = 0.4 \times 10^{-4}s^{-1}$, with a height difference of 15 cm over this narrow current of 50 km width, $v = 75 \, cm \, s^{-1}$. These are typical examples of how this simple idea can be used to quantify and monitor different current systems in the world. The satellites carrying altimeters can provide the sea surface height fields across a number of high-speed currents in real time and can allow us to use Geostrophic Balance to infer surface velocities. Of course, success of such applications depends on time scale, the quality of the data, and location of the current with respect to the satellite track, and other factors.

3.3 DYNAMIC HEIGHT AND GEOSTROPHIC VELOCITY

The geostrophic equation lets you quantify the motion of the fluid (air or water) parcel on a rotating earth to the first degree considering its rotation. Yes, if you can measure the temperature and salinity of the water masses at a point (x, y) at various depths (z), then you can obtain the density profile using the equation of state. Assuming hydrostatic balance (equation 3.4), one can then relate the density to the surface pressure. So, if you have two nearby vertical profiles of temperature and salinity, you can then quantify the pressure gradient between these two stations. And, that pressure gradient force (which is normal to the plane occupied by the two profiles) should be balanced by the Coriolis force acting along that plane due to the fluid motion normal to the plane. Using the above equations one can quantify "v."

3.3.1 A SIMPLE CONSTRUCT OF THE GEOSTROPHIC VELOCITIES

The geostrophic balance can be used to construct a two-dimensional velocity section in the $x - z$ or $y - z$ plane, which is very useful to numerical model initialization. Let us look at a simple construct here, which will then be expanded later in the modeling Chapters 8 and 10.

The beauty of the geostrophic balance is that it connects the velocity in one direction to the pressure gradient across the flow (or the differential pressure) on the plane perpendicular (normal) the direction of flow. For an oceanic front, eddy or gyre, the along-stream flow is supported by the cross-stream pressure gradient. The along-steam flow is accompanied by the Coriolis force (normal to the flow direction) on a rotating earth; the Coriolis force is balanced by the pressure gradient force, being directed to this normal direction opposing the Coriolis force – keeping the flow in a balanced condition.

Now, in a three-dimensional ocean, velocity and pressure both vary with depth (z), so it is useful to see if we can use the geostrophic relationship (which is a horizontal balance) to describe the vertical variation of horizontal velocity or to deduce the horizontal velocity profile.

3.3.2 SPECIFIC VOLUME ANOMALY

Note that the pressure gradient term $\alpha\frac{\partial p}{\partial x}$ is primarily dependent upon density distribution ($\alpha = \frac{1}{\rho}$) and the pressure at any depth is related to the density as $dp = \rho g dz$, via the hydrostatic balance. Note that the density itself is again a function of the pressure at a depth. So, the equation of state is non-linear and the hydrostatic balance is a piece-wise (depth-dependent) balance of gravity and pressure-gradient in the z-direction.

At this time, it is important to think about the similarities and differences of the dependence of density of air and water on the mass constituents or the fundamental properties of the respective fluids. For air, $\rho = \rho(T, p)$, where T is absolute temperature. The inversion of density or the specific volume ($\alpha = \frac{1}{\rho}$), which has the unit of m^3/kg can be expressed by $\alpha = \alpha(T, p)$. For a perfect gas, $\alpha = RT/p$ (Note Boyle's Law (See Holton, 1973), $pV = nRT$, which leads to $V/n = RT/p$; p and V are the pressure and volume of n units of gas at temperature T; so, V/n becomes the specific volume), where R is the gas constant.

For seawater, the temperature is generally in Celcius, and $\alpha = \alpha(S, T, p)$ is generally expressed as a polynomial. The specific volume anomaly is generally written as

$$\alpha(S, T, p) = \alpha(35, 0, p) + \delta \qquad (3.12)$$

where $\alpha(35, 0, p)$ is the specific volume anomaly at $S = 35$ psu, $T = 0°C$ and at a pressure p at the depth of the parcel. The term δ is the part of the specific volume anomaly which accounts for additional contributions to the density due to variations of the specific volume (density) on temperature, salinity, and pressure and their combined effects from $S = 35$, $T = 0$, and surface pressure.

3.3.3 VELOCITY ACROSS A SECTION

Let us now consider two stations, A and B, at a distance L from each other, as in Figure 3.2. The surface pressure is constant along line AB with a value of "p."

The difference in Geostrophic balances between the two stations yields,

$$2\Omega \sin\phi \, (V_1 - V_2) = g \, (\tan i_1 - \tan i_2) \qquad (3.13)$$

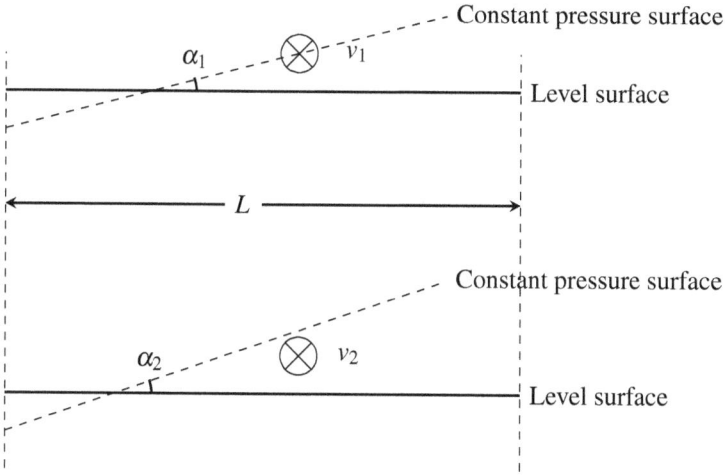

Figure 3.2: Using geostrophy to get velocities with depth.

Note that $\tan i_1$ and $\tan i_2$ can be expressed as the ratio of depth differences (in terms of 'z's) between the constant pressure surfaces (Φ_1 and Φ_2) and the distance between two stations, L. Algebraic manipulations (see Pond and Pickard, 1983, Chapter 8 for details) lead to

$$(V_1 - V_2) = \frac{1}{L\,2\Omega\sin\phi}\left[\int_{p_1}^{p_2}\delta_B dp - \int_{p_1}^{p_2}\delta_A dp\right] \qquad (3.14)$$

The last two terms in the RHS are the integrals of the specific volume anomaly at the two stations B and A over depth ranges between the constant pressure surfaces. These terms are called "dynamic height" at these stations. Mathematically, we can now define the dynamic heights as:

$$\Delta\Phi_B = \int_{p_1}^{p_2}\delta_B dp \qquad (3.15)$$

$$\Delta\Phi_A = \int_{p_1}^{p_2}\delta_A dp \qquad (3.16)$$

The dynamic height has the unit of $m^2 s^{-2}$, similar to that of "gz." As the value of g is 9.8 ms^{-2}, we can approximately set $g = 10 ms^{-2}$, and this is often used to convert the dynamic height in a unit called "dynamic meter."

3.3.4 ABSOLUTE VELOCITIES

Equation 3.14 expresses the differences between the velocities at two levels – but how do you get the actual values of these velocities at each level? Unfortunately, this is the limitation of the geostrophic balance. Well, one way to get the absolute velocities at different depths is to assume a zero-velocity layer or surface at some depth – called the level of no motion (LNM) and then integrate the specific volume anomaly from that depth upward to get velocity structure. This is often a very good first approximation to initialize numerical models for real-time forecasting. Often times, you collect CTD (conductivity- temperature-depth) observations at two stations by using CTD probes. These are typical instruments in a cage or in a small casing (even handheld units are available now) to be lowered from the side of a boat. If you have these instruments to provide the temperature and salinity profiles – $T(z)$ and $S(z)$, then you can get the density profile $\rho(z)$ and subsequently get the specific volume anomaly at all depths and use the above technique to get the geostrophic velocity at both stations with respect to an LNM, which will be very close to absolute velocity, if the assumption of LNM holds. An extraordinary book on how the ocean works and what are the limitations to constructing the absolute velocity profiles in the real ocean is lamented by Henry Stommel to the Chief Engineer of a cruise ship is worth a good read for anyone who wants to be more curious of the ocean (Stommel, 1987).

Some recent works allow for using a *level of known motion* (as opposed to the level of no motion) from ADCP or current meters at a depth and then use the same idea of integrating the specific volume anomaly to get the absolute velocity profile. While such methodologies work for a vast majority of flows (because they are geostrophic, to a large degree), many flows in the coastal oceans and equatorial regions happen in layers and there are multiple LNMs, with counter-currents and energetic eddy fields; and one has to augment the geostrophic flow field solutions with consideration of ageostrophic (other than geostrophic) and time-dependent flow solutions. Some of these considerations will be discussed later in Section 6.

In summary, geostrophic velocity sections are very useful to a first approximation and can be interpreted as a representation of the velocity shear in the vertical being balanced by the pressure gradient in the cross-stream direction. The balance is between the Coriolis force and the pressure gradient force.

3.4   LARGE-SCALE GEOSTROPHIC BALANCE

It is instructive to think about a few known phenomena in our weather system (atmosphere) to understand the effect of the earth's rotation and the resulting geostrophic balance in the atmospheric motions.

An example of geostrophic balance is evident in the large-scale wind directions over the earth's surface. In Section 1.5, it was shown how the surface winds tend to come from north to south near the equator in the northern hemisphere. As they flow toward the equator, they would veer to their right of the flow direction or toward west. This is typical of the equatorial trade winds. Similarly, in the southern hemisphere, near equator, the flow coming from south on the surface (note that they are part of the direct cell in the tropics in Figure 1.9) would feel the earth's rotation and veer to the left, again toward west. Thus the hemispheric equatorial trade winds coming closer to the equator converge with each other in a westward flow and creates this region called the inter tropical convergence zone (ITCZ).

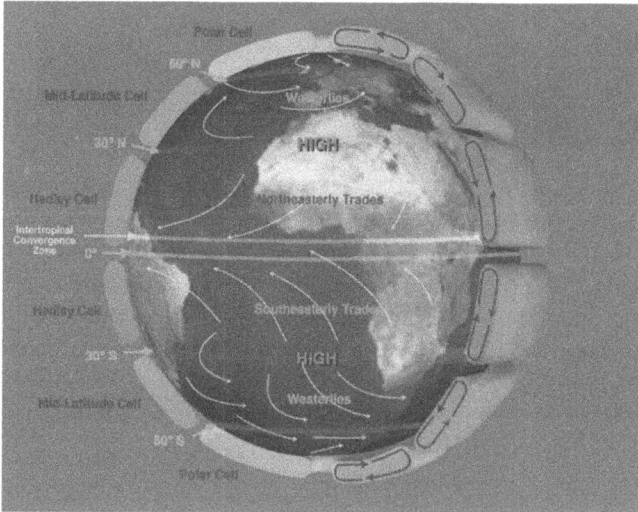

Figure 3.3: Depiction of global atmospheric circulation, showing major convection cells at various latitudes, including Hadley, mid-latitude, and polar cells, and showing the resulting surface winds. (NASA/File: Earth Global Circulation.jpg)

It is apparent in Figure 3.3 that the polar surface winds veer to the right (left) in the northern (southern) hemisphere. Again they are part of the direct cells. The winds in the subtropics behave differently and are subject to more complex dynamics.

Finally, let us think about a common weather phenomenon. Hurricanes in the Atlantic, Typhoons in the Pacific, and Cyclones in the Indian Ocean – they all rotate in anti-clockwise circulation within the vortex in the northern hemisphere. Why? The eye of the vortex or the center is a low pressure which attracts the air from the high-pressure regions around it. As the air flows from high to low (radially inward to a core of low depression), it will feel the rotation of the earth. In the northern hemisphere, they will thus tend toward right of their inward motion and gradually form the anti-clockwise circulation (or cyclonic circulation).

Figure 3.4: This image from EUMETSAT's METEOSAT-7 satellite over the Indian Ocean shows co-occurring typhoons, Thane in the northern hemisphere on shore in India and Benilde over the open ocean in the southern hemisphere. This rare lineup of storms shows that storms in the northern hemisphere rotate counterclockwise, while southern hemisphere storms rotate clockwise. This image was taken by METEOSAT-7 at 0530Z on December 30, 2011. (NOAA/https://www.nnvl.noaa.gov/MediaDetail2.php?MediaID=918&MediaTypeID=1)

In the southern hemisphere, the inward moving parcels from high to low will tend toward left and form a clock wise vortex – thus cyclones in the southern hemisphere rotate in the opposite direction to the cyclones in the northern hemisphere! A cyclone in the northern and another in the southern hemisphere is shown in Figure 3.4 for the same day (December 30, 2011).

3.5   INERTIAL MOTION

It is natural to ask, "Well, what happens if the pressures are similar on the sea surface, i.e., the surface of the ocean is flat?" How does the ocean adjust to pure rotation of the earth and no pressure gradient?

Let us look at the equation of motion 3.2 again. Let us set $\frac{\partial p}{\partial x}$ and $\frac{\partial p}{\partial y}$ to be zero, as there is no pressure gradient. However, we need to keep the acceleration terms on the LHS now, as they will balance the Coriolis term on the right-hand side.

Then the equations of motion become:

$$\frac{du}{dt} = 2\Omega \sin \phi \, v \qquad\qquad (3.17)$$

$$\frac{dv}{dt} = -2\Omega \sin \phi \, u \qquad\qquad (3.18)$$

Which leads to a simple solution of

$$u = V_H \sin(2\Omega \sin \phi \, t) \qquad\qquad (3.19)$$
$$v = V_H \cos(2\Omega \sin \phi \, t) \qquad\qquad (3.20)$$

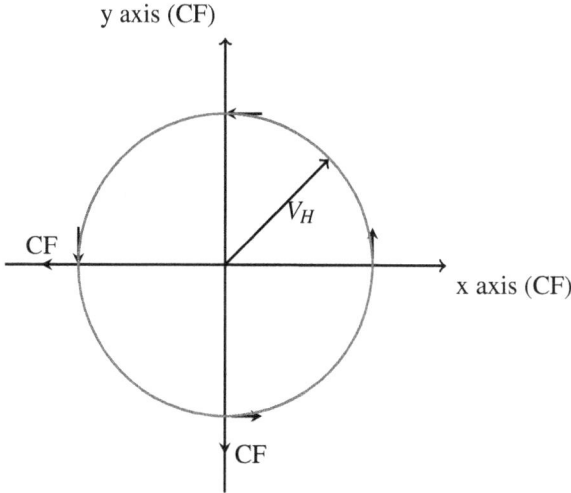

Figure 3.5: A particle on a rotating earth would follow the inertial circle in the absence of any other horizontal forces, the Coriolis force (CF) being to the right of the motion (black arrows). The motion would be counter clockwise in the northern hemisphere as shown and will be clockwise in the southern hemisphere.

This solution is shown in Figure 3.5. The locus of the particle on the rotating earth without any effect of pressure gradient would trace a circle with radius $V_H = \sqrt{(u^2 + v^2)}$, and the components (u, v) are proportionately distributed by the sin and cosine of the angle defined by the product of the Coriolis frequency (f, the inertial frequency) $(2\Omega \sin \phi)$ at that latitude and the time (t). This circle is appropriately called the "inertial circle," the motion being inertial with the inertial frequency. So, the adjustment time period for the parcels is on the order of the inertial period. The inertial period, the period of an inertial oscillation, is $2\pi/f$, where f is the Coriolis parameter. A key feature of this rapid adjustment of the fluid – in a time on the order of the rotation period – to an equilibrium that is not a state of rest. The equilibrium satisfies continuity exactly, and the momentum equations are degenerate.

3.6 THERMAL WIND

As we have seen above, the horizontal geostrophic equations and the hydrostatic equation can be used to calculate the velocity difference between two layers – or the velocity shear in the vertical direction.

For the atmosphere, density is directly related to its temperature, and thus changes in the temperature in the horizontal direction could directly lead to vertical variation of velocity. This idea of linking the temperature gradient across the atmospheric front with its velocity shear in the vertical along the front has led to understanding and forecasting weather for a long time. Hence, these velocities are often referred to as those of the "Geostrophic Winds" and the relationship that connects Geostrophy with hydrostatics is also known as the "Thermal Wind Balance." It is easier to see this in the following way.

Let us consider the x-momentum equation:

$$fv = \frac{1}{\rho}\frac{\partial p}{\partial x} \tag{3.21}$$

$$\rho fv = \frac{\partial p}{\partial x} \tag{3.22}$$

Differentiating with respect to z yields:

$$\rho f\frac{\partial v}{\partial z} = \frac{\partial}{\partial z}\left(\frac{\partial p}{\partial x}\right) \tag{3.23}$$

$$= \frac{\partial}{\partial x}\left(\frac{\partial p}{\partial z}\right) \tag{3.24}$$

using change of variables for p.

Using hydrostatic balance, $\frac{\partial p}{\partial z} = -g\rho$, we get,

$$\rho f\frac{\partial v}{\partial z} = -g\frac{\partial \rho}{\partial x} \tag{3.25}$$

$$\frac{\partial}{\partial z}(\rho fv) = -g\frac{\partial \rho}{\partial x} \tag{3.26}$$

$$\frac{\partial}{\partial z}(\rho fu) = +g\frac{\partial \rho}{\partial y} \tag{3.27}$$

So, from the density field distribution, one can obtain the vertical shear of velocity, NOT the velocities themselves! However, using a known velocity somewhere within the atmosphere, one can get a very good estimate of the whole three-dimensional wind field.

CONCLUSION

In this chapter, we considered two of the major forces, pressure-gradient and the Coriolis force, and assumed steady state. This balance between the pressure gradient and Coriolis due to Earth's rotation is called *the geostrophic equilibrium*.

A key concept within the geostrophic balance is the idea of a specific volume anomaly (SVA) which allows vertical variation of velocity linkage with density of a water column using the hydrostatic balance. The water column-integrated expression of the SVA is known as the dynamic height. The gradient of the dynamic height across any two locations provides a measure of the cross-sectional velocity. We have seen how given a level of no (or known) motion. The geostrophic equations can be used to obtain velocity sections. The geostrophic balance helps explain major set-ups of oceanic gyres, strengths of boundary currents, atmospheric fronts, and eddies. We have also looked into two types of simpler balances known as the inertial motion and thermal wind balance closely related to the Geostrophic Balance situation, but with key differences.

FURTHER READING

Cushman-Roisin, B. (1994). *Introduction to geophysical fluid dynamics*. Englewood Cliffs, NJ, Prentice-Hall.

Gill, A. E. (1982). *Atmosphere-ocean dynamics (international geophysics series)*. Academic Press.

Holton, J. R. (1973). An introduction to dynamic meteorology. *American Journal of Physics*, *41*(5), 752–754.

Knauss, J. A., & Garfield, N. (2016). *Introduction to physical oceanography*. Waveland Press.

Pedlosky, J. (2013a). *Geophysical fluid dynamics*. Springer Science & Business Media.

Pond, S., & Pickard, G. L. (1983). *Introductory dynamical oceanography*. Gulf Professional Publishing.

Rossby, C.-G. (1936). *Dynamics of steady ocean currents in the light of experimental fluid mechanics*. Massachusetts Institute of Technology; Woods Hole Oceanographic Institution.

Stommel, H. M. (1965). *The Gulf Stream: A physical and dynamical description*. University of California Press.

Stommel, H. M. (1987). *A view of the sea: A discussion between a chief engineer and an oceanographer about the machinery of the ocean circulation*. Princeton University Press.

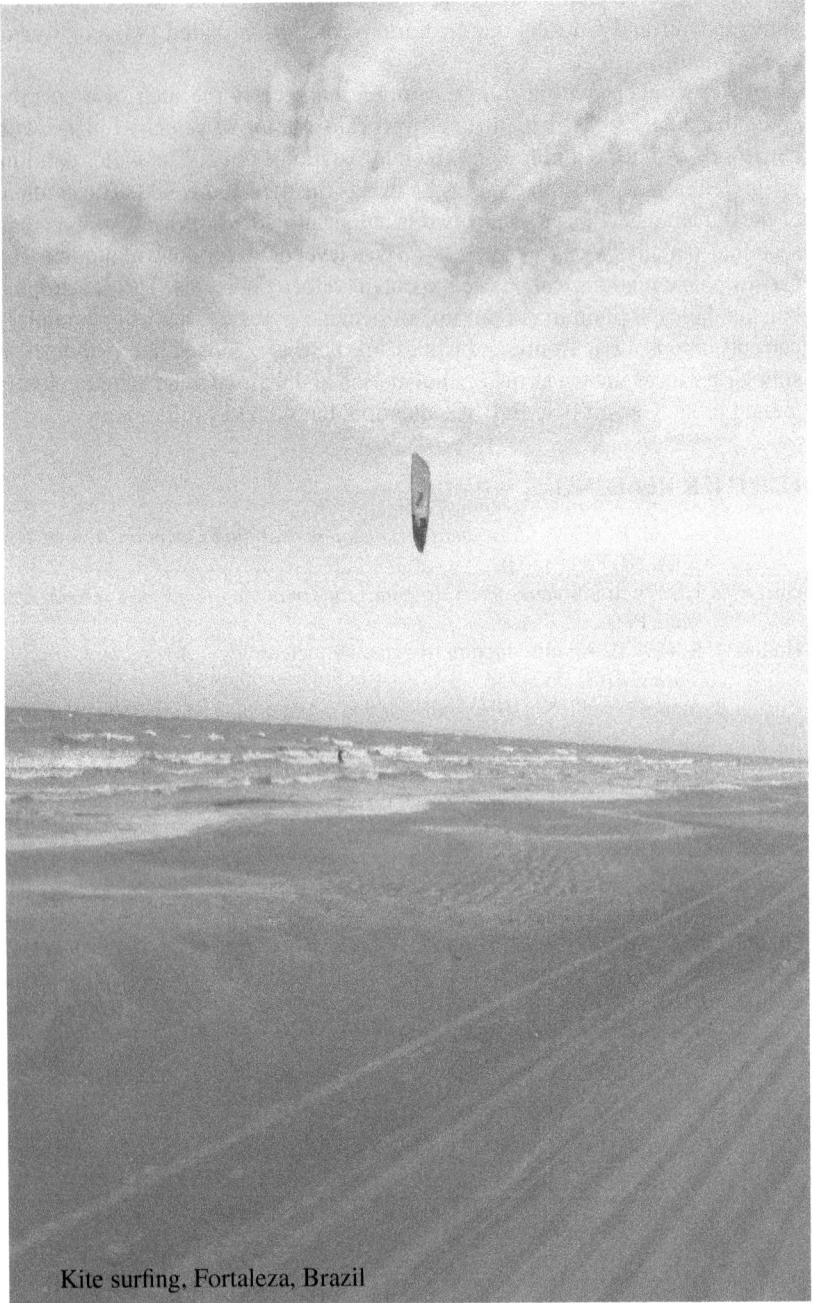

Kite surfing, Fortaleza, Brazil

4 Wind-Driven Circulation

> The answer is blowing in the wind.
>
> —————————————————————
> Bob Dylan (1941 -)*

OVERVIEW

In the second chapter, we derived the governing equations of motion and discussed how the four different kinds of forces (pressure, rotation, gravity, and friction) act on a fluid element. In the last chapter, we have shown how two of these forces, namely the pressure and Coriolis (due to Earth's rotation), balance each other in a steady-state flow situation, which is called the Geostrophic Balance.

In this chapter, we will focus on the effects of wind forcing on the fluid element along with pressure and Coriolis force. We will discuss the development of our understanding of the wind-driven circulation using the historical milestones. Specifically, we will first conceptualize the friction applied by the wind-stress on the water, as it differs from molecular friction counterpart, and develop the idea of "eddy viscosity." This idea was first put forward by Ekman in 1905 while he was looking at the effect of rotation on moving icebergs (see Chapter 1). We will then extend the wind-forcing as a body force on the water column and develop the integrated flow relationship called the Sverdrup balance. Finally, we will add bottom friction and discuss the Stommel solution which explained the westward intensification of subtropical wind-driven gyres. We will end this chapter with a brief outline of the Munk model with additional lateral friction as this is the end of the analytical solution of the theories of wind-driven ocean circulation.

4.1 INTRODUCTION

The fundamental question of understanding ocean circulation is to ask "What forces the ocean to circulate?" Or, what forces the ocean currents? What forces the oceanic eddies? What forces the fronts? The upwelling regions? The abyssal flows? To a large extent, "The answer, my friend, is blowing in the winds!" As we understand it, most of the circulation of the upper ocean is imparted by the winds blowing over it; and the circulation in the deeper ocean is driven by density-differences.

———————————————————

*Credit/Source: Bob Dylan. (n.d.). AZQuotes.com. Retrieved September 29, 2021, from AZQuotes.com Web site: https://www.azquotes.com/quote/851800

DOI: 10.1201/9780429347221-4

The wind pattern changes seasonally over the globe and it is important to realize the setup. See Figure 4.1 for the typical wind pressure and flow system. In the boreal winter (January), the Aleutian Low and the Icelandic Low are dominant at high northern latitudes. The Hawaiian High and the Azores High are on the Pacific and Atlantic subtropics, while there is a predominantly strong Siberian High over the Asian subcontinental region. In the southern hemisphere, you can see the South Pacific High, the South Atlantic High, and the Indian Ocean High. In the boreal summer (July), the Icelandic Low and the Aleutian Low are weaker than in winter, while the lows over India, Sahara, and North America become dominant. The Azores High and the Hawaiian High are stronger than their winter counterparts. The Indian Ocean High and the South Atlantic high become stronger. Such typical wind patterns affect the underlying oceanic currents and their ecosystems in various ways. In this chapter, we will dive into the basics of how the winds affect the motion in the water first and then see if typical wind distribution can lead us to some analytical solutions of ocean circulation.

We can look at the ocean starting at the equator and move towards the pole in each hemisphere. See Figure 1.10 and look for the different currents. Near the equator, where the Coriolis forces are negligible or very small, the generally westward winds pile up water to the west in the three basins. This creates a large anomalous pool of warm water to move to the west and brings evaporation and large-scale precipitation to place like Indonesia in the Western Pacific and Amazon in the Western Atlantic. On the other hand, on the eastern sides of these two oceans, upwelling sets in as a response to the westward movement of water from the surface layers by the winds. Such east-west SST difference forced by winds in the Equatorial Pacific is the backdrop for the well-known phenomenon called "El Niño" and "La Niña," which will be discussed later in Chapter 7. As we look to the subtropics in this chapter, the asymmetric anticyclonic gyre circulation in all mid-latitude oceans can be explained by a balance of Coriolis, winds, and pressure gradient forces. The polar and subpolar regions are more driven by the thermohaline changes imposed by heat loss in the winter and buoyancy driving (resulting from ice melting in the summer) and will be discussed in the next chapter. Sinking of heavy water from winter cooling (due to strong cold winds blowing over the polar waters) results in the slow and steady density-driven flow along the topographic contours in the abyssal depths.

In other words, the features of circulation in the upper ocean is driven by winds and called the "wind-driven circulation" and the circulation in the deep ocean (below the main thermocline) is driven by the density difference between water masses of different temperature and salinity constituents and thus called "thermohaline circulation" or the "abyssal circulation." Of course, the picture in the coastal ocean is a mixed scenario because of its shallowness, which allows for a vigorous competition of wind driving and thermohaline effects within a

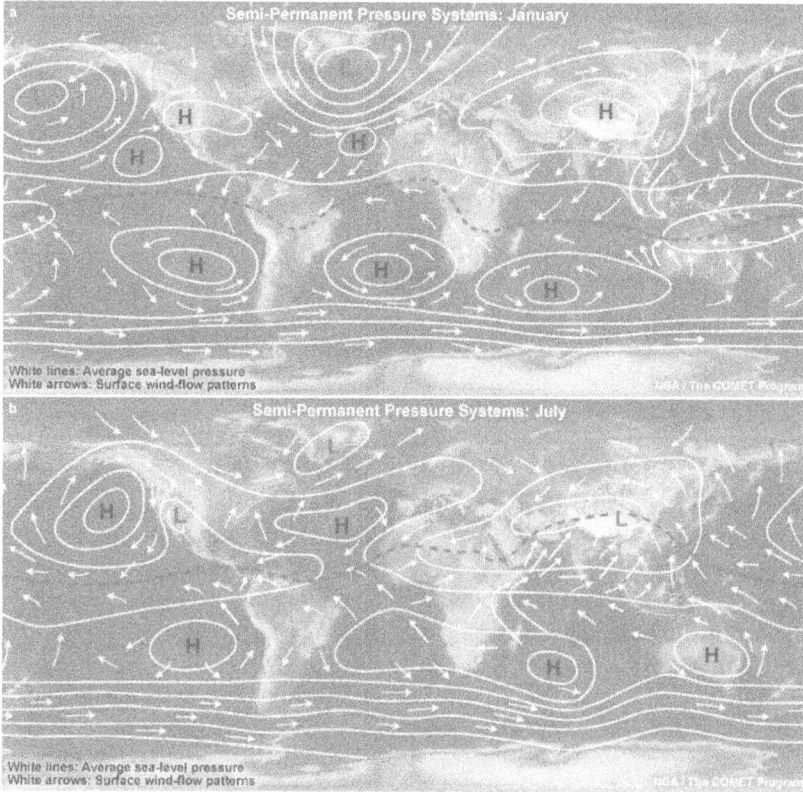

Figure 4.1: Mean pressure systems and representative wind vectors at the surface for (a) January and (b) July. The red dashed lines represent the equatorial trough (and monsoon trough over Asia). "The source of this material is the COMET Website at http://meted.ucar.edu/ of the University Corporation for Atmospheric Research (UCAR), sponsored in part through cooperative agreement(s) with the National Oceanic and Atmospheric Administration (NOAA), U.S. Department of Commerce (DOC). 1997–2017 University Corporation for Atmospheric Research. All Rights Reserved."

shallow water column. In this chapter, we will uncover a few of the important concepts of wind-driven ocean circulation. In the next chapter, we will consider a few of the novel theories that have been put forward to explain the nature of the slow thermohaline circulation that then binds the total oceanic circulation in a four-dimensional space-time dynamical system.

4.2 WIND STRESS AND EDDY VISCOSITY

As shown in Section 2.6.2, the viscous force or shear stresses on a fluid element can be represented by Newton's Law of friction which states that the shear stress (τ) is proportional to the vertical shear of the horizontal velocity ($\frac{\partial u}{\partial z}$). Or,

$$\tau = \mu \frac{\partial u}{\partial z} \tag{4.1}$$

where μ is the molecular viscosity; which can also be written as $\mu = \rho v$, where v is the kinematic viscosity.

At 20°C: $\mu = 10^{-3}$ kgm^{-1}s^{-1}, $v = 10^{-6}$ m^{2}s^{-1}, and $\rho = 10^3$ kgm^{-3}.

The above holds true for fluids for Reynolds number $Re = \frac{UL}{v} < 1000$ or for pre-turbulent (laminar) flow fields. Here U and L are the characteristic velocity and length scales of the flow. For pipe flow, U is about $0.1 - 1$ ms^{-1}, $L = 1 - 10$ m; so, $Re \approx 10^4$. Typically, the fluid is really turbulent at Reynolds Number, $Re >> 10^6$. See Batchelor and Batchelor, 2000 for details on laminar versus turbulent flow characteristics. For a typical oceanic flow, where $L = 100$ km; $U = 1$ ms^{-1}; $Re = 10^9$, and the stress-velocity relationship given by Newtonian molecular characterization (equation 4.1) yields a very small magnitude of shear stress to be effective in the oceanic momentum equations. Or, in other words, molecular shear stresses are too small to account for the viscous forces in the ocean. This limitation was overcome by Ekman while working with the winds and rotation (see Chapter 1 for the story of Nansen and Ekman). Ekman introduced the idea of "eddy viscosity." He proposed that the oceanic friction works on eddying scales and not on molecular scales. It is the turbulent eddies that provide the viscosity of viscous forces in the ocean and the atmosphere. So, he reworked the viscosity coefficients at the eddy level (much larger length scales than the molecular levels) and called it "eddy viscosity." Ekman introduced different eddy viscosity coefficients for the horizontal and vertical directions, which are:

$$A_H = 10^5 \text{ m}^2\text{s}^{-1} \tag{4.2}$$

$$A_z = 10^{-1} \text{ m}^2\text{s}^{-1} \tag{4.3}$$

Before deriving the Ekman balance, let us now introduce the forcing by the wind stress on the top surface of the fluid element, as shown in Figure 4.2. The setup is similar to the pressure and mass conservation setups discussed in Chapter 2.

The wind stresses in the horizontal (x) direction on the two surfaces (top -1; and bottom -2) are designated by τ_{zx1} and τ_{zx2}, respectively, as these surfaces are perpendicular to the z-axis following the tensor notation in Chapter 2. Assuming a vertical gradient of wind stress as $\frac{\partial \tau_{zx}}{\partial z}$, it is easy to see that,

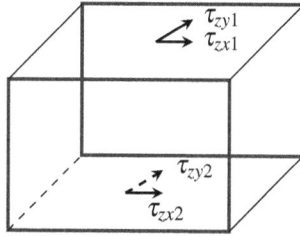

Figure 4.2: Wind stress on an element of fluid. The top surface is denoted by '1' and the bottom surface is denoted by '2'.

$$\tau_{zx2} = \tau_{zx1} + \frac{\partial \tau_{zx}}{\partial z} \Delta z$$

Thus, the effective wind stress-derived force per unit mass on the element in the x-direction (after dropping the "1" for the surface stress):

$$\frac{\left(\tau_{zx} + \frac{\partial \tau_{zx}}{\partial z} dz - \tau_{zx}\right) \Delta x \Delta y}{\rho \, \Delta x \, \Delta y \, \Delta z} = \frac{1}{\rho} \frac{\partial \tau_{zx}}{\partial z} \tag{4.4}$$

The above derivation is very similar to the other force-balance derivations shown in Chapter 2. Similarly, for the y-direction, the force per unit mass due to wind stress τ_{zy} can be given by:

$$\frac{1}{\rho} \frac{\partial \tau_{zy}}{\partial z}$$

4.3   THE EKMAN BALANCE

The horizontal equations of motion in the presence of winds can be written as follows:

$$\frac{du}{dt} = fv - \alpha \frac{\partial p}{\partial x} + \frac{1}{\rho} \frac{\partial \tau_x}{\partial z} \qquad \text{x-momentum equation} \tag{4.5}$$

$$\frac{dv}{dt} = -fu - \alpha \frac{\partial p}{\partial y} + \frac{1}{\rho} \frac{\partial \tau_y}{\partial z} \qquad \text{y-momentum equation} \tag{4.6}$$

Let us simplify the above equations by first assuming steady state (no time-dependence and thus LHS = 0); and then by considering that the ocean surface has a constant pressure, and thus no pressure-gradient ($\frac{\partial p}{\partial x} = \frac{\partial p}{\partial y} = 0$). Under these assumptions, the horizontal equations of motion reduce to a simple balance of Coriolis force against the wind-forcing and can be written as follows:

$$fv + \frac{1}{\rho}\frac{\partial \tau_x}{\partial z} = 0 \tag{4.7}$$

$$-fu + \frac{1}{\rho}\frac{\partial \tau_y}{\partial z} = 0 \tag{4.8}$$

Now, note that the eddy friction force can be re-written in terms of the vertical shear of the horizontal velocity of the fluid element as below (using the equation in Section 2.6.2).

$$\frac{1}{\rho}\frac{\partial \tau_x}{\partial z} = \frac{1}{\rho}\frac{\partial}{\partial z}\rho A_z \frac{\partial u}{\partial z} \tag{4.9}$$

$$= A_z \frac{\partial^2 u}{\partial z^2} \tag{4.10}$$

and similarly,

$$\frac{1}{\rho}\frac{\partial \tau_y}{\partial z} = \frac{1}{\rho}\frac{\partial}{\partial z}\rho A_z \frac{\partial v}{\partial z} \tag{4.11}$$

$$= A_z \frac{\partial^2 v}{\partial z^2} \tag{4.12}$$

So, the equations of motion become:

$$fv = A_z \frac{\partial^2 u}{\partial z^2} \tag{4.13}$$

$$-fu = A_z \frac{\partial^2 v}{\partial z^2} \tag{4.14}$$

The above two equations are known as the Ekman Equations. They represent a balance of the wind stress supported by the Coriolis in the absence of any pressure-gradient in the steady state. The absence of pressure-gradient implies that this balance would work independently of the geostrophic balance. Thus, the total velocity of the fluid can now be decomposed into two parts when acted upon by winds and pressure-gradient. A part of the velocity (U_E and V_E) that is supported by wind-driven Ekman balance, and called Ekman Velocities; and another part which is purely geostrophic (U_G and V_G) which is the balance of the pressure gradients and the Coriolis force resulting from these velocity components.

The solution to the Ekman Equations are straightforward and the reader is referred to his original paper (Ekman, 1905) for details. The solutions are given by:

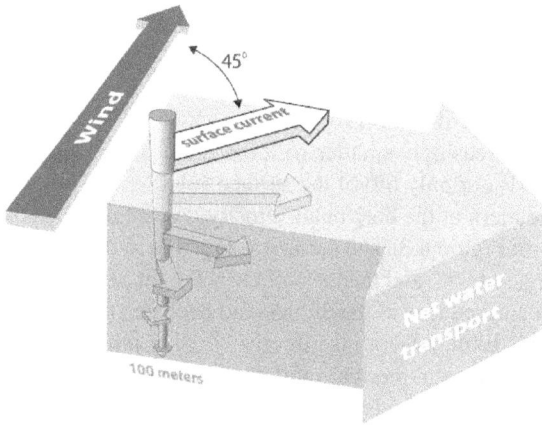

Figure 4.3: The Ekman spiral occurs as a consequence of the Coriolis effect. When surface water molecules are moved by the wind, they drag deeper layers of water molecules below them. Like surface water, the deeper water is deflected by the Coriolis effect – to the right in the Northern Hemisphere and to the left in the Southern Hemisphere. As a result, each successively deeper layer of water moves more slowly to the right or left, creating a spiral effect (NOAA, oceanservice.noaa.gov).

$$u_E = \pm V_0 \cos\left(\frac{\pi}{4} + \frac{\pi Z}{D_E}\right) e^{\left(\frac{\pi Z}{D_E}\right)} \qquad (4.15)$$

$$v_E = V_0 \sin\left(\frac{\pi}{4} + \frac{\pi Z}{D_E}\right) e^{\left(\frac{\pi Z}{D_E}\right)} \qquad (4.16)$$

$$V_0 = (\sqrt{2}\pi\tau_{yn})/(D_E\,\rho\,|f|) \qquad (4.17)$$

$$D_E = \pi\left(\frac{2A}{|f|}\right) \qquad (4.18)$$

where τ_{yn} is the magnitude of the wind stress and D_E is the Ekman Depth down to which the effect of the wind stress at the surface is realized by the velocity fields.

At the sea surface, $Z = 0$,

$$u_E = \pm V_0 \cos 45° \tag{4.19}$$

$$v_E = V_0 \sin 45° \tag{4.20}$$

The general solution above is proportional to $V_0 e^{(\pi Z/D_E)}$, which makes the velocity becomes increasingly smaller with increasing depth and at $Z = -D_E$, $V = V_0 e^{-\pi} = 0.06\, V_0$, or only 6% of the surface velocity.

Note that the pattern of the horizontal velocity is a spiral, called the "Ekman Spiral" as shown in Figure 4.3. The surface velocity is to the right of the wind direction at an angle of 45°, and then it veers to the right as it descends through the water column and reduces in magnitude and becomes negligible below the Ekman Depth. This Ekman Depth is effectively the depth of the wind-mixed layer and is often used in numerical models for use in heat and salt balance equations and calculating budgets in the upper layers.

Since D_E is on the order of 50–100 meters (using typical values of A and mid-latitude "f"), we limit the Ekman solution to the upper 50–100 meters and divide the total velocity into two parts (geostrophic and Ekman) as indicated above.

An example of the application of Ekman velocity in conjunction with the geostrophic currents is the generation of total surface velocities as provided by the reanalysis called "OSCAR (Ocean Surface Currents Analyses Real-time)." Satellite Altimeter data provides the geostrophic currents and reanalysis or real-time atmospheric models provide the wind speed components. Both of these are used separately to give geostrophic and Ekman velocities at the surface, which are then added and a combined field is the product for the measure of an average velocity field representative of the upper layer. This is often used for validation of average surface layer velocity fields from different numerical models.

4.4   THE INTEGRATED EKMAN TRANSPORT

If we now let the Coriolis force balance the wind-stress in an integrated sense throughout the water column, then the horizontal equations of motion simply become:

$$f\, V_E + \frac{1}{\rho} \frac{\partial \tau_x}{\partial z} = 0 \tag{4.21}$$

$$-f\, U_E + \frac{1}{\rho} \frac{\partial \tau_y}{\partial z} = 0 \tag{4.22}$$

If $\tau_x, \tau_y = 0$ at the bottom of this element in a homogeneous ocean, and if we integrate from the surface $(z = 0)$ to a depth $(z = -H)$, where there is no wind or the bottom, we get:

$$\int_{-H}^{0} fV_E dz + \int_{-H}^{0} \frac{1}{\rho} \frac{\partial \tau_x}{\partial z} dz = 0 \tag{4.23}$$

$$\int_{-H}^{0} -fU_E dz + \int_{-H}^{0} \frac{1}{\rho} \frac{\partial \tau_y}{\partial z} dz = 0 \tag{4.24}$$

Setting the limits,

$$fM_{yE} + \frac{1}{\rho}[\tau_x|_{sfc} - \tau_x|_{bottom}] = 0 \tag{4.25}$$

$$-fM_{xE} + \frac{1}{\rho}[\tau_y|_{sfc} - \tau_y|_{bottom}] = 0 \tag{4.26}$$

Using $\tau_x, \tau_y = 0$ at the bottom, its a simple balance of integrated mass transport with the wind stress in its normal direction, given by:

$$fM_{yE} + \frac{1}{\rho}[\tau_x|_{sfc}] = 0 \tag{4.27}$$

$$-fM_{xE} + \frac{1}{\rho}[\tau_y|_{sfc}] = 0 \tag{4.28}$$

What does this integrated Ekman balance physically signify? It is important to realize that the transport in one direction is balanced by the wind-stress in its normal direction due the Coriolis effect on the transport. Let us look at a couple of practical application of this simple balance next, called coastal upwelling and equatorial upwelling.

4.4.1 UPWELLING (COASTAL AND EQUATORIAL)

The simple Ekman balance described above has huge implication near the coast. Imagine a coast along which the wind blows parallel to the coast for a sustained period of time. This happens in many coastal regions. Now, in the northern hemisphere, the setup of the atmospheric circulation in the mid-latitudes are such that in anticyclonic wind gyres (see Figure 4.1), the winds flow from North to south along the eastern boundaries of the Pacific and Atlantic Oceans. For example, if you think about the coast of California the equator-ward winds will induce an Ekman transport to the right of the wind direction, or away from the coast. If the water in the surface layers move out of the coast, what happens to the void created by this Ekman drift? Well, due to continuity, mass has to be conserved, and so water from below will come up and fill the void created by the Ekman drift due to wind! This phenomenon is known as coastal upwelling. In the Southern hemisphere, for an eastern boundary, when wind blows equatorward (from south to north), again coastal upwelling will happen. This is what happens along the coasts of Peru and Chile in the south Pacific. An example is shown in Figure 4.4d for a sequence of this process for the southern hemisphere (SH).

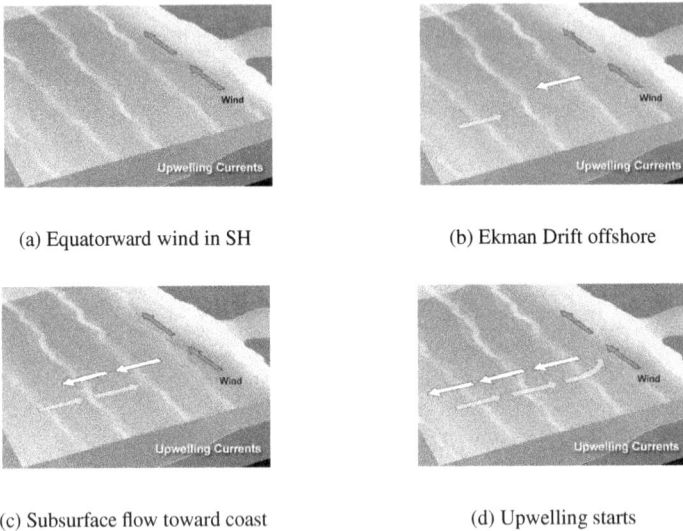

(a) Equatorward wind in SH

(b) Ekman Drift offshore

(c) Subsurface flow toward coast

(d) Upwelling starts

Figure 4.4: Progressive nature of Coastal Upwelling in the southern hemisphere (NOAA Oceanservice).

It is important to realize that upwelling is a phenomenon that incorporates the Ekman balance in the horizontal (between Coriolis and wind stress) and the continuity equation in three-dimension given by the equation 2.33. The Ekman drift (transport) creates a void of water or divergence at the surface, and the fluid from below upwells to fill the empty space to conserve mass. The coast acts as the support or the wall through which there could be no flow horizontally to replenish the void created by the wind drift. A similar wall is the Equator due to that fact that, $f = 0$ at the equator, rendering no Ekman drift on the Equator. However, right outside of the Equator, the Ekman drift is to the north and to the south of the prevailing westward wind on the Equator (see Figure 4.1) in the northern and southern hemispheres, respectively. This divergence of flow at the surface from the equator creates a void right on the Equator and the fluid from below has to come up and replenish the space above. This results in a permanent setup of "Equatorial Upwelling" on the equator, which is shown in Figure 4.5.

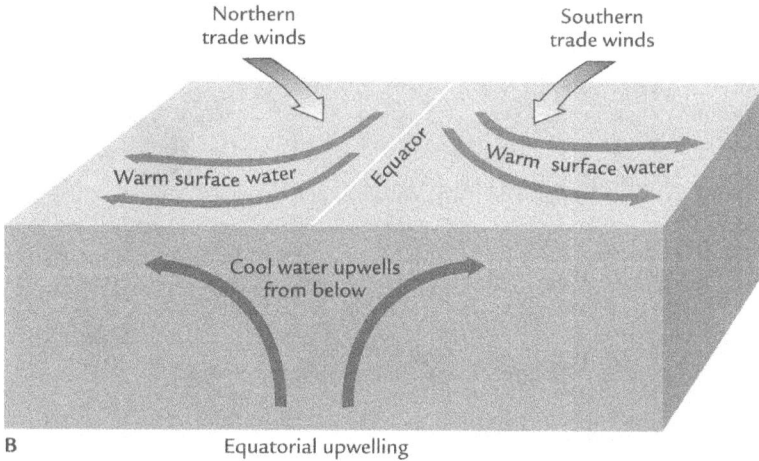

Figure 4.5: A schematic of Equatorial Upwelling (from Climate.gov).

About one-half of the world's population leave within 20 miles of a coast in various continents. So, they depend on the coastal ecosystem which is supported by the colder nutrient-rich waters being continuously or intermittently upwelled from below as the surface near coastal waters are moved offshore due to upwelling-favorable winds. A typical upwelling index considers the Ekman flow based on wind-speed and direction, and model such phenomenon (Bakun, 1973). Such upwelling indices are used for monitoring and predicting for fisheries and other related purposes in the coastal zones (M. Jacox & Edwards, 2012; M. G. Jacox et al., 2018). These coastal upwelling zones in the world are shown in Figure 4.6. Sea surface reflectance data from Satellite (see Chapter 11) are used to monitor, understand and predict the variability in these zones by many agencies and industries for various purposes.

4.5 SVERDRUP DYNAMICS

Ekman derived the balance between the wind stress and the Coriolis force. The importance of the rotation of earth on fluid motion via the geostrophic balance was known to both meteorological and oceanographers for some time. Let us recapitulate these two separate balances for a moment in the background of the overarching equations of motion.

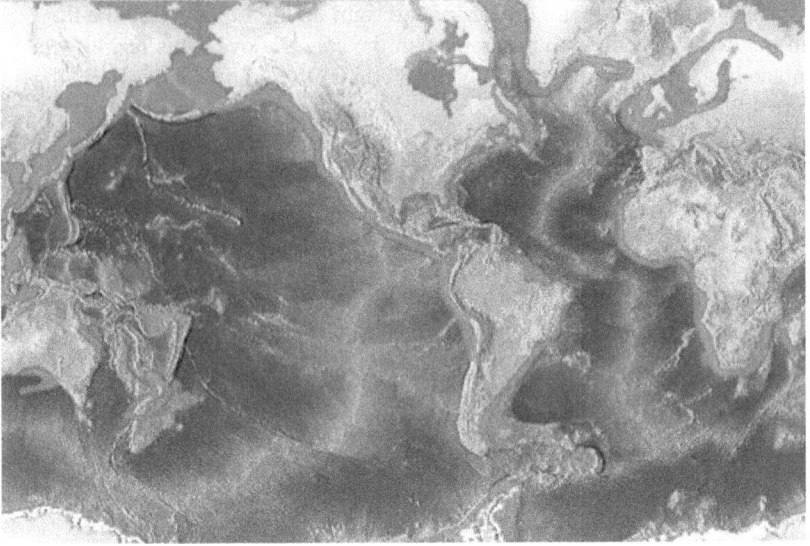

Figure 4.6: World-wide distribution of Coastal Upwelling regions. Source (NOAA).

The vector notation (for brevity) is useful to describe the three-dimensional equations of motion as follows:

$$\underbrace{\frac{du}{dt}}_{\text{accln.}} = \underbrace{-\alpha\nabla p}_{\text{pressure}} \underbrace{-2\Omega\times V}_{\text{Coriolis}} + \underbrace{g}_{\text{gravity}} + \underbrace{F}_{\text{Forcing}} \qquad (4.29)$$

In the steady-state the LHS of the above equation 4.29 is zero. So, the horizontal equations of motion (without gravity) in the presence of three factors, namely, pressure gradient, Coriolis and winds can be written as:

$$\overbrace{-\alpha\frac{\partial p}{\partial x} + fv}^{\text{Geostrophy}} + \frac{1}{\rho}\frac{\partial \tau_x}{\partial z} = 0 \qquad (4.30)$$

$$-\alpha\frac{\partial p}{\partial y} \underbrace{-fu + \frac{1}{\rho}\frac{\partial \tau_y}{\partial z}}_{\text{Ekman}} = 0 \qquad (4.31)$$

Look at the above two equations 4.30 and 4.31, and think about what we did so far in the derivations of geostrophy in Chapter 3 and Ekman balance in this chapter. If you use the Ekman balance, you get a solution for the upper level flow down to the Ekman layer which is about 50–100 m deep. If you use the geostrophic balance, you get a velocity distribution throughout the water column which balances the pressure gradient which supports the flow. They are part of the solution of the whole system, but not yet the whole solution!

A natural evolution would be to look into a circulation scheme that would encompass a balance of all three forces – Coriolis, pressure gradient and wind stress. However, a closed form solution was almost obtained by Sverdrup in 1942 (Sverdrup et al., 1942), and successfully developed by Henry Stommel in 1948 (H. Stommel, 1948), and extended by Walter Munk in 1950 (Munk, 1950; Munk & Carrier, 1950). Let us look at these solutions in the following.

Sverdrup started with the steady-state ocean and the Coriolis + pressure gradient + wind stress balance as follows for a homogeneous ocean of constant density ρ and constant depth H. He then looked for a depth-integrated solution using the conservation equation to eliminate the pressure terms. Let's explore this process in detail.

Now, integrating down through the water column, the x-momentum equation becomes,

$$\int_{-H}^{0} f v dz - \int_{-H}^{0} \alpha \frac{\partial p}{\partial x} dz + \frac{1}{\rho} \int_{-H}^{0} \frac{\partial \tau_x}{\partial z} dz = 0$$

$$\int_{-H}^{0} \rho f v dz - \int_{-H}^{0} \frac{\partial p}{\partial x} dz + \int_{-H}^{0} \frac{\partial \tau_x}{\partial z} dz = 0$$

$$f M_y - \int_{-H}^{0} \frac{\partial p}{\partial x} dz + \tau_x|_{sfc} - \tau_x|_{-H} = 0$$

$$f M_y - \int_{-H}^{0} \frac{\partial p}{\partial x} dz + \tau_x|_{sfc} = 0 \qquad (4.32)$$

Similarly, for the y-momentum equation,

$$f M_x - \int_{-H}^{0} \frac{\partial p}{\partial y} dz + \tau_y|_{sfc} = 0 \qquad (4.33)$$

where the depth-integrated mass transport in x- and y-direction is represented by $M_x = \int_{-H}^{0} \rho u dz$ and $M_y = \int_{-H}^{0} \rho v dz$.

Now, it is easy to see that if we cross differentiate the x and y-momentum equations and subtract one from the other, we can eliminate the pressure terms. Let us do exactly that first.

So, by applying $\frac{\partial}{\partial y}(4.32) - \frac{\partial}{\partial x}(4.33)$, we obtain

$$\frac{\partial}{\partial y}(fM_y) + \frac{\partial \tau_x}{\partial y} + \frac{\partial}{\partial x}(fM_x) - \frac{\partial \tau_y}{\partial x} = 0 \qquad (4.34)$$

$$\underbrace{\frac{\partial f}{\partial y}}_{\beta} M_y + \frac{\partial M_y}{\partial y}f + \frac{\partial \tau_x}{\partial y} + \frac{\partial f}{\partial x}M_x + \frac{\partial M_x}{\partial x}f - \frac{\partial \tau_y}{\partial x} = 0 \qquad (4.35)$$

$$\beta M_y + f(\frac{\partial M_x}{\partial x} + \frac{\partial M_y}{\partial y}) = \frac{\partial \tau_y}{\partial x} - \frac{\partial \tau_x}{\partial y} \qquad (4.36)$$

Here, $\beta = \frac{\partial f}{\partial y}$, or the variation of Coriolis frequency with latitude. You can think about a plane on the earth's surface where $f = f_0 + \frac{\partial f}{\partial y} \times y = f_0 + \beta y$. Such linearization leads to a simpler mathematical construct and often referred to as the "beta-plane" approximation. We will use this often in subsequent derivations without explicitly mentioning it.

The continuity equation $\frac{\partial u}{\partial x} + \frac{\partial v}{\partial y} = 0$ implies that the depth-integrated continuity holds as

$$\frac{\partial M_x}{\partial x} + \frac{\partial M_y}{\partial y} = 0 \qquad (4.37)$$

and we are left with

$$\beta M_y + \frac{\partial \tau_x}{\partial y} - \frac{\partial \tau_y}{\partial x} = 0 \qquad (4.38)$$

From the continuity equation, we can define a mass transport stream function ψ, which is usually given by: $M_x = \frac{\partial \psi}{\partial y}$ and $M_y = -\frac{\partial \psi}{\partial x}$.

Substituting ψ in the above equation 4.38, we get the Sverdrup solution for the mass-transport stream function for the homogeneous ocean in the steady state under the action of winds and pressure gradient on a rotating earth as:

$$-\beta \frac{\partial \psi}{\partial x} = \frac{\partial \tau_y}{\partial x} - \frac{\partial \tau_x}{\partial y} = curl \ \tilde{\tau} \qquad (4.39)$$

or,

$$\beta M_y = \text{wind stress } curl \qquad (4.40)$$

What is this wind-stress curl? Well, let us think about how the winds interact with the ocean. Other than pushing the water near the coasts (Ekman drift), it can also impart momentum into the ocean. This transfer of momentum is a complex process and we will not get into the details; however, we can appreciate the idea that the rotational effect (or torque) by the wind stress might impact the oceanic motion. This rotational tendency of the wind is captured in its 'vorticity' or the wind-stress curl. We will discuss the vorticity aspect of a fluid in the next subsection.

The Sverdrup equation connects the stream function, a measure of the oceanic circulation to the wind stress curl. Simply put, a negative wind stress curl would result in a southward velocity (integrated transport) in the homogeneous ocean. Most of the subtropical North Atlantic, subtropical North Pacific, subtropical South Atlantic, Subtropical South Pacific, Springtime Bay of Bengal and Spring time Arabian Sea have a predominantly negative wind stress curl distribution, which would result in large-scale equatorward flow in these regions. Sverdrup then assumed that since there is no strong boundary current to the eastern side of these ocean basins, we can put a boundary condition of $\psi = 0$ at the eastern side of the basin, add up all the Equatorward (southward in the northern hemisphere and northward in the southern hemisphere) transport as we integrate along any latitude westward. Then by continuity, we need to invoke a strong western boundary current (WBC) which would need to flow poleward. See Figure 4.7 for a schematic of the Sverdrup solution setup.

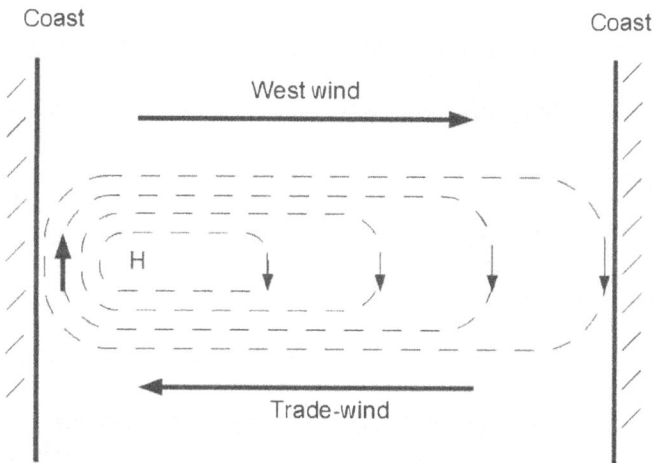

Figure 4.7: Sverdrup solution. Continuity and integration from East to West leads to the WBC. Figure Credit: Schwedenhagen at German Wikipedia.This file is licensed under the creative commons attribution-share Alike 3.0 Germany.

Examples of such strong and organized poleward currents in the Northern Hemisphere are the Gulf Stream in Western North Atlantic, the Kuroshio in the Western North Pacific, the springtime Western Boundary Current in the Bay of Bengal and the Somali Current in the Arabian Sea. Note that, in the Sverdrupian sense, the East India Coastal Current (EICC), is more of a density-driven boundary current (and remotely forced by basin-scale Kelvin Waves) in the autumn/winter and is dynamically different than the springtime WBC. For a deeper discussion on these modes in the Bay of Bengal see Gangopadhyay, Bharat Raj, et al. (2013). Similar WBCs in the Southern hemisphere are the Brazil Current in the western South Atlantic, the East Australian Current in the western South Pacific and the Agulhas current in the South Indian Ocean. See Figure 1.10 for the locations of these different WBCs (see WBC).

Note that the Sverdrup Equation is NOT a momentum equation (unlike the Geostrophic and Ekman Equations), it is a *vorticity balance equation*. It is a balance between the relative vorticity (curl of the velocity) of the fluid in response to the wind forcing (wind vorticity/torque) represented by the wind stress curl. Mathematically, the momentum equations are cross-differentiated and subtracted to eliminate the pressure terms first; continuity is then used to reduce the u and v components of the velocity field in one representative stream function to represent the circulation. This series of operations converts the momentum equations into a single vorticity equation which can describe the oceanic circulation. Physically, the vorticity equation is balance of vorticities in a force-response system, one order more complex than the simpler momentum balances. The momentum balances are direct representation of Newton's second law of motion. The Vorticity equation blends the rotational tendency of the fluid with continuity equation and balances with the torque imposed on the fluid by the wind. The vorticity equation is a statement about the circulation of the fluid in one integrated expression. In fact, circulation and vorticity are well linked – mathematically, it can be shown that the former is the integral of the latter along a streamline. We will leave the reader to more advanced texts (Gill, 1982; J. Holton, 2016) for those discussions.

4.6 VORTICITY

What is vorticity? Think of three different flow fields as shown in the top left panel of Figure 4.8. The flow field in the middle has a constant velocity $u(y)$. So, its gradient field ($\frac{\partial u}{\partial y}$) is non-existent or zero. Any fluid parcel within this flow will feel no tendency to deviate or rotate out of the straight path.

The flow in the bottom of the three-panel set has a distribution of velocity where the velocity is increasing as one approaches the axis of the flow from the south. So, in this case any parcel within the flow will feel an anticyclonic rotational tendency, due to the positive flow gradient and negative vorticity ($\frac{\partial u}{\partial y} > 0; \xi < 0$). In contrast, for the flow with gradually decreasing velocities to the north (top of the three-panel set) will feel a cyclonic rotational tendency, due to the negative flow gradient and positive vorticity ($\frac{\partial u}{\partial y} < 0; \xi > 0$). Similarly, if we consider a north-south flow field, the gradients $\frac{\partial v}{\partial x}$ can generate cyclonic or anticyclonic rotational tendency depending on if $\frac{\partial v}{\partial x}$ being negative or positive.

Flow

Cyclonic

Zero

Anticyclonic

Relative (to earth)
Vorticity

$$\zeta = \underbrace{(\frac{\partial v}{\partial x} - \frac{\partial u}{\partial y})}_{\text{curl } \vec{V}}$$

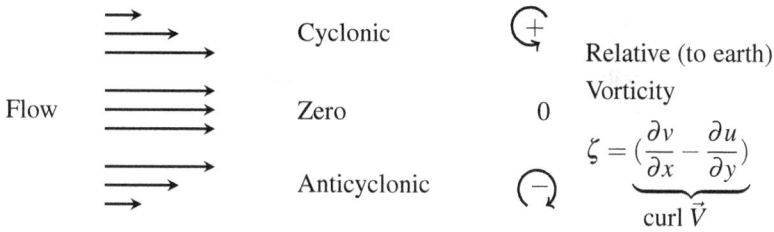

Planetary Vorticity

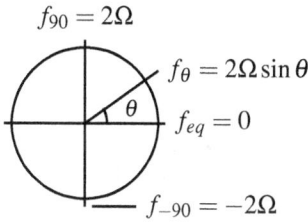

$f_{90} = 2\Omega$

$f_\theta = 2\Omega \sin\theta$

$f_{eq} = 0$

$f_{-90} = -2\Omega$

Potential Vorticity

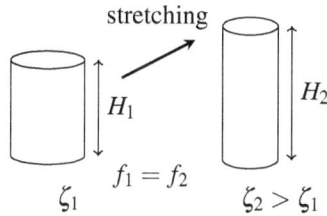

stretching

H_1

H_2

$f_1 = f_2$

ζ_1

$\zeta_2 > \zeta_1$

Figure 4.8: Different ways a fluid experiences vorticity.

4.6.1 RELATIVE VORTICITY

Thus, the relative vorticity, ζ, for a general flow field with components $u(x,y)$ and $v(x,y)$ can be given by adding the relevant rotational tendencies as follows:

$$\zeta = \frac{\partial v}{\partial x} - \frac{\partial u}{\partial y}$$

$\zeta > 0$ leads to cyclonic (anti-clockwise) or positive vorticity, and
$\zeta < 0$ leads to anticyclonic (clockwise) or negative vorticity.
Note that the vorticity has a unit of frequency (s^{-1}).
In vector notation, $\vec{\zeta}$ is the curl of the velocity $\bar{u}$, and is given by:

$$\vec{\zeta} = curl \ \bar{u} = \bar{\nabla} \times \bar{u}$$

4.6.2 PLANETARY VORTICITY

The rotational tendency of any parcel on the surface of the earth is simply given by the local Coriolis frequency, or,

$$f = 2\Omega \sin\phi$$

Its range varies from -2Ω at the South Pole to $+2\Omega$ at the North Pole passing through a zero value on the Equator (see the bottom left quadrant of Figure 4.8.

4.6.3 ABSOLUTE VORTICITY

On a rotating earth, the sum of relative vorticity and planetary vorticity is called the Absolute Vorticity and is given by

$$\zeta + f$$

4.6.4 POTENTIAL VORTICITY

The potential vorticity of a fluid element incorporates the absolute vorticity and the height of the fluid column or the thickness of density layers. In a rudimentary sense, it is defined mathematically as follows:

$$q = \frac{\zeta + f}{H} \tag{4.41}$$

For a fluid element with a depth H, it preserves a quantity called potential vorticity (q) following its path and this conservation of potential vorticity is given by:

$$\frac{dq}{dt} = 0 \tag{4.42}$$

$$\frac{d}{dt}\left(\frac{\zeta + f}{H}\right) = 0 \tag{4.43}$$

The bottom right panel in Figure 4.8 shows an example of the conservation of potential vorticity. Consider a fluid element with initial relative vorticity of ζ_1 and height H_1 is moved to a different location along the same latitude $(f = constant)$ and gains relative cyclonic vorticity (say, from wind) and its new relative vorticity is then ζ_2, which is greater than ζ_1. To preserve the potential vorticity, the fluid element will have to stretch $(H_2 > H_1)$, so that

$$\frac{\zeta_1 + f}{H_1} = \frac{\zeta_2 + f}{H_2} \tag{4.44}$$

In a general sense, the potential vorticity of a fluid parcel is proportional to its Absolute vorticity and stratification (which is proportional to the vertical density gradient, $\frac{\partial \rho}{\partial z}$). We will discuss this later in Chapter 7 and section 7.2.1. For more on vorticity and how it plays a crucial role in fluid dynamics, please refer to the excellent texts such as Batchelor and Batchelor (2000), Cushman-Roisin and Beckers (2011), Kundu et al. (2012), Pedlosky (2013a), and others.

4.7 STOMMEL'S SOLUTION

Let us recap the Sverdrup Solution derived in Section 4.5. The final solution was given by

$$\beta M_y + \frac{\partial \tau_x}{\partial y} - \frac{\partial \tau_y}{\partial x} = 0$$

Now, substituting the stream function formulation $M_x = \frac{\partial \psi}{\partial y}; M_y = -\frac{\partial \psi}{\partial x}$, the continuity equation $\frac{\partial M_x}{\partial x} + \frac{\partial M_y}{\partial y} = 0$ is automatically satisfied.

Now let us look at relative vorticity of the fluid column, which could be given by

$$\frac{\partial M_x}{\partial y} - \frac{\partial M_y}{\partial x} = \frac{\partial^2 \psi}{\partial y^2} + \frac{\partial^2 \psi}{\partial x^2} = \nabla^2 \psi$$

So, the Laplacian of the stream function field is the relative vorticity field of the fluid circulation. This is a very important relationship between the circulation represented by streamfunction and vorticity. They are integrally linked by the Laplacian relationship.

Now let us look into the next development of the wind-forced analytical solution as obtained by Stommel in 1948 (H. Stommel, 1948).

The limitation of the Sverdrup solution was in its implicit assumption that the boundary condition of $\psi = 0$ had to be imposed on the Eastern side of the ocean to generate the western boundary current and thus close the solution. The LHS of the Sverdrup equation is a velocity component, not a stream function that describes a two-dimensional field, which could uniquely define and describe the circulation of the homogeneous ocean basin. Mathematically thinking, Stommel was looking for a way to arrive at a closed solution expressed by using the Laplacian of the stream function that would then yield a closed form solution.

So, let's take another look at the momentum equations 4.29 with wind-forcing, pressure-gradient and the Coriolis force in a steady state. We have to cross-differentiate, subtract, use continuity and then we can get to one equation that describes our circulation! That is all good, but how can we get a Laplacian ($\nabla^2 \psi$) at the end? This was the precise problem that Stommel solved so elegantly.

While Sverdrup assumed bottom friction ($\tau_x|_{bottom} = 0$), Stommel introduced a new idea of the bottom friction being proportional to the velocity. As long as there is some velocity, there would be some friction at the bottom, however small it is, and he introduced a so-called "bottom friction coefficient". So, the first thing Stommel did was to introduce a new term (ru), where r is the bottom friction coefficient and u is the velocity. Clearly, such a term would lead

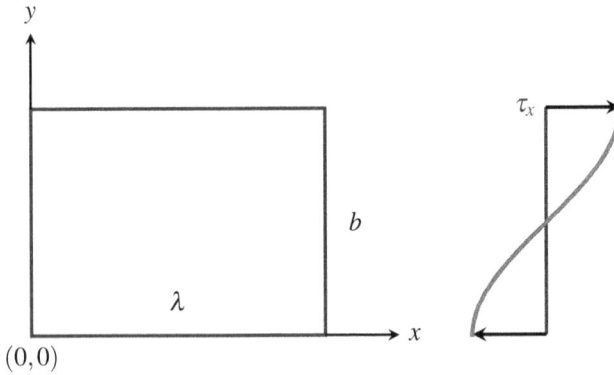

Figure 4.9: Stommel's rectangular basin; right panel shows the simple sinusoidal wind stress distribution.

to cross-derivatives through the mathematical operations of reaching the final solution and end up in a Laplacian for the stream function.

The Second thing Stommel did was to set the top of the water level at a variable height $h(x,y)$, i.e., as a function of x and y, instead of being at $Z = 0$. This allows the equations to have a solution for the pressure field as well in terms of $h(x,y)$.

The third thing Stommel did was to simplify the wind-stress distribution and assumed the meridional wind-stress to be negligible and the zonal wind stress to be sinusoidal in nature, which allowed for some comprehensible mathematical expressions to appear in the solutions. The sinusoidal distribution was justified as they closely resembled the mid-latitude zonal wind stresses on both the Pacific and Atlantic oceans.

Let us now briefly follow Stommel's model below.

Consider a homogeneous ocean with a length λ and width b.

$$\tau_x = F_0 \quad \text{at} \quad y = b$$

and

$$\tau_x = -F_0 \quad \text{at} \quad y = 0$$

So, it is possible to assume a sinusoidal distribution $\tau_x = -F_0 \cos(\frac{\pi y}{b})$. Thus

$$\frac{\partial \tau_x}{\partial y} = \frac{F\pi}{b} \sin(\frac{\pi y}{b})$$

Note that bottom friction terms are ∞ velocity $\longrightarrow -Ru, -Rv$.

Integrating the x-momentum equation like in Section 4.5 and putting the boundary condition, one can easily arrive at the following depth-integrated equations in the two directions.

$$0 = f(D+h)v - F_o \cos(\frac{\pi y}{b}) - Ru - g(D+h)\frac{\partial h}{\partial x} \quad (4.45)$$

$$0 = f(D+h)u - Rv - g(D+h)\frac{\partial h}{\partial y} \quad (4.46)$$

and, the continuity equation is given by:

$$\frac{\partial}{\partial x}[(D+h)u] + \frac{\partial}{\partial y}[(D+h)v] = 0 \quad (4.47)$$

Cross-differentiating the momentum equations, subtracting, and using continuity, one can obtain a single equation describing the solution:

$$(D+h)v\frac{\partial f}{\partial y} + \frac{F\pi}{b}\sin(\frac{\pi y}{b}) + R(\frac{\partial v}{\partial x} - \frac{\partial u}{\partial y}) = 0 \quad (4.48)$$

Substitute $\alpha = \frac{D}{R}\frac{\partial f}{\partial y}$ and $\gamma = \frac{F\pi}{Rb}$ to yield the well-known Stommel's solution:

$$\underbrace{-\nabla^2\psi}_{\text{Relative Vorticity}} + \underbrace{\alpha v}_{\text{Planetary Vorticity}} + \underbrace{\gamma\sin(\frac{\pi y}{b})}_{\text{Wind Stress Curl}} = 0 \quad (4.49)$$

Using $v = -\frac{\partial\psi}{\partial x}$, we get

$$\nabla^2\psi + \alpha\frac{\partial\psi}{\partial x} - \gamma\sin(\frac{\pi y}{b}) = 0 \quad (4.50)$$

Now α contains $\beta = \frac{\partial f}{\partial y}$, which is the Coriolis parameter. So, in Stommel's equation, α represents the change of the Coriolis frequency with latitude.

The solution to the above equation is given in various books and in Stommel's paper as a sum of polynomials. The mathematically skilled students can write the code for the solution for two cases: $\beta = 0$ and $\beta = 10^{-11}$.

For $\beta = 0$, $f = $ constant, which means a non-rotating ocean, the solution is a symmetric circulation gyre as shown in the top panel of Figure 4.10. For, $\beta \neq 0$ this solution yields an asymmetric gyre with the streamlines crowding in the western side of the basin. See the bottom panel of Figure 4.10. This clearly shows that the westward intensification of the wind-driven ocean circulation is caused by the parameter β, or the change of Coriolis frequency with latitude, called the "Coriolis parameter."

(a) No rotation or β

(b) Rotating Basin

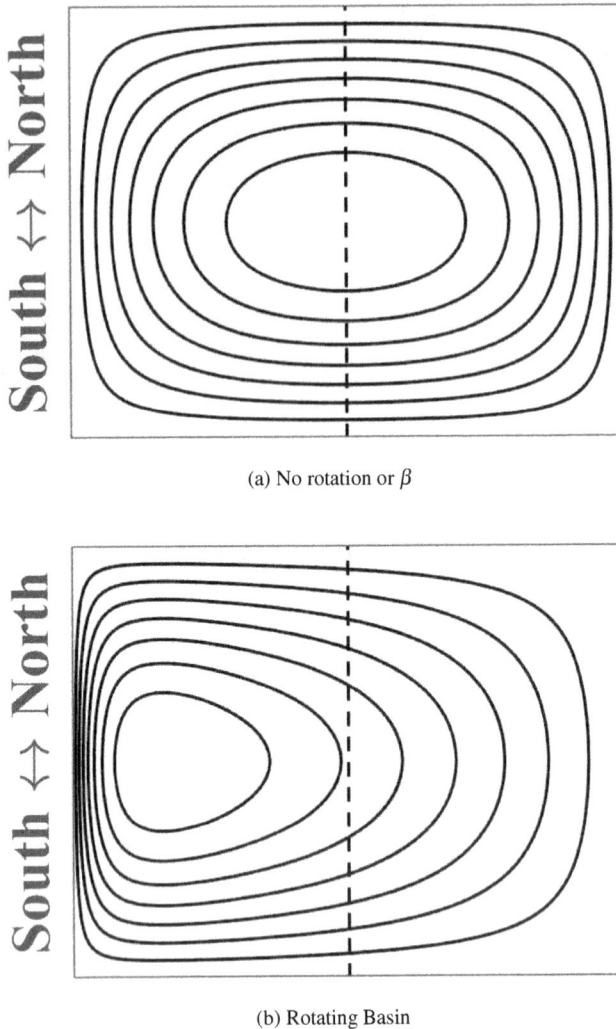

Figure 4.10: Two solutions for Stommel's Model explaining the western boundary current formation. Upper panel: No rotation or $\beta = 0$ with symmetric gyre (no WBC) and lower panel for $\beta \neq 0$ with WBC intensification and an asymmetric gyre (After: H. Stommel, 1948).

4.8 MUNK'S MODEL AND FUTURE DIRECTIONS

Further modifications of the analytical work were done by Munk in 1952, who introduced another factor as lateral friction in addition to bottom friction and came up with a fourth-order (∇^4) solution for the final equation. His solution is as follows:

$$A \nabla^4 \Psi - (\beta \frac{\partial \Psi}{\partial x}) - curl\tau = 0 \qquad (4.51)$$

where A is the eddy viscosity coefficient for lateral friction within the fluid. The biharmonic operator ∇^4 is simply given by

$$\frac{\partial^4}{\partial x^4} + 2 \frac{\partial^4}{\partial x^2 \partial y^2} + \frac{\partial^4}{\partial y^4}$$

The first term in Equation (4.51) is the vorticity generated by the lateral stress, the second term is the planetary vorticity and the third term is the wind stress curl. The last two terms are similar to the terms in the Sverdrup (4.39) or Stommel (4.50) equations.

Because of the fourth-order equation in Munk's model, it is possible to satisfy four boundary conditions – no-flow through and no-slip conditions across and along the eastern and western boundaries. The solutions then allow for a narrow western boundary current regime with a counter-current region to its eastern side, similar to observed recirculation cells to the southeast of the Gulf Stream and the Kuroshio.

Additionally, the Munk solution also quantifies the width of the narrow boundary current, within which the first two terms of Equation 4.51 would approximately balance each other.

Later in the sixties, in 1962, two-gyre models by Carrier and Robinson (1962) were proposed, which were based on the observation that the wind-stress fields provide for two oceanic gyres (subtropical and subpolar) and the western boundary currents flow along the boundary between these oppositely circulating gyres. This was more realistic as the western boundary currents do not end at the northern boundary of the subtropical gyre, but they do separate from the coast at a particular latitude and flows to the northeast as a narrow swift current. The Gulf Stream leaves the coast at Cape Hatteras (35°N, 75°W); the Kuroshio leaves the coast at around 35°N in North pacific, the springtime WBC in the Bay of Bengal leaves the coast at around 18°N (Gangopadhyay, Bharat Raj, et al., 2013); the Brazil Current leaves the Brazilian coast at around 35°S in the South Atlantic (Silveira et al., 2008).

In the seventies, Gill showed how the wind-stress over the Atlantic generates westward propagating waves (called Rossby Waves, see Chapter 6.2.2). Upon reflection from the western boundary, these waves become very fast and dissipate within a narrow width creating the western boundary current. Credence to this time-integrated propagating Rossby waves generating the western boundary current was found in the seminal work by Parsons (1969) and Veronis et al. (1973), who connected the steady-state Sverdrup, Stommel and Munk models with the time-dependent Gill model to define an upper limit of the steady-state balance, which is the latitude of separation of the Western Boundary Currents from the coast. The Parsons-Veronis model of wind-driven circulation and generation of the Western Boundary current and its separation from the coast was then validated by Gangopadhyay et al. (1992) and by A. H. Taylor and Stephens (1998). The path of the Gulf Stream after separation can also be related to the overlying wind-stress curl maximum in a two-gyre model (Carrier & Robinson, 1962; Gangopadhyay & Chao, 2000). Some recent work (Chi et al., 2019; Sanchez-Franks et al., 2016) also connected the Icelandic Low as the subpolar gyre's signature to the path of the Gulf Stream. Current research on a combined model of basin-wide wind forcing and the Icelandic Low movement to the formation and separation of the Gulf Stream is ongoing (Silver et al., 2021b).

CONCLUSION

In this chapter, we discussed and developed the analytical theories of wind-driven ocean circulation. We learned how Ekman designed a concept of eddy viscosity and applied Newtonian friction theory to solve for the wind-driven motion under rotation, with no pressure-gradient and in steady-state. This simple "water flows to the right of the wind in northern hemisphere" solution is key to understand many different physical phenomena including coastal and equatorial upwelling. However, the impact of the wind is limited to the wind-driven mixed layer (the Ekman layer). Sverdrup extended the force-response balance to include the pressure gradient term with rotation and wind stress. He integrated the momentum equations vertically and derived the well-known Sverdrupian balance which explained the basin-wide meridional flow under subtropical wind stress curl. He then used a heuristic argument of boundary condition and continuity to explain the existence of the WBCs. It was Henry Stommel who added the bottom friction term in the momentum equations to arrive at a closed form solution for the gyre circulation. Stommel showed that the variation of Coriolis frequency with latitude or the "beta $(\frac{\partial f}{\partial y})$" term was responsible for the WBCs. Without beta, the oceanic gyres would be symmetric! In the process of developing an analytical equation for the circulation (in terms of stream function); realize that the final equation for the solutions in Sverdrup, Stommel and Munk formulations is a "vorticity balance equation." The concept of vorticity was briefly discussed here. The different terms being relative, planetary, absolute, and potential vorticity for the flow and wind-stress curl for the wind-forcing.

FURTHER READING

Ekman, V. W. (1905). *On the influence of the earth's rotation on ocean-currents.* Almqvist & Wiksells Boktryckeri, A.-B.

Gill, A. E. (1982). *Atmosphere-ocean dynamics (international geophysics series)*. Academic Press.

Holton, J. R. (1973). An introduction to dynamic meteorology. *American Journal of Physics*, *41*(5), 752–754.

Munk, W. H. (1950). On the wind-driven ocean circulation. *Journal of Meteorology*, *7*(2), 80–93.

Pond, S., & Pickard, G. L. (1983). *Introductory dynamical oceanography*. Gulf Professional Publishing.

Robinson, A. R. (1963). *Wind-driven ocean circulation: A collection of theoretical studies*. Blaisdell Publishing Company.

Stommel, H. (1948). The westward intensification of wind-driven ocean currents. *Eos, Transactions American Geophysical Union*, *29*(2), 202–206.

Stommel, H. M. (1965). *The Gulf Stream: A physical and dynamical description*. University of California Press.

Perito Moreno Glacier, Argentina

5 The Abyssal Connection

> What we know is not much. What
> we don't know is enormous.
>
> ---
>
> Pierre-Simon Laplace (1749–1827)

OVERVIEW

In the previous chapter, we have developed a number of analytical closed-form solutions for the wind-driven ocean circulation for a homogeneous ocean. We have explored how upper layers of the ocean respond to the combined effects of rotation and wind forcing. It is possible to describe the dynamical balances as combined effects of pressure, Coriolis, hydrostatics, and wind-stress for overall understanding of the first-order vertically integrated behavior of the currents, eddies, and gyres.

In this chapter we will dive into the deep ocean. The abyssal circulation is similarly impacted by rotation, now the difference being that the sources of flow coming from sinking of water-masses in the polar regions and upwelling in the subtropics. Eventually, the abyssal circulation is connected to the wind-driven upper layer via what is known as *the conveyor belt*. A water-mass distribution is presented to connect the two layers in different oceans. A conceptual analytical model of the thermohaline circulation is presented to illustrate the dynamical balance. An example of a combined upper-layer and abyssal flow in the Atlantic is presented in the final section of this chapter.

5.1 THE BASICS

Let us think about the other kind of circulation – the circulation in the abyss – the thermohaline circulation. The word thermohaline is a mix of two words – "thermal (temperature)" and "haline (salinity)." It is the effect of different water-masses having different temperatures and salt contents that creates a density gradient in the fluid. This density difference then drives the flow. The abyssal flow experiences the effect of rotation and creates circulation features in certain patterns and interacts with other water-masses (including the surface wind-driven current) in a fully coupled three-dimensional inter-oceanic balanced system. Let us explore this three-dimensional behavior of ocean circulation here.

DOI: 10.1201/9780429347221-5   **91**

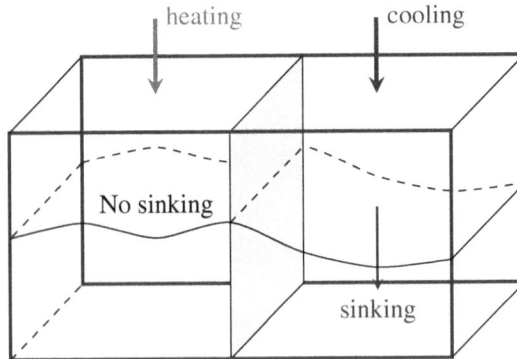

Figure 5.1: Effect of heating versus cooling to crease density difference in the vertical.

Consider a simple example of how a density difference might happen in the ocean. Think about how the fluid will behave when it is a layered fluid with two different cases of heating, as in Figure 5.1.

In the top panel in Figure 5.1, the water is being heated from above, so the density in the upper layer of the fluid will decrease and will not sink down as the density below is already higher than above. However, in the case of the lower panel, when the water is being cooled from top, the temperature decrease will result in increasing density at the surface layers – which will at some point become heavier than the water below and will sink!

This simple mechanism of sinking can become a little more complex in the polar regions of the Earth. Let us see how that happens. As we know, the Earth is getting excess heat on the equatorial regions and a deficit of heat on the poles, which results in the icecaps of the Arctic and Antarctica. We also know that the air sinks down on poles and blow equatorward near the poles at the surface (in the Hadley/Direct cell system) – see Figure 1.9. Now think about the seasonality of the winds. During winter, these sinking air-masses are really cold, and the surface cold air on the poles will start freezing the seawater adjacent to the ice already in existence in the Arctic and Antarctica for years and months.

Now, as the seawater freezes, ice forms and the salt is ejected out of the ice-forming seawater to the surrounding cold water (which is already becoming cooler due to cold air blowing over it). Either direct cooling or indirect salting or a combined effect (temperature decrease and salinity increase) will make the seawater near Arctic heavier at the surface than the water below (like in 5.1 bottom panel). This leads to winter convection and the new water (low temperature, high salinity) forms and sinks to an appropriate density level for it to start flowing following the isobaths (constant bathymetric contours) and preserving potential vorticity (See Section 4.5) in its layer thickness.

5.2   THE SINKING REGIONS

So, the thermohaline circulation begins in the Earth's polar regions. Upper surface water increases its density either directly by cooling or indirectly when ice freezes out to eject salt or due to a combined effect of these two, and thus, increasing the density of the remaining water. To repeat again, in short, when ocean water in these areas gets very cold, sea ice forms. The surrounding seawater gets saltier, increases in density, and sinks.

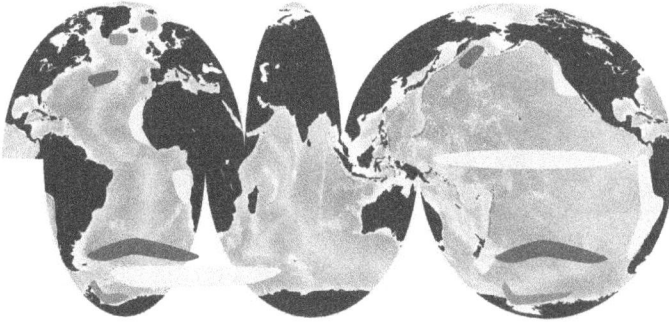

Figure 5.2: The deep water formation regions around the world oceans. Purple – deep sinking; blue – intermediate level sinking; and yellow – upwelling.

The Earth's present equilibrium has four different sinking regions. The major sinking happens in the northern North Atlantic where the North Atlantic Deep Water (NADW) forms in Labrador Sea Greenland Sea, and Norwegian Sea. See Figure 5.2.

The Mediterranean Deep Water (MDW) forms off of the Mediterranean sea. The Antarctic Bottom Water (AABW) forms in the southern ocean on the margins of the Antarctic continent in the Wedell and Ross seas. For a full description of these different types of water-masses please see the book on regional oceanography by Tomczak and Godfrey (2013).

There is not much sinking happening in the other oceans (Indian or Pacific) except in the Bering Sea, where the intermediate waters at mid-depth, called the Pacific Intermediate Water (PIW) sink and form bottom water of the Pacific. See Figure 5.2.

5.3 THE CONVEYOR BELT

Don't you want to know what happens to water that sinks every year in the Labrador sea during winter convection period? How does the NADW spread or move through the abyss? Scientists have been looking for answers to that question for a number of years until recently. And even after you know how it flows through the deep abyssal pathways, they would have to connect to the upper-level wind-driven circulation patterns discussed in Chapter 4. Right?

After years and years of observations and analysis of chemical tracers which are very important and longer lasting-markers of the water's age (and thus very helpful in tracking them along a pathway), a global picture has evolved. A unified theory of thermohaline circulation was put forward in the late 1980s and early 1990s by (Broecker, 1997; Gordon, 1986). Eventually we have begun to understand how the thermohaline circulation works and how it connects to the upper layer circulation in a global sense.

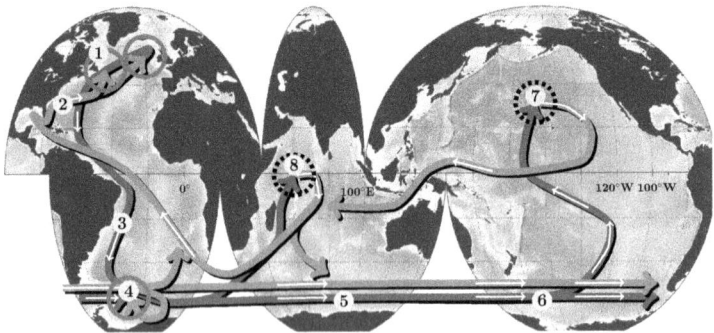

Figure 5.3: A simple depiction of the Thermohaline Circulation. Note the upper and lower layers of circulation. The sinking regions (solid circles) and upwelling regions (dashed circulations) are located at different parts of the world ocean connecting the abyssal circulation with the wind-driven upper ocean circulation. Figure Credit: Adrienne Silver.

We now know that the thermohaline circulation drives a global-scale system of currents called the "global conveyor belt." The conveyor belt begins on the surface of the ocean near the pole in the North Atlantic. Here, as mentioned in Sections 5.1 and 5.2, the water is chilled by arctic air temperatures. It also gets saltier because when sea ice forms in winter, the salt does not freeze and is left behind in the surrounding water. The cold water is now more dense due to the added salts, and sinks toward the ocean bottom. Surface water moves in to replace the sinking water, thus creating a current.

Let us follow the water using their water-mass (t-s) characteristics. See Figure 5.3 for the locations marked from 1 through 8 in yellow. Starting from the Labrador convection region (#1) the water flows in a slow circulation around the three different basins and comes back following the conveyor belt route. See Figure 5.4a for the isothermal near-zero temperature with increasing salt for deeper waters in the Labrador water t-s signature. This deep water moves south, between the continents, past the equator, and down to the ends of Africa and South America. See Figure 5.3. This water-mass is called the NADW, which can be traced along the pathway of the Deep Western Boundary Current (DWBC) and observed to flow under the Gulf Stream (Figure 5.4b) in the subtropical North Atlantic, then across the equator near the Caribbean, and then flow along the coast of Brazil southward under the multiple layers of opposing flows in the Brazil Current system (Figure 5.4c).

This DWBC flow is refueled with new water as it travels along the coast of Antarctica and picks up more cold, salty, dense water from this secondary sinking region. See the region before the middle branching in Figure 5.3 (Atlantic sector). The main current splits into two sections: one traveling northward into the Indian Ocean, while the other flows into the western Pacific, a part of which goes all around the Antarctica to become part of some of the deeper layers of the Antarctic Circumpolar current. See the middle branching in Figure 5.3 (ACC splitting in the Atlantic sector).

The water-masses along the ACC remain very similar in their character along its circumpolar path (along 58-60°S) in all three sectors. See panels d (Atlantic sector at 30°E), e (Indian sector at 90°E), and f (Pacific sector at 170°W) in Figure 5.4 for locations #4, #5, and #6, respectively, in Figure 5.3.

Next, see how these branches in the Indian and Pacific Oceans flows rise and turn back in Figure 5.3. The two branches of the current warm and rise as they travel northward, then loop back around southward and westward. The equatorial region of the Indian Ocean and the polar regions of the Pacific Ocean are thought to be the rising (upwelling) regions for these modified waters where they turn southward and westward, mixing with the wind-driven upper layer warm surface currents. See the modified water-mass of the North Pacific (Figure 5.4g) and notice how the deeper waters are connected back to the waters of ACC in the three previous panels (d, e, f) of Figure 5.4. Finally, realize that the upwelled waters in the Indian Ocean mixed with surrounding waters from the western equatorial and tropical Pacific make the waters in the tropical Indian Ocean (Figure 5.3h) warmer and saltier than those in the north pacific (Figure 5.3g).

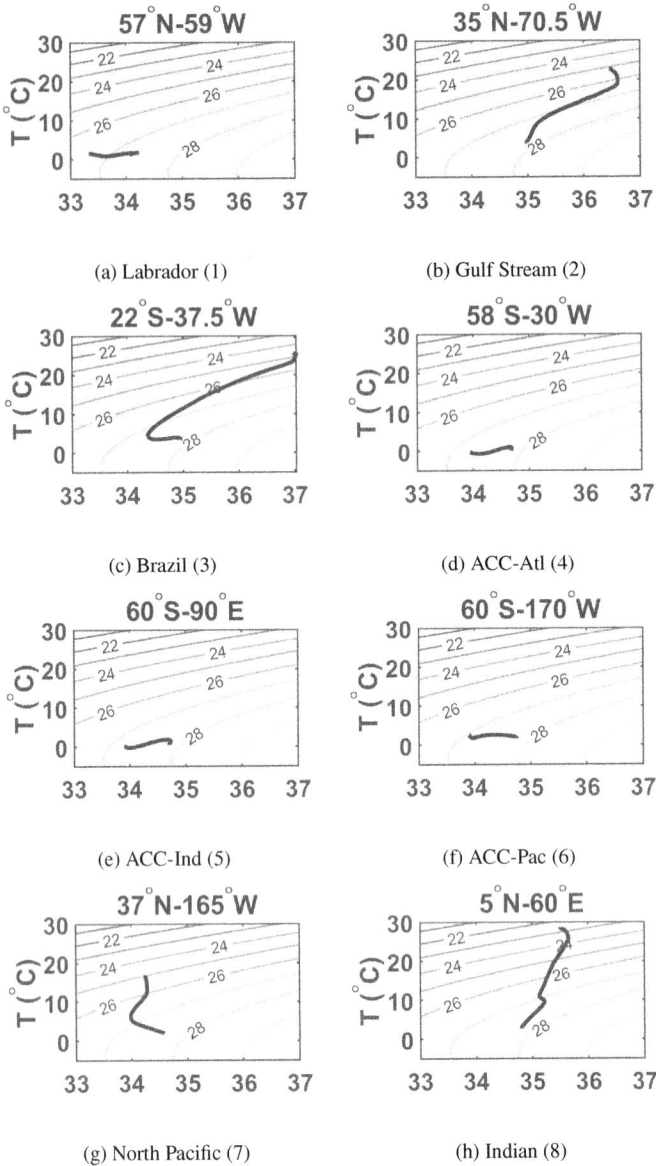

Figure 5.4: Typical climatology water-mass distribution in the eight locations identified by numbers (1-8) in the Thermohaline Circulation schematic shown in Figure 5.3. The x-axis is salinity (psu). The contour lines are for $\sigma_\theta = \rho - 1000$ in each panel. Figure Credit: Adrienne Silver.

The now-warmed surface waters continue circulating around the globe. They eventually return to the North Atlantic, where the cycle begins again.

In summary, Figure 5.3 shows a conceptual schematic diagram of the global ocean circulation pathway known as "conveyer" belt due to its northward transport at the ocean surface and southward return flow in the abyss in the Atlantic. The ocean circulation system is a slow, three-dimensional pattern of flow involving the surface and deep oceans around the world. The water-masses shown in Figure 5.4 connects the waters of different regions through the different segments of the conveyor belt.

Important elements of the thermohaline circulation are the Labrador Current, the Deep Western Boundary Current. Other branches of the circulation is less well-defined – however, it is better expressed as a collection of water-masses such as NADW (found in the South Atlantic), Antarctic Bottom water, and a number of intermediate-level water-masses – the understanding of which are still evolving in the first quarter of the twenty-first century. A somewhat less obvious fact that is ignored in the whole scheme of things is the role that is played by the Eastern Boundary Current systems.

5.4   MERIDIONAL OVERTURNING CIRCULATION

So, there you have it. Three oceans – three different kinds of circulation with commonalities. Upper ocean Sverdrup flow and Western Boundary Currents in every ocean. Lower thermocline DWBC primarily in the sinking region of North Atlantic and other deep water slow flows.

Look at the conveyor belt in Figure 5.5 carefully. The sinking regions of the northern North Atlantic and the Antarctic region continuously replenish the deep water formation region through atmospheric cooling from the top. The key to the conveyor belt circulation is the upwelling regions (black circles) in the middle of the Pacific and Indian Oceans. How does it happen that waters reaching these two regions do not go further and starts to upwell? One possible explanation is that the ambient water in the northern Arabian Sea and equatorial Indian Ocean always remain dense enough due to supply of very saline intermediate Red Sea Water. So, the relatively lighter water coming from Antarctica rises up and mixes with the other surface flows coming through the Indonesian Throughflow region and collectively they flow back southward in the upper layer. See Figure 5.5.

The situation in the pacific is similar with relatively fresh waters in the northern latitudes are not dense enough to sink (leaving the dense water below) and the northward flow from the Antarctica region upwells to mix with the southward upper layer flow in the North Pacific and flow further south and then westward with the equatorial flow system.

Putting it all together, we get a complex realization of the three-dimensional circulation of the world ocean in a global context. This is what is called the

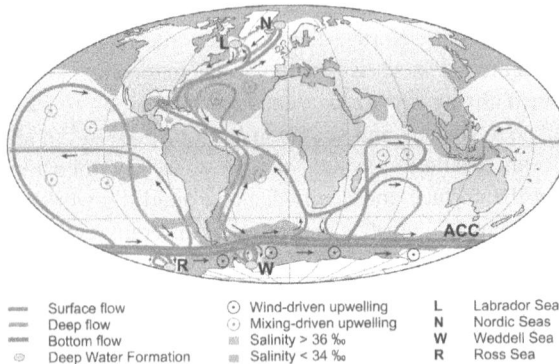

Figure 5.5: Schematic of the Global Overturning Circulation. Strongly simplified sketch of the global overturning circulation system. In the Atlantic, warm and saline waters flow northward all the way from the Southern Ocean into the Labrador and Nordic Seas. By contrast, there is no deep water formation in the North Pacific, and its surface waters are fresher. Deep waters formed in the Southern Ocean become denser and thus spread in deeper levels than those from the North Atlantic. Note the small, localized deepwater formation areas in comparison with the widespread zones of mixing-driven upwelling. Wind-driven upwelling occurs along the Antarctic Circumpolar Current (ACC). After *Rahmstorf* [2002]. Republished with permission of John Wiley & Sons, from Kuhlbrodt et al. (2007). On the driving processes of the Atlantic meridional overturning circulation. *Reviews of Geophysics*, 45(2); permission conveyed through Copyright Clearance Center, Inc.

"Overturning circulation." Note the meridional balance that is brought about by the flow system. See Figure 5.5.

The meridional overturning circulation is a system of surface and deep currents encompassing all ocean basins. It transports large amounts of water, heat, salt, carbon, nutrients, and other substances around the globe, and connects the surface ocean and atmosphere with the huge reservoir of the deep sea.

This is a rather simple picture. More complicated understanding of the Conveyer belt and its probable morphing to the idea of a more general concept of a "Global Meridional Overturning Circulation" or the "Meridional Overturning Circulation" (MOC) is found in Talley, 2011 and Talley, 2013 and other recent studies. For the Atlantic ocean, this spans over both south and north Atlantic and is called the "Atlantic MOC" or "AMOC." Let us discuss this a bit now.

5.4.1 ATLANTIC MERIDIONAL OVERTURNING CIRCULATION (AMOC)

The theoretical ideas of Stommel (1957) and the conceptual conveyor belt model by (Broecker, 1987, 1991) helped our understanding of the connection between the abyssal circulation and the upper-layer circulation to a great extent. This conceptual understanding is based on temperature, salinity, and oxygen signatures of the intermediate and deep water-masses and tracing them all over the world from their source regions (Richardson, 2008 and references therein). The flow pathways were inferred (see Figure 5.6) by assuming a primarily advective system of flow for the tracers. Formation of deep water and its role as the source water and driver for the abyssal circulation was a critical element for our understanding of the conveyor belt (Lozier, 2012). We will discuss the analytical model of Stommel and Arons, 1960 on the set up of the abyssal flow using such source terms in the next section.

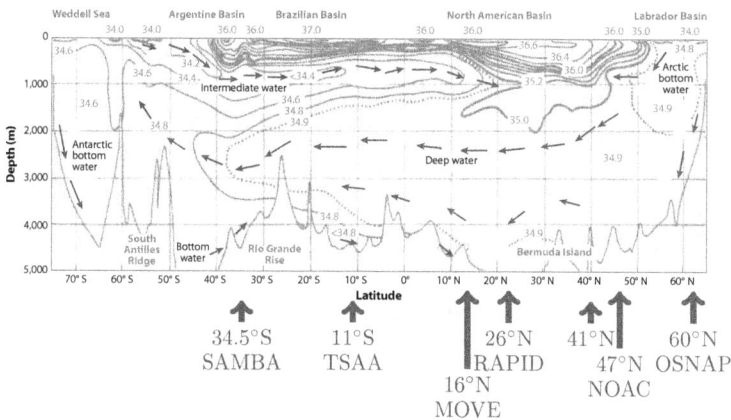

Figure 5.6: Meridional cross-section of salinity in the western Atlantic Ocean. Figure modified from Lozier (2012) with permission from Annual Reviews. (Figure originally adapted from Tchernia 1980, where it was reprinted from Merz 1925.) Observing arrays for monitoring the MOC in the Atlantic with AMOC transport estimates are shown along the bottom meridional axis with arrows. These are (from north to south): OSNAP at $\approx 55 - 60°$N (from 2014), NOAC 47°N (still to be produced), RAPID 26°N (from 2004), MOVE 16°N (from 2000), TSAA 11°S (still to be produced), and SAMBA 34.5°S (from 2009).

Naturally, much of the details of the workings of the conveyor belt are still under investigation. Some of the interesting questions that have been asked in last two decades are related to the observations that (i) the DWBC flow in not continuous and is eddy-dominated at certain latitudes, (ii) there exist other interior pathways than the DWBC for abyssal transport – an observed difference between mean and synoptic pathways, (iii) the upper limb of the overturning is also discontinuous. See Lozier, 2010 and references therein for detail discussion on these aspects. A number of studies have recently suggested that the impact of buoyancy and wind forcing on the MOC transport is different over different time-scales; wind-forcing dominating the seasonal, interannual, and decadal variability while buoyancy forcing dominates over the longer, centennial time-scales (Biastoch et al., 2008; Mielke et al., 2013; Zhao & Johns, 2014a, 2014b). Such findings bring up the natural question: What forces the overturning? How do wind and buoyancy impact this complex multiscale process of the ocean's three-dimensional circulation from synoptic eddy scale to the climatic scales?

Starting in 2004, a series of monitoring mooring arrays have been placed to measure the volume and heat transport across a selected set of latitudes in the north and south Atlantic. These locations are shown in Figure 5.6 along the bottom meridional axis. New results are being analyzed from the data from these mooring which highlights that the picture of overturning is far from being complete. It is a story of evolution of our understanding of the complex and intricate pathways and transport of the flow and heat of the overturning circulation that connects the upper limb to the abyssal flows. For a recent review of results from our observational and modeling understanding of AMOC variability based on data from 2004 until 2020, see the works by Frajka-Williams et al. (2019) and by Caesar et al. (2021).

5.5   A SIMPLE MODEL FOR THE THC

It is instructive to understand the basic dynamical reason behind the formation of the deep currents, much like the asymmetric setup of the wind-driven ocean currents where the forcing by the wind stress curl and the beta-effect result in the western boundary current there is a similar setup for the abyssal circulation. Here, the abyss is dominated by multiple deep western boundary currents. These are forced by the deep water formation and sinking near the poles in the Labrador/Greenland/Norwegian Seas and the Wedell Seas. See Figure 5.2. The original complete theory of the abyssal circulation was put forward by Stommel and Arons (1960). Their solution is shown in Figure 5.7.

Let us see how the DWBC occurs in the Stommel and Arons, 1960 analytical model setup. We assume that the deep ocean is a shallow layer of homogeneous density, with depth H. The circulation is set up by the source of cold and dense water at the poles and upwelling of such waters in a very distributed manner over the ocean interior, as explained above. In this idealized model, the ocean bottom is flat, there is no dynamical coupling with the upper ocean (where the upwelled water needs to mix with the surface flow), and of course there is no stratification.

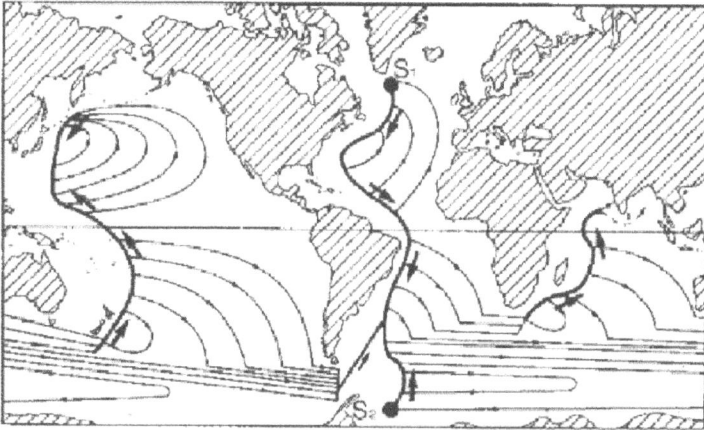

Figure 5.7: Stommel-Arons solution for abyssal circulation. Republished with permission of Elsevier Science and Technology Journals, from Stommel, H. (1958). The Abyssal circulation. *Deep Sea Research, 5*; permission conveyed through Copyright Clearance Center, Inc.

Following Stommel and Arons original paper, we write the equations of motion in spherical coordinates. We encourage you to jump to Chapter 8 and see the details about the setup. Let us consider a simple basin geometry, which is a hemispheric sector (much like North Atlantic) bounded by the equator and north pole ($\phi = 0, \pi/2$), and the longitudes going from an arbitrary 0 to a λ_1. The sectorial difference in longitude is λ_1 and the sector could start at any particular longitude (the zero longitude). See Figure 5.8.

So, the volume flux that down wells at the pole is given by

$$S_0 = Q_0 \dot{A}rea = Q_0 a^2 \lambda_1 \tag{5.1}$$

Using geostrophy, we can write the equations of motion balancing the pressure gradient with the Coriolis force and the continuity equation as follows:

$$-fv = -\frac{1}{a\cos\phi}\frac{\partial p}{\partial \lambda} \tag{5.2}$$

$$+fu = -\frac{1}{a}\frac{\partial p}{\partial \phi} \tag{5.3}$$

$$\frac{1}{a}\frac{\partial u}{\partial \lambda} + \frac{1}{a}\frac{\partial}{\partial \phi}(v\cos\phi) + \frac{\partial}{\partial z}(w\cos\phi) = 0 \tag{5.4}$$

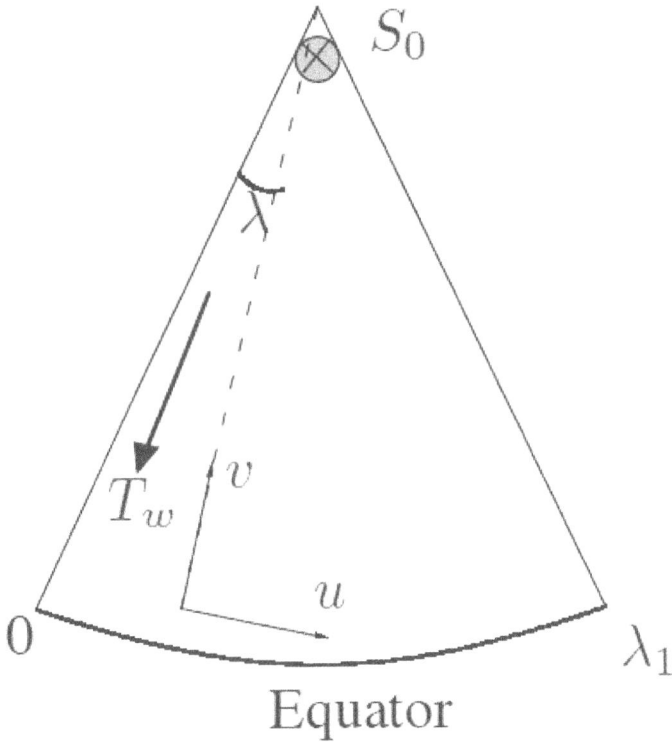

Figure 5.8: Sectorial domain for abyssal circulation analytical model. The volumetric source S_0 is meant to represent dense water formation at high latitudes. Republished with permission of Elsevier Science and Technology Journals, from Cushman-Roisin, B. (2009). *Introduction to geophysical fluid dynamics: Physical and numerical aspects*; permission conveyed through Copyright Clearance Center, Inc.

Here, a is the radius of the Earth, λ is the longitude, ϕ is the latitude, z is the local vertical coordinate, and $f = 2\Omega \sin \phi$, as before. We can use the usual procedure of cross-differentiation to eliminate the pressure term from the two momentum equations, which results in

$$\frac{\partial}{\partial \lambda}(fu) + \frac{\partial}{\partial \phi}(fv \cos \phi) = 0. \qquad (5.5)$$

Now, we can use the conservation of volume equation (5.4), and we reach the abyssal equivalence of Sverdrup solution for wind-driven circulation given by,

$$\beta v = f \frac{\partial w}{\partial z} \qquad (5.6)$$

where β is $f = 2(\Omega/a)\cos\phi$.

Following Sverdrup and Stommel's development for the wind-driven solutions, we can now integrate this equation over depth with boundary conditions $w(z=0) = w_0$, and $w(z=-H) = 0$. This leads to the meridional velocity being,

$$\int_{-H}^{0} v\,dz = \mathbf{V} = aw_0 tan\phi \qquad (5.7)$$

Assumption of a barotropic abyssal layer (from equation 5.2) leads to even a simpler form for the northward velocity,

$$v = \frac{aw_0}{H} tan\phi \qquad (5.8)$$

which means, $v = 0$ on the equator ($\phi = 0$) and $v = \inf$ at the poles ($\phi = 90°$). It also means that, with $w_0 > 0$ everywhere, v is northward everywhere, between the equator and the pole in the northern hemisphere and v is southward everywhere between the equator and the pole in the southern hemisphere.

Now, using the above expression for v in the conservation equation 5.4, we can get the following expression describing the zonal velocity,

$$\frac{\partial u}{\partial \lambda} = -\frac{\partial}{\partial \phi}(\cos\phi\, v) - \frac{aw_0}{H}\cos\phi \qquad (5.9)$$

Now, we can use the eastern boundary condition for $u = 0$, at $\lambda = \lambda_1$, and integrate between any longitude (λ) and the eastern boundary to get the zonal velocity at any longitude (λ), which is given by,

$$u = -\frac{a}{H\sin\phi} \frac{\partial}{\partial \phi} \left(\sin^2\phi \int_{\lambda_1}^{\lambda} w_0\,d\lambda \right) \qquad (5.10)$$

Since the upwelling w_0 is assumed uniform over the domain, this reduces to,

$$u = 2\frac{a}{H} w_0(\lambda_1 - \lambda)\cos\phi \qquad (5.11)$$

So, the zonal velocity is always positive, i.e. eastward.

So, this idea is that the solution of the abyssal circulation points to a northward and eastward flow field around the domain. This is counter-intuitive in the first place, as we have a huge source at the pole which is shedding flow southward to begin with!

Well, the only solution would be to allow for that source-driven flow to flow southward in a very narrow western boundary current towards equator and on its way recirculate back to the eastwards and northward flows of a slow abyssal gyre. See the right panel of Figure 5.8 for a possible schematic solution. This is how the Deep Western Boundary Current in the deep oceans was explained by Stommel and Arons, 1960. Of course, the later models have verified and modified the complex system of thermohaline circulation as we know it now. However, the above exposition helps set the stage for the basic dynamics of the THC.

Finally, let us go one step forward with this formulation and write the transport budget at a particular latitude ϕ. For this we follow Cushman-Roisin and Beckers, 2011 and consider the northern hemisphere. The Source S_0 and the northward transport integrated over the domain (between ϕ and the pole) should balance the DWBC transport at latitude ϕ and the upwelling water at the top for the pole-to-ϕ domain. We can write this as,

$$S_0 + \int_0^{\lambda_1} Va\cos\phi\, d\lambda = T(\phi) + \int_\phi^{\pi/2} \int_0^{\lambda_1} w_0 a^2 \cos\phi\, d\lambda\, d\phi \qquad (5.12)$$

For an uniform w_0, and using equation 5.7 to eliminate V, we get,

$$S_0 + \sin\phi\, \lambda_1 a^2 w_0 = T(\phi) + (1 - \sin\phi)\lambda_1 a^2 w_0 \qquad (5.13)$$

On further simplification, the transport in the western boundary current is given by,

$$T(\phi) = S_0 + (2\sin\phi - 1)\lambda_1 a^2 w_0 \qquad (5.14)$$

Three cases can be conceptualized for three different scenarios of relative magnitude of the source versus the upwelling intensity.

Scenario 1: If $S_0 = \lambda_1 a^2 w_0$, then the upwelling from the equator to the pole matches exactly with the source. The transport $T(\phi) = 2S_0 \sin\phi$, which is twice the Source transport at the pole and reduced to zero on the equator. Effectively, this scenario has the hemispheres decoupled from each other in their abyssal circulation!

Scenario 2: If $S_0 > \lambda_1 a^2 w_0$, then the strength of the convective source exceeds the return distribution through upwelling. In this case, the excess southward flow will cross the equator and flow into the southern hemisphere. This is the thermohaline circulation situation that currently prevails in the North Atlantic.

Scenario 3: If $S_0 < \lambda_1 a^2 w_0$, then the source is unable to sustain the required upwelling; and in this case, an additional northward boundary-layer flow across the equator is needed to feed the system to make up for this difference. The current circulation in the North Pacific is an example of this scenario.

5.6   THE COMBINATION OF WIND-DRIVEN AND THERMOHA-LINE CIRCULATION

Finally we want to close this chapter with an example of how the thermohaline circulation described in this chapter interacts with the wind-driven circulation in Chapter 4 in a three-dimensional circulation on the western boundary of the Atlantic Ocean. Let us think about the North Atlantic basin with the wind-driven Gulf Stream, which flows along the coast of North America from 25°N to 35°N before separating out to the East. The wind-driven transport is about 25Sv at around 27°N with a seasonal cycle of about 2Sv. The transport increases to about 85-90Sv at 35°N due to recirculation in its offshore side.

In contrast, with a similar set up, in the Southwest Atlantic, the Brazil Current, which is the wind-driven western boundary current along the coast of Brazil, flows southward with only 4Sv at 25°S, which increases to about 14Sv at 32°S (Majumder et al., 2016).

The question is, "why is this difference between the two currents which are similarly set up by the subtropical wind-driven Sverdrupian gyres?"

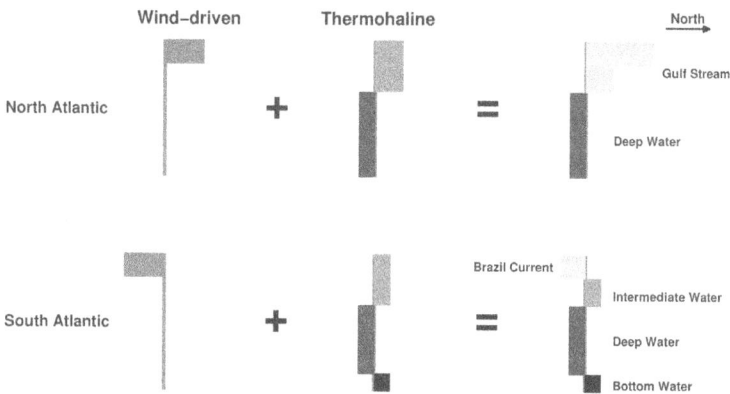

Figure 5.9: Gulf Stream and Brazil Current – an example of combining wind-driven and thermohaline flows in the WBCs. Figure Credit: Professor Dr Ilson Silveira

The answer lies in the fact that ocean circulation is three-dimensional and both effects of wind-driven and thermohaline circulation combine to provide a broader picture. See Figure 5.9. A simpler two-layer (wind-driven upper layer and thermohaline lower layer) was originally proposed by Stommel (1957). Later observations in the 80s and 90s have exposed the nature of complicated three-dimensionality of the system with multiple limbs interconnecting the abyssal circulation with the upper ocean circulation with upwelling and sinking regions in different regions of the world ocean. A four-layer system conceived by Schmitz (1995) was able to explain major circulation pathways of the world ocean and is a great synthesis.

However, the quest is still on as we have a lot of questions on how the ocean works through its interbasin exchanges between the Atlantic and Pacific, between Atlantic and Indian Oceans, and through the Indonesian throughflow between the Pacific and the Indian Ocean in the tropics. New discoveries are eagerly awaiting the future generation of oceanographers.

CONCLUSION

Much like the surface circulation, the abyssal circulation in the ocean is governed by the Earth's rotation, the driving force being the difference in density. Polar regions of sinking and upwelling regions in the subtropics and tropics connect the abyssal circulation to the upper layers. The global representation of a combined surface and abyssal circulation, known as the *Global conveyor belt* is linked with the evolving idea of the Meridional Overturning Circulation (MOC) in different ocean basins. A conceptual analytical model of the thermohaline circulation was presented to illustrate the dynamical balance. A combined analysis of upper-layer and abyssal flow for the Gulf Stream and the Brazil Current illustrated the observed differences in their transports while being at similar latitudinal bands in the north and south Atlantic.

FURTHER READING

Broecker, W. S. (1991). The great ocean conveyor. *Oceanography*, *4*(2), 79–89.

Broecker, W. S. (1997). Thermohaline circulation, the Achilles heel of our climate system: Will man-made CO_2 upset the current balance? *Science*, *278*(5343), 1582–1588.

Cushman-Roisin, B., & Beckers, J. (2011). *Introduction to geophysical fluid dynamics: Physical and numerical aspects*. Elsevier.

Gordon, A. L. (1986). Interocean exchange of thermocline water. *Journal of Geophysical Research: Oceans*, *91*(C4), 5037–5046.

Hogg, N. G., & Huang, R. X. (1995). *Collected works of Henry M. Stommel* [in 3 volumes]. American Meteorological Society.

Stommel, H. (1957). A survey of ocean current theory. *Deep Sea Research*, *4*, 149–184.

Stommel, H. M. (1965). *The Gulf Stream: A physical and dynamical description*. University of California Press.

Stommel, H., & Arons, A. B. (1960). On the abyssal circulation of the world ocean – I: Stationary planetary flow patterns on a sphere. *Deep Sea Research*, *6*, 140–154.

Talley, L. D. (2013). Closure of the global overturning circulation through the Indian, Pacific, and Southern Oceans: Schematics and transports. *Oceanography*, *26*(1), 80–97.

Tomczak, M., & Godfrey, J. S. (2013). *Regional Oceanography: An introduction.* Elsevier.

Van Aken, H. M. (2007). *The oceanic thermohaline circulation: An introduction* (Vol. 39). Springer Science & Business Media.

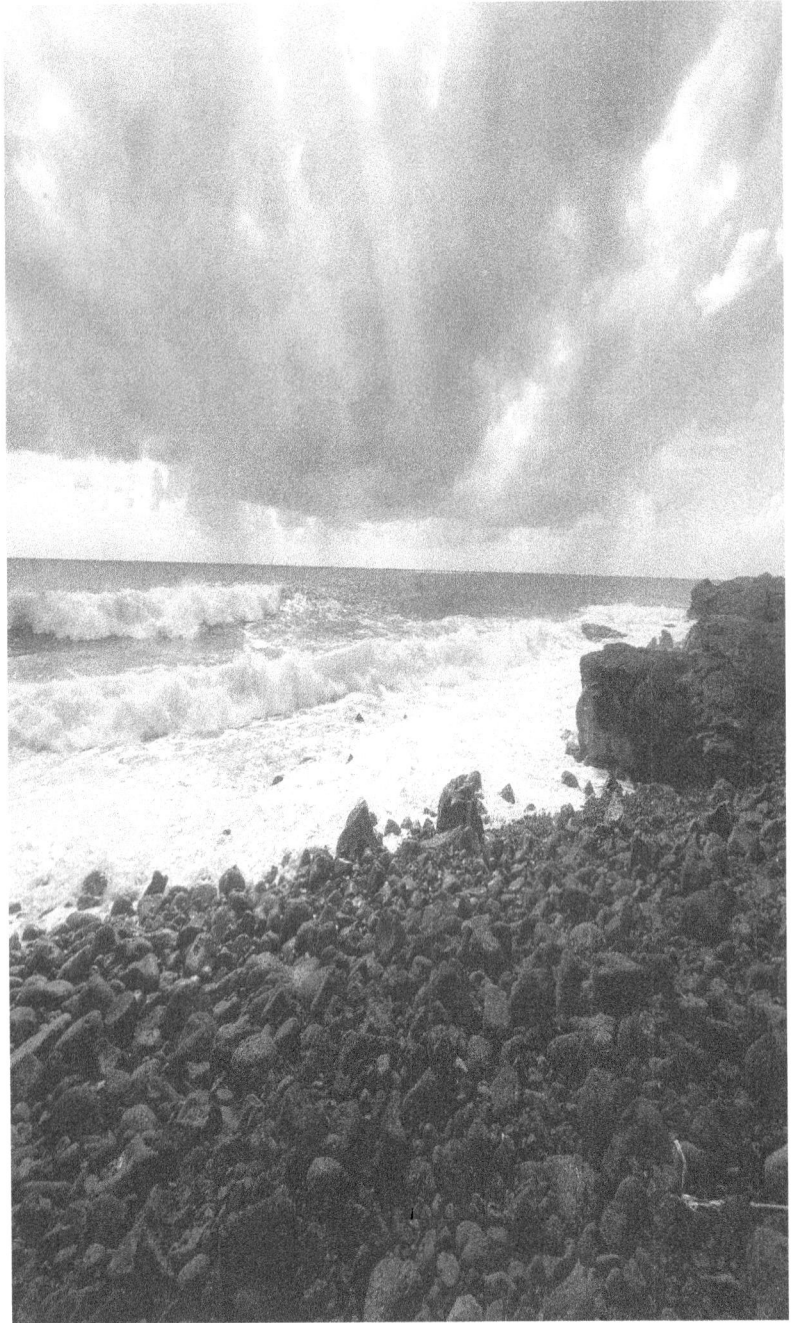

Laupahoehoe Beach Park, Hawai'i

6 Time–Dependent Circulation

> Because there's nothing more beautiful than the way the ocean refuses to stop kissing the shoreline, no matter how many times it's sent away.

> Sarah Kay (1988 -)*

OVERVIEW

So far, we have discussed ocean circulation from a steady-state perspective. The geostrophic equilibrium in Chapter 3, the wind-driven circulation in Chapter 4 and the thermohaline circulation in Chapter 5 – all of these balances were developed assuming that the acceleration is zero ($\frac{du}{dt} = 0$) on the left side of the governing equations. Further dynamical assumptions resulted in a closed analytical solution for the integrated stream function, which defined the circulation patterns in a steady-state.

In this chapter, we will investigate the possible solutions when we allow time-dependence for the forcing fields and response variables. These solutions are typically in the form of *waves* of multiple frequencies and wavelengths. We discuss the characteristics of a single traveling wave. The idea of Fourier Transform is presented next which allows us to decompose or reconstruct an irregular time-series (of force or response) into a set of useful harmonics (single-frequency/wavelength components). Then we focus on a number of simple geophysical settings with time-dependent equations of motion whose solutions are geophysical waves such as the Kelvin waves and Rossby waves. Finally we discuss the unique characteristics (balances and properties) of these waves as can be recognized in typical oceanic circulation.

6.1 TIME-DEPENDENCE AND WAVES

It was important to look at the ocean and how it works in steady-state, or essentially, with no time-dependence. Now, we want to introduce a bit of time-dependence. How do we do this?

---

*Credit/Source: From B by Sarah Kay, © 2015. Reprinted by permission of Hachette Books an imprint of Hachette Book Group, Inc.

DOI: 10.1201/9780429347221-6

109

Consider the simplest set up with acceleration being resulting from a combination of Coriolis, pressure-gradient and friction forces:

$$\frac{du}{dt} = +fv - \frac{1}{\rho}\frac{\partial p}{\partial x} + F_x \tag{6.1}$$

$$\frac{dv}{dt} = -fu - \frac{1}{\rho}\frac{\partial p}{\partial y} + F_y \tag{6.2}$$

In the steady-state, the first term is zero and the rest of the equation is then applied to different balances described earlier in Chapters 2, 3, and 4.

$$\underbrace{\frac{du}{dt}}_{\text{steady state}} \!\!\!\!\! \overset{0}{=} \underbrace{+fv}_{\text{Coriolis}} \underbrace{-\frac{1}{\rho}\frac{\partial p}{\partial x}}_{\text{PGF}} + \underbrace{F_x}_{\text{Friction}} \tag{6.3}$$

$$\text{Sverdrup/Stommel/Munk}$$

$$\text{Coriolis + PGF} \Rightarrow \text{Geostrophy}$$

$$\text{Coriolis + Friction} \Rightarrow \text{Ekman}$$

Here PGF is the pressure gradient force.

Let us go back to the basic equation $F = ma$ and look at the total acceleration term again. The acceleration a has those four components, local derivative and three advection terms, remember? See section 2.5 and the Taylor series expansion equation 2.3.

Let us now look at the total derivative (the a term) mathematically for the x-component of the velocity:

$$\frac{du}{dt} = \frac{\partial u}{\partial t} + u\frac{\partial u}{\partial x} + v\frac{\partial u}{\partial y} + w\frac{\partial u}{\partial z} \tag{6.4}$$

All of the balances considered so far (in Chapters 2, 3, and 4) assumed a steady state or $\frac{du}{dt} = 0$, i.e. no time-dependence.

To allow for time-dependence, but still not consider non-linearity, we can ignore or neglect the non-linear advective terms but keep the local derivative on the right-hand side of equation 6.4 as follows:

$$\frac{du}{dt} = \frac{\partial u}{\partial t} + \underbrace{u\frac{\partial u}{\partial x} + v\frac{\partial u}{\partial y} + w\frac{\partial u}{\partial z}}_{\overset{0}{\diagup}} \tag{6.5}$$

And now we can allow for dynamic evolution of the velocity with time-dependence using Equations 6.1 and 6.2, which then become,

$$\frac{\partial u}{\partial t} = +fv - \frac{1}{\rho}\frac{\partial p}{\partial x} + F_x \qquad (6.6)$$

$$\frac{\partial v}{\partial t} = -fu - \frac{1}{\rho}\frac{\partial p}{\partial y} + F_y \qquad (6.7)$$

As soon as we introduce such time-dependence in the equations, the solutions for u and v will be functions of time (t) as well as of x and y. A typical time-dependent solution to such evolution is "waves". There are no waves in the pure steady-state solution! It is the time-dependence that allows us to realize solutions in terms of waves.

Waves carry energy from one place of the ocean to another. They transfer energy. The stored potential energy in the ocean gets perturbed – waves are generated – they propagate – when they come in contact with other oceanic features or bathymetry, or lateral boundary; the waves interact, reflect, dissipate, they transfer their energy to other forms of kinetic energy. Currents are created. The winds disturb the ocean– this cycle goes on and on. Where does the potential energy come from? It's from the sun – the heat (creating temperature difference within the fluid, creating density differences across and through the fluid medium) – where does this energy finally go? Diffusion, other processes, friction, cyclones hitting land and losing energy, precipitation, currents carrying heat to northern latitudes, and then sinking and then flowing through the abyssal planes. Wind-driven circulations to thermohaline circulations and back through vertical motion. Excess heat near equatorial regions is carried by WBCs poleward and then cold currents flow around the eastern boundaries and through the abyssal planes. Energy gets transferred from scale to scale – through what is called multiscale processes. Generation of waves, their propagation, their eventual dissipation, and distribution, of energy – it is the life cycle of nature.

A wave breaking is a great example. Swells are created by winds perturbing the ocean far away from the coast. These waves carry their energy toward coast. As they come closer, they feel the shoaling bathymetry. The speed of wave is given by a simple formula, $c = \sqrt{gH}$, that we will investigate in the next section. Here g is gravity and H is the water depth. As a number of waves travel together, the wave that leads becomes slower than the one that follows. So, the one that was behind can now overcome the one that was in front and overtakes the other by moving ahead. However, the one that moves ahead does not have the support of the water column anymore and cannot support itself, thus breaks. This is wave breaking, in short. Ready to surf?

6.2   THE INERTIA-GRAVITY WAVE EQUATIONS

Let us develop the wave equation now. We start from Equations 6.6 and 6.7 in a modified form. Assume no friction, so waves are free to travel and they maintain their character, i.e. non-dispersive. The friction terms F_x, and F_y are set to zero. Following the development of Stommel, we also assume that the pressure is given by the small perturbation at the sea surface, or $p = \rho g \eta$, where, $\eta(x, y, t)$ expresses the sea surface elevation above the mean sea level and effectively becomes the signature of the wave height and amplitude.

Under these assumptions, Equations 6.6 and 6.7 become the following:

$$\frac{\partial u}{\partial t} = + fv - g\frac{\partial \eta}{\partial x} \tag{6.8}$$

$$\frac{\partial v}{\partial t} = - fu - g\frac{\partial \eta}{\partial y} \tag{6.9}$$

And the associated continuity equation to following Chapter 2 (equation 2.23) and Chapter 4 (Section 4.7, eq 4.47) is given by

$$\frac{\partial \eta}{\partial t} + H\left(\frac{\partial u}{\partial x} + \frac{\partial v}{\partial y}\right) = 0 \tag{6.10}$$

This system of equations, 6.8, 6.9, and 6.10 (three equations with three unknowns, u, v, η) are called the free wave equations on the rotating earth. The balance is between the time-dependent evolution of the free surface (η), the Coriolis force, and the pressure gradient. To the lowest order when time-dependence ceases to exist, the solution returns to a geostrophic equilibrium (pressure gradient balancing the Coriolis term).

6.2.1   SURFACE GRAVITY WAVES

It is instructive to set up the time-dependent horizontal equations of motion in a non-rotating frame first. We can also think of this as a balance when gravity dominates the dynamical balance over rotation. The resulting dynamical equations are simply obtained by letting $f = 0$ in the system of equations, 6.8, 6.9, and 6.10. These are:

$$\frac{\partial u}{\partial t} = - g\frac{\partial \eta}{\partial x} \tag{6.11}$$

$$\frac{\partial v}{\partial t} = - g\frac{\partial \eta}{\partial y} \tag{6.12}$$

The phenomenon that is governed by the above momentum equations 6.11 and 6.12 with the continuity equation 6.10 is essentially a balance of gravity and vertical motion of the sea surface, representing the surface gravity waves.

Now we can eliminate u and v to obtain a single equation in η in two steps. First, we can take partial derivatives of equations 6.11 and 6.12 with respect to x and y, respectively, and add them to yield,

$$\frac{\partial^2 u}{\partial x \partial t} + \frac{\partial^2 v}{\partial y \partial t} = -g \left[\frac{\partial^2 \eta}{\partial x^2} + \frac{\partial^2 \eta}{\partial y^2} \right] \tag{6.13}$$

$$\frac{\partial}{\partial t} \left(\frac{\partial u}{\partial x} + \frac{\partial v}{\partial y} \right) = -g \left[\frac{\partial^2 \eta}{\partial x^2} + \frac{\partial^2 \eta}{\partial y^2} \right] \tag{6.14}$$

In the second step, use equation 6.10 to substitute the divergence term $(\frac{\partial u}{\partial x} + \frac{\partial v}{\partial y})$ by $-\frac{1}{H}\frac{\partial \eta}{\partial t}$ to result in the following well-known wave equation:

$$\frac{\partial}{\partial t} \left(\frac{1}{H} \frac{\partial \eta}{\partial t} \right) = g \left(\frac{\partial^2 \eta}{\partial x^2} + \frac{\partial^2 \eta}{\partial y^2} \right) \tag{6.15}$$

$$\frac{\partial^2 \eta}{\partial t^2} = gH \nabla^2 \eta \tag{6.16}$$

$$\frac{\partial^2 \eta}{\partial t^2} = c^2 \nabla^2 \eta \tag{6.17}$$

Here, $\nabla^2 = \frac{\partial^2}{\partial x^2} + \frac{\partial^2}{\partial y^2}$ is the second-order operator.

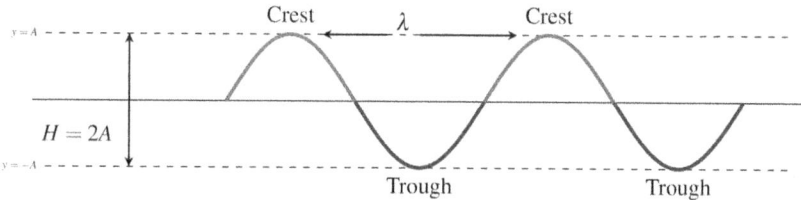

Figure 6.1: A typical standing waveform and its parameters. The distance between two consecutive crests is the wavelength, λ. The height (H) of the wave is twice the amplitude (A). Any displacement y is a sinusoidal function of x given by equation 6.18.

6.2.2 WHAT EXACTLY IS A WAVE?

Now we can seek solutions in terms of waves, which are generally given by sine and cosine functions. It is the repetitive nature of these functions that makes them ideal wave solutions. See Figure 6.1.

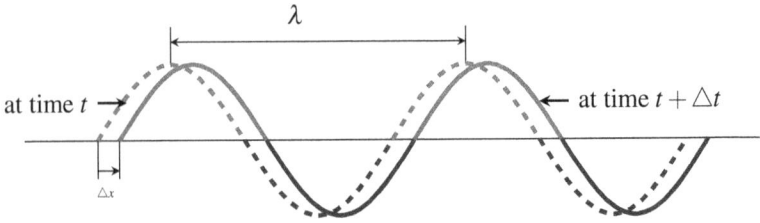

Figure 6.2: A wave in progression. It moves by $\triangle x$ from a time t to another time $(t + \triangle t)$. The wave moves as a whole entity. The wave speed is thus $c = \frac{\triangle x}{\triangle t} = \frac{\lambda}{T} = \frac{w}{l}$.

A standing wave can be expressed mathematically by a sine function, as seen in Figure 6.1. The displacement y at any location x can be given by a sine function of nature,

$$y(x) = A \sin lx \qquad (6.18)$$

Now, what is "l"? Since x is in a physical space (distance) and the argument of the sine function is degrees or radians (from 0 to 2π), we need to convert the distance to an equivalent angle with repeating nature after 2π.

Clearly, @$x = 0$, $y = 0$; The nature of the sine function allows us to define a wavelength λ, the distance from one crest to the next, or from one trough to the subsequent trough. So, @$x = \lambda$, $y = A\sin(l\lambda) = 0$, which then leads to $l\lambda = 2\pi$, which yields $l = \frac{2\pi}{\lambda}$. Note that @$x = \lambda/4$; $y = A\sin(\pi/2) = A$; and @$x = \lambda/2$; $y = A\sin(\pi) = 0$ and @$x = 3\lambda/4$; $y = A\sin(3\pi/2) = -A$.

So, l is called the "Wavenumber," it is $2\pi/\lambda$ and has units of radians/km. Similarly, for a two-dimensional wave, there is a zonal (l) and a meridional (m) wavenumber. Many times, a horizontal wave number (k) is invoked in describing general horizontally propagating waves, which is given by $k = \sqrt{l^2 + m^2}$.

Now, let us consider this wave in a propagating context. Imagine the wave to be propagating in time and repeating itself after a period T. See Figure 6.2.

This progressive wave can then be captured in the following form using separation of variables (x and t) and using sine function for both space and time variables. Correct? Simply put, a progressive wave can then be written as,

$$y(x,t) = A \sin\left(\frac{2\pi}{\lambda}x\right) \cos wt \qquad (6.19)$$

where $w = \frac{2\pi}{T}$. At $t = 0$, this waveform reduces to the standing wave in Equation 6.18.

Now, let us use a little trigonometric identity to simplify equation 6.19. For any two angles, θ and ϕ, we have the identity,

$$\sin(\theta + \phi) + \sin(\theta - \phi) = 2\sin\theta\cos\phi \tag{6.20}$$

or,

$$\sin\theta\cos\phi = \frac{1}{2}[\sin(\theta + \phi) + \sin(\theta - \phi)] \tag{6.21}$$

Using the above identity, equation 6.19 can be written as

$$y(x,t) = \frac{A}{2}[\sin(lx - wt) + \sin(lx + wt)] \tag{6.22}$$

The two terms in the above equation actually represent two traveling waves – one in the positive x-direction and one in the negative x-direction. How?

Let us focus our attention to the first term in equation 6.22. Imagine that time is frozen at a time t, and the waveform would be that similar to a standing wave in Figure 6.1 which is redrawn in Figure 6.2 with the dashed waveform. Let us now think about a single value of $y(x,t)$ and when will that same value occur for the wave at a later time $t + \triangle t$. Over this small time $\triangle t$, the wave progresses by a distance $\triangle x$, or the point would be at a location $x + \triangle x$. So, we must have,

$$y(x,t) = y(x + \triangle x, t + \triangle t) \tag{6.23}$$

or

$$\sin(lx - wt) = \sin(l(x + \triangle x) - w(t + \triangle t)) \tag{6.24}$$

To satisfy the above, one must have

$$l\triangle x - w\triangle t = 0$$

or, $\frac{\triangle x}{\triangle t} = \frac{w}{l}$.

Over the small time $\triangle t$, the wave progresses by a distance $\triangle x$, so its speed is given by $c = \frac{\triangle x}{\triangle t}$ – this is also called the phase speed.

So, equation 6.24 is automatically satisfied if the full waveform progresses with a speed of $c = \frac{w}{l}$ as shown in the solid waveform in Figure 6.2.

Over the period T, the wave travels by one full wavelength, or λ. So, the speed of the wave is nothing but, $c = \lambda/T = (2\pi/l)/(2\pi/w) = w/l$.

Similarly, you can take the second term in equation 6.19 and see that the other form with $\sin(lx + wt)$ translates in the negative x-direction.

Obviously, there is a meridional phase speed $c_y = w/m$. And, then a horizontal effective phase speed c_p can be defined as $c_p = w/k$, where w is the frequency and k is the horizontal wavenumber. Thus, $c_p = w/\sqrt{l^2 + m^2}$. So, the full two-dimensional traveling wave can be expressed in terms of its wavenumber, frequency, and amplitude as follows

$$y(x,y,t) = A\sin(lx + my - wt) \tag{6.25}$$

6.2.3   EXAMPLES OF WAVES IN OTHER MEDIA

The wave equation described above applies to many other processes and many other mediums. Their mathematical expression is very similar to equation 6.17, but their interpretation is specific to the process – to the wave that carries the energy through a particular medium. It is worthwhile to appreciate this simple fact. Here are some examples of unidirectional waves. You can extrapolate those for two- or three-dimensions.

- *Acoustic waves* The acoustic wave equation is used to describe the propagation of sound waves. Both sound pressure (p) and particle velocity (u) follow the similar (and standard) wave equation (in x-direction)

$$\frac{\partial^2 p}{\partial x^2} = \frac{1}{c^2}\frac{\partial^2 p}{\partial t^2} \tag{6.26}$$

$$\frac{\partial^2 u}{\partial x^2} = \frac{1}{c^2}\frac{\partial^2 u}{\partial t^2} \tag{6.27}$$

  where c is the speed of sound propagation which depends on the medium and is given by the Newton-Laplace equation

$$c = \sqrt{\frac{K}{\rho}} \tag{6.28}$$

  where K is the coefficient of stiffness or the bulk modulus of the medium, and ρ is the density of the medium. See the similarity of this with $c = \sqrt{gH}$ for surface gravity or Kelvin waves!

- *Waves on a stretched string*
  The displacement of a massless string can similarly be represented by waves governed by similar equations, where the speed is a function of the tension (T) and mass per unit length (μ) and given by

$$c = \sqrt{\frac{T}{\mu}} \tag{6.29}$$

- *Electromagnetic waves, light waves, etc.* Similarly, for electromagnetic waves, light waves, longitudinal and transverse waves, plasma waves, and P-waves (earthquakes), similar wave equations apply. However, each phenomenon is represented by its own characteristics or response variable and the wave speed of propagation is dependent upon some of the properties of the medium and/or forcing parameters that cause those waves.

6.2.4 A SINGLE WAVE IS A KEY REPRESENTATION OF MANY OF THEM – THE FOURIER DECOMPOSITION

As was shown in Section 6.2.2, the phase speed is a function of wavenumber and frequency as follows.

$$c_p = \frac{w(l,m)}{\sqrt{l^2 + m^2}} = c_p(w,l,m) \qquad (6.30)$$

This means that waves with different wavenumbers and frequencies will have a different speed.

Now let us talk about another idea – the idea of 'Fourier decomposition' of the forcing function and looking at the response of the system to each component of the forcing function, and then summing up the individual responses to realize the full response.

So, what is Fourier decomposition?

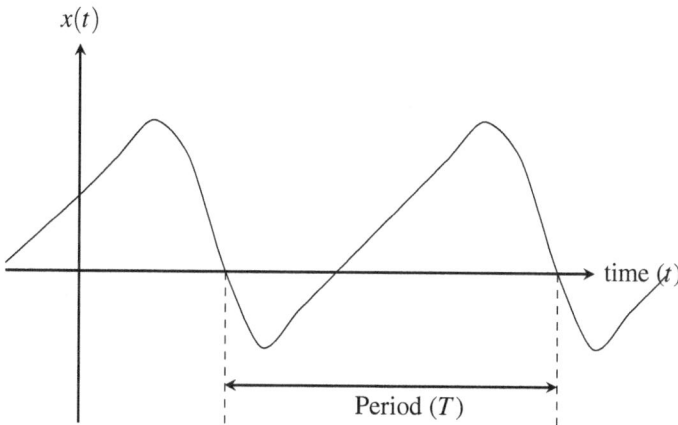

Figure 6.3: Any irregular forcing time-series.

In a time-dependent framework, the force applied to the system is generally an irregular time-series like in Figure 6.3. The famous French mathematician J. B. Fourier (1768–1830) in 1807 proposed a fundamental theorem to understand such time-series. Fourier proposed that *any* disturbance that repeats itself regularly with a period T can be made up from a set of pure sinusoidal vibrations of periods $T, T/2, T/3, T/4$, etc., with some amplitudes. These amplitudes can be determined from the disturbance time-series itself using the so-called "Fourier Integrals".

Mathematically, the series $x(t)$ in Figure 6.3 can be expressed as an infinite sum of sines and cosines as

$$x(t) = a_o + a_1 \cos \frac{2\pi t}{T} + a_2 \cos \frac{4\pi t}{T} + a_3 \cos \frac{6\pi t}{T} + \dots$$
$$+ b_1 \sin \frac{2\pi t}{T} + b_2 \sin \frac{4\pi t}{T} + b_3 \sin \frac{6\pi t}{T} + \dots \quad (6.31)$$

The Fourier coefficients $a_0, a_1, \dots$ and $b_0, b_1, b_2, \dots$ can be determined using the time-series $x(t)$ itself as follows.

$$a_0 = \frac{1}{T} \int_{-T/2}^{T/2} x(t) dt$$

$$a_k \atop {k \geq 1} = \frac{2}{T} \int_{-T/2}^{T/2} x(t) \cos \frac{2\pi k t}{T} dt$$

$$b_k \atop {k \geq 1} = \frac{2}{T} \int_{-T/2}^{T/2} x(t) \sin \frac{2\pi k t}{T} dt \quad (6.32)$$

Furthermore, realizing that each of the coefficients is linked to a harmonic of the fundamental frequency ($w_k = 2\pi k/T$), Fourier developed and showed that the time-series $x(t)$ is indeed a sum of Fourier Integrals with amplitudes linked with those harmonics.

Or, with a little more algebra, one can see that,

$$A(w) = \frac{1}{2\pi} \int_{-\infty}^{\infty} x(t) \cos wt \, dt$$

$$B(w) = \frac{1}{2\pi} \int_{-\infty}^{\infty} x(t) \sin wt \, dt \quad (6.33)$$

yields

$$x(t) = 2 \int_0^{\infty} A(w) \cos wt \, dw + 2 \int_0^{\infty} B(w) \sin wt \, dw \quad (6.34)$$

In the language of spectral analysis, $A(w)$ and $B(w)$ are the components of the *Fourier transform* of $x(t)$, and equation 6.34 is a representation of the time-series $x(t)$ by its *Fourier Integral* (in spectral domain) or *inverse Fourier transform*.

So, there we have it. Any irregular time-series (of sufficiently long duration) can then be thought to be repeating itself after the observation period and then be represented by a fundamental frequency and its harmonics. Tides, wind forcing, surface gravity waves, tsunamis, earthquakes, Gulf Stream responses, ecosystem responses – any forcing or any response can be thought of as a sum of an infinite set of frequency and corresponding amplitudes. And, most of the time, a finite number of these frequencies would dominate the process and explain most of the energy distribution (or variance, in statistical terms) in the time-series. So, in a force-response system, it is often that we try to understand the response of the system as a sum of responses to a number of components of the forcing function. These forcing components are composed of (according to Fourier decomposition) a fundamental and a few leading order harmonics of different amplitudes. So, once you know the response of the system from one representation of the forcing component (frequency-wavenumber), you are on your way to build the whole response system for the process. However, a word of caution. Most of our processes in the ocean are non-linear and multiscale. So, one has to be careful about applying the Fourier decomposition idea for complex processes. Having said that, this idea of Fourier decomposition and understanding one wave response to understand the whole system response is extremely fruitful and appealing for both periodic and aperiodic phenomena.

6.2.5   AN EXAMPLE OF FOURIER DECOMPOSITION

Let us look at the decomposition of the time-series presented in Figure 6.4. This time-series is actually a tidal signal in one of the ports on the eastern shores of India (Visakhapatnam, at $83.32°E, 17.69°N$) the hourly time-series is for the month of April (1–30) 2014. As we know the tidal signals are generally dominated by the M2 tidal frequency at around 12-hour period. In addition there is a daily period of 24 hours in the signal.

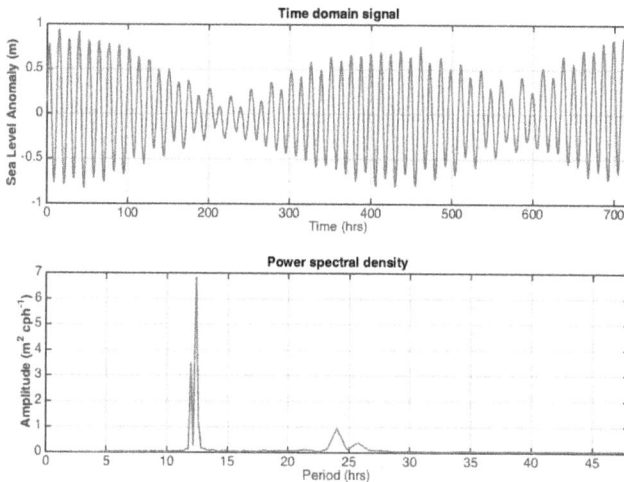

Figure 6.4: A time-series of tides on April 2014 at Visakhapatnam, India. Image and Data from Dr. Sourav Sil.

Using any available FFT analysis package (MATLAB, R, etc.) one can compute a spectrum of the time-series shown in the top panel of Figure 6.4, and this spectrum is shown in the bottom panel of Figure 6.4. The spectra clearly shows two nearby peaks: one at 12 hours and the other is at 12.41 hours. The question remains is that are they significant? How do we know if one is different from the other when they are so close together? We describe the process of recognizing two important parameters of spectral analysis below, which will help guide your interpretation of any spectrum. Two important parameters of the spectral analysis are: (i) the maximum frequency of analysis and (ii) the resolution interval in the frequency domain. The first one, i.e. the maximum frequency of analysis, is called the Nyquist frequency and is given by $f_{max} = 1/(2.\delta t)$, where δt is the sampling interval of the time-series. The value of f_{max} is the maximum frequency or the minimum period ($1/f_{max}$) that you will be able to get out of a time-series. So, if the sampling interval is one month, you cannot get periods less than two months from spectral analysis. If the sampling interval is 1 hour, you cannot resolve any period less than two hours. If δt is 1 day, then the minimum period in the frequency spectrum is 2 days.

The second one is a bit tricky. Suppose the time-series has N intervals with sampling intervals of δt, with a total time-length of T. So, $T = N \times \delta t$. Theoretically, you can only resolve two peaks whose frequencies are apart by a $\delta f = 1/T$ in the frequency domain (axis). This really depends on the length of the time-series! The FFT function determines the spectrum after computing the Fourier transforms carried out with the spacing $\delta f = 1/(N \times \delta t)$; or, $\delta f = 1/T$. Thus, the frequency axis for the FFT spectrum should go from "Zero" to "f_{max}" (given by the Nyquist value, or $1/(2\delta t)$) with an interval of δf (given by $1/T$). Given the above time-series presented in Figure 6.3, you have enough resolution to get 12 hours and 12.41 hours in the spectral peaks depending on how long of a time-series you have.

Let us now see how this signal is a superposition of multiple sine waves, which is what Fourier's theorem is. Let us consider two sine waves. One for a period of 24 hours and another for a period of 12 hours. The two time-series are shown in the top panels of Figure 6.5 (a,b). If we compute the power spectra of these two single-frequency pure sine waves, we will get solitary peaks as shown in Figure 6.5 lower panels (a,b). Now look at Figure 6.5c. The top panel is the superposition of two sine waves (24 and 12 hours). The bottom panel is the FFT spectrum, which clearly resolves both the peaks! Similarly, if you add another time-series of 12.41 hours you will get the time-series in Figure 6.4 and the FFT will give its spectra with all three peaks like it did in the bottom panel of 6.4.

(a)

(b)

(c)

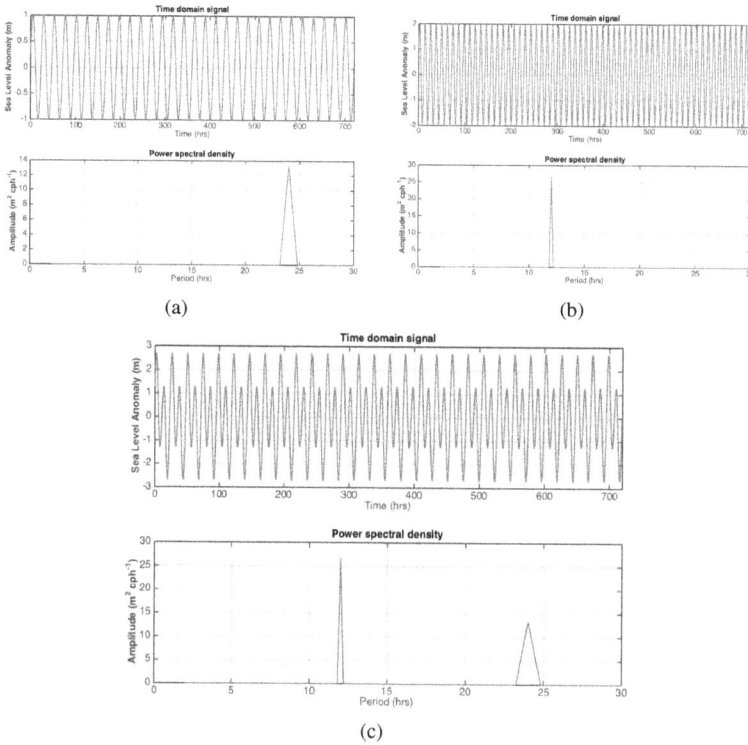

Figure 6.5: Decomposition and reconstruction of tidal series according to Fourier components. Image and Data from Dr. Sourav Sil of SEOCS, IIT, Bhubaneswar, India.

6.3 THE DISPERSION RELATIONSHIP

Let us apply a more general solution of the waveform to the velocity components and the sea surface elevation. We can represent the wave form with sine and cosine functions as a complex function ($e^{ix} = \cos x + i \sin x$). See Strang and Freund, 1986 for details. The general solution for $u \neq 0$, and $v \neq 0$ can be written as:

$$u = u_0 e^{i(lx+my-wt)} \tag{6.35}$$

$$= u_0 [\cos(lx+my-wt) + i \sin(lx+my-wt)] \tag{6.36}$$

$$v = v_0 e^{i(lx+my-wt)} \tag{6.37}$$

$$\eta = \eta_0 e^{i(lx+my-wt)} \tag{6.38}$$

Using the above forms in Equations 6.8, 6.9, and 6.10, we get the following three equations.

$$-iwu - fv = -igl\eta \tag{6.39}$$
$$-iwv + fu = -igm\eta \tag{6.40}$$
$$-iw\eta + H(ilu + imv) = 0 \tag{6.41}$$

Or in matrix form,

$$\underbrace{\begin{bmatrix} -iw & -f & igl \\ f & -iw & igm \\ iHl & iHm & -iw \end{bmatrix}}_{\mathbf{A}} \begin{bmatrix} u \\ v \\ \eta \end{bmatrix} = 0 \tag{6.42}$$

This is a linear system of equations and has a trivial solution of $u = v = \eta = 0$ unless the determinant is null. Since u, v, or η cannot be zero, the determinant of this (3×3) matrix ($\mathbf{A}$) on the left has to be zero. This means that for the wave solution to exist, the following equation must hold.

$$\text{Det}[\mathbf{A}] = 0 \tag{6.43}$$
$$w[w^2 - f^2 - gH(l^2 + m^2)] = 0 \tag{6.44}$$

This equation 6.44 is called the "Dispersion Relation." It characterizes the relationship between the wavenumbers (l and m) and the frequency w.

From equation 6.44, if we set the frequency, $w = 0$, which is one of the solutions, then we recover the steady-state solution or no time-dependence. With the time-dependence being absent, our balance resorts back to Geostrophy (Equations 6.8 and 6.9 without the time-dependent term on the left).

The wave solution is inside the parenthesis in equation 6.44 and these are generally called the inertia-gravity waves. These are represented by the remaining two roots given by

$$w = \pm\sqrt{f^2 + gH(l^2 + m^2)} = \pm\sqrt{f^2 + gHk^2} \tag{6.45}$$

These correspond to traveling waves, called the Poincare Waves, after the famous French mathematician, who first derived the above relationship.

Consider two situations when the first term can be neglected within the square-root on the RHS of equation 6.45, i.e., $f \approx 0$ or $f \ll \sqrt{gHk}$. The first situation is when the inertia (f) can be ignored (gravity rules). The second situation is when the wavenumber (k) is large or short-wavelength gravity waves compared to the radius of deformation ($\sqrt{gH}$). In such cases, $w = \sqrt{gH}k$; and then the phase speed of these traveling waves is simply

$$c_p = w/k = \sqrt{gH}/f$$

These are classical gravity waves. In the second situation, the wavelength is shorter than the radius of deformation and they don't feel the effect of rotation! Again, gravity rules.

On the other hand, when $f > (\sqrt{gH}k)$, i.e., for extremely low wavenumbers (k), when the wavelengths are much larger than the radius of deformation, $k^2 << \frac{f^2}{gH}$, and the rotational effects dominate gravity, and $w \approx \pm f$. The waves follow the inertial period at that latitude ($w \approx \pm f = 2\Omega \sin \phi$, where ϕ is the latitude).

So, in summary, for intermediate wavenumbers, the frequency w is generally greater than f and the waves are called "superinertial." As the Poincare waves exhibit a mixed behavior between gravity waves and inertial oscillations, they are also called the "inertia-gravity waves."

6.4 THE DISPERSION DIAGRAM

Let us look into the details of the dispersion relationship for the inertia-gravity waves a bit more carefully.

$$w^2 - f^2 - gH(l^2 + m^2) = 0 \tag{6.46}$$

let $k^2 = l^2 + m^2$, where k is the horizontal wavenumber.
So, Equation 6.46 becomes

$$w^2 - f^2 - gHk^2 = 0 \tag{6.47}$$

Then,

$$w^2 = f^2 + gHk^2 \tag{6.48}$$
$$w = \sqrt{f^2 + gHk^2} \tag{6.49}$$

So, the phase speed is given by

$$c = \frac{w}{k} = \frac{\sqrt{f^2 + gHk^2}}{k}$$

and group speed is given by

$$c_p = \frac{\partial w}{\partial k}$$

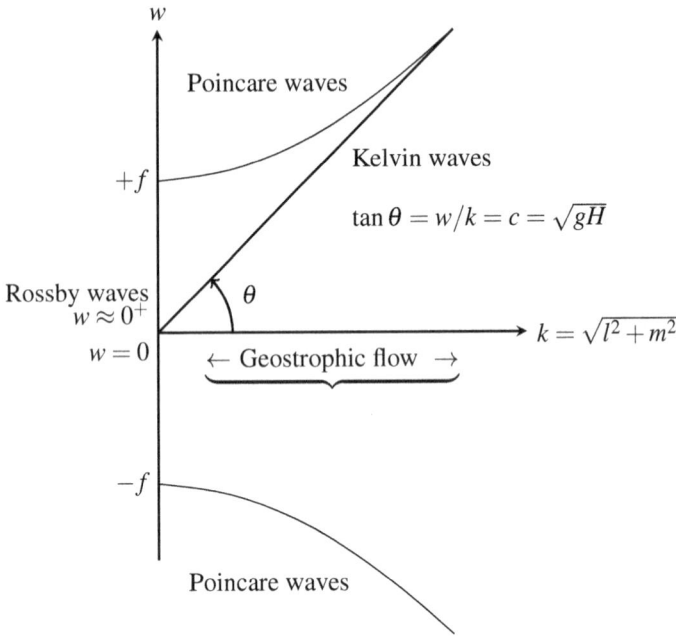

Figure 6.6: Figure for dispersion diagram.

If we ignore the inertial or rotational effect (i.e. $f = 0$), we are left with $w = \sqrt{gH}\,k$, which yields a wave speed $c = \frac{w}{k} = \sqrt{gH}$. Clearly, these are the surface gravity waves discussed in Section 6.2.1.

One can represent these different waves on a so-called "dispersion diagram" shown in Figure 6.6. There are two special cases, which are of extreme importance to ocean circulation. These are Kelvin waves and Rossby waves.

Kelvin waves happen when one of the velocity components is set to vanish but the rotational effects remain. We will discuss these next in Section 6.5.

Then there is an interesting class of waves when the Coriolis frequency is very small – So, these are very long period waves which are still waves on the rotating earth, but close to the Geostrophic limit of $w \approx 0$. These are called "Rossby waves." We will discuss these waves in Section 6.6.

6.5 KELVIN WAVES

The equations 6.7, 6.8, and 6.9 capture the basic set of time-dependent equations for inertia-gravity waves. They have the effects of gravity and rotation to support the dynamical circulation of the ocean in terms of various waves, as shown in the dispersion diagram 6.6.

One particular curious case is when there is a flow along the coast. In this case, we can assume either u or v to vanish if we set up the coordinate axes appropriately.

Let's set $v = 0$ in the equations 6.7, 6.8, and 6.9. we also assume that f is constant.

So, the two horizontal momentum and the continuity equations become:

$$\frac{\partial u}{\partial t} = -g\frac{\partial \eta}{\partial x} \tag{6.50}$$

$$fu = -g\frac{\partial \eta}{\partial y} \tag{6.51}$$

$$\frac{\partial \eta}{\partial t} = -H\frac{\partial u}{\partial x} \tag{6.52}$$

It is interesting to note that the three equations above have two independent variables u and η. In fact, if we consider $f \approx 0$, or a non-rotating ocean, equations 6.50 and 6.52 describe the shallow water motion for $v = 0$. And, on a rotating earth, very close to the equator, where $f \approx 0$, and if there is no meridional velocity ($v = 0$), then the solution can be exactly the same as given by equations 6.50 and 6.52; while the additional equation 6.51 can be an expression of geostrophy supporting the pressure gradient ($\frac{\partial \eta}{\partial y}$) in the meridional direction by the zonal flow u, close to the equator ($f \approx 0$).

In both situations, the solution is similar to the shallow water gravity waves discussed earlier but are called Kelvin waves. One is called the coastal Kelvin waves, the other is called the equatorial Kelvin waves. The equations are very similar; their solutions are similar too. Their wave characteristics are defined in a similar manner; – BUT *their manifestations in processes are different*!

Using the x-momentum equation and the continuity equation to eliminate η, we arrive at a simplified wave equation –

$$\frac{\partial^2 u}{\partial t^2} = gH\frac{\partial^2 u}{\partial x^2} = c^2\frac{\partial^2 u}{\partial x^2} \tag{6.53}$$

This wave solution is called the Kelvin wave. It propagates along the coast (x-axis) and is supported by the geostrophic pressure gradient in the y-direction (equation 6.51).

Non-dispersive Kelvin waves have a speed of $\sqrt{(gH)}$. The solution can be written as a superposition of 2 non-dispersive waves traveling in opposite directions. These are given by

$$u = u_1(x - ct, y) + u_2(x + ct, y) \tag{6.54}$$

$$\eta = F_1(x - ct, y) + F_2(x + ct, y) \tag{6.55}$$

There are two waves in the general solution, one traveling in the +ve x-direction, the other traveling in -ve x-direction.

Note that,

$$\frac{\partial^2 u}{\partial t^2} = c^2 \left(u_1(x+ct,y) + u_2(x-ct,y) \right) \tag{6.56}$$

$$\frac{\partial \eta}{\partial x} = -\frac{c}{g} \left(u_1(x+ct,y) - u_2(x-ct,y) \right) \tag{6.57}$$

Substituting these in equation 6.53 satisfies the equation nicely. Well, what about the structure of u_1 and u_2? What do they look like? Maybe we can use the other equation, the Geostrophic equation 6.51, to look at those.

Using the expression for u and η, and omitting the functional expression of $(x-ct,y)$,

$$f[u_1 + u_2] = -g\frac{c}{g}[\frac{\partial u_1}{\partial y} - \frac{\partial u_2}{\partial y}] \tag{6.58}$$

Using separation of variables, and collecting terms for u_1 and u_2,

$$\frac{\partial u_1}{\partial y} = -\frac{f}{\sqrt{gH}} u_1 \tag{6.59}$$

$$\frac{\partial u_2}{\partial y} = +\frac{f}{\sqrt{gH}} u_2 \tag{6.60}$$

These lead to the following exponential solutions for u_1 and u_2.

$$u_1 = U_{10}(x-ct)e^{-y/R} \tag{6.61}$$

$$u_2 = U_{20}(x+ct)e^{+y/R} \tag{6.62}$$

where $R = \frac{\sqrt{gH}}{f} = \frac{c}{f}$, is called the Rossby Radius of Deformation or just the radius of deformation. It is the *length scale* at which effects of rotation and gravity become comparable.

For equatorial regions, of the two independent solutions, one wave decays exponentially in the positive y-direction, whereas the other decays exponentially in the negative y-direction. This exponential behavior is very distinctive of Kelvin waves, away from the coast or from the equator.

Among the two solutions, one of them (the one with the positive exponential power of $(+y/R)$ increases exponentially with increasing distance away from the shore (proxy for x-axis if $v = 0$) and is thus declared physically unrealistic.

So, we are left with the solution for u, v, and η as

$$v = 0 \tag{6.63}$$

$$u = \sqrt{gH}F(x-ct)e^{-\frac{y}{R}} \tag{6.64}$$

$$\eta = -HF(x-ct)e^{-\frac{y}{R}} \tag{6.65}$$

There are two things here. First, in the northern hemisphere, when $f > 0$, the Kelvin waves travel in the positive x-direction because the solution with $F(x-ct)$ is the only real solution. Second, the wave amplitude and current will decrease as one moves away from the coast in the positive y-direction because of the multiplier $e^{-\frac{y}{R}}$.

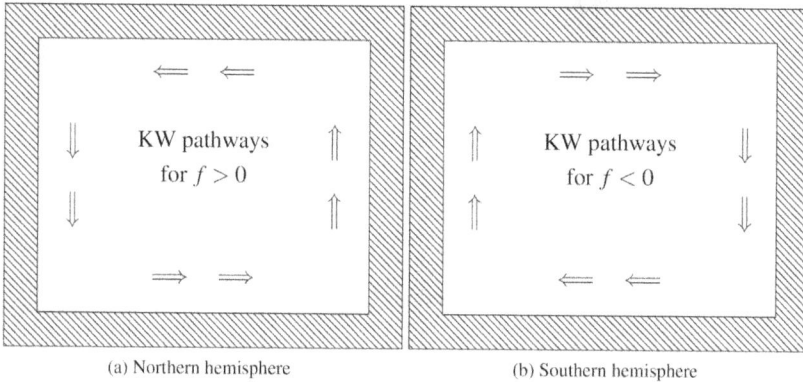

(a) Northern hemisphere   (b) Southern hemisphere

Figure 6.7: Propagation of Kelvin waves along the coast.

In the northern hemisphere, if the coast is zonal (East-West) and the ocean's southern boundary, the Kelvin wave would propagate eastward; for a northern boundary as the coast, the Kelvin wave would propagate westward. Similarly, for a coast along the western (eastern) boundary, the Kelvin wave would propagate southward (northward). Generalizing this, rotating the axes, it is easy to realize that the Kelvin waves travel with the coast to their right in the northern hemisphere and to their left in the southern hemisphere (because of negative f). See Figure 6.7 for typical propagation directions.

It is for this reason of the exponential decay away from the coast that the Coastal Kelvin waves are also called coastally trapped waves (CTW). The Rossby radius, R ($\sqrt{gH}/f$), is the measure of the trapping distance.

Note that because of the interrelationship between η and u, the Kelvin waves have upwelling ($\eta > 0$) and downwelling ($\eta < 0$) signatures as they propagate. We defer those discussions for more advanced texts on Geophysical Fluid Dynamics such as Cushman-Roisin and Beckers (2011) and Gill (1982) and The Sea, Volume 10 (K. H. Brink & Robinson, 2005).

An example of coastal Kelvin waves traveling along the coast for upwelling and downwelling conditions in the northern hemisphere is shown in Figure 6.8.

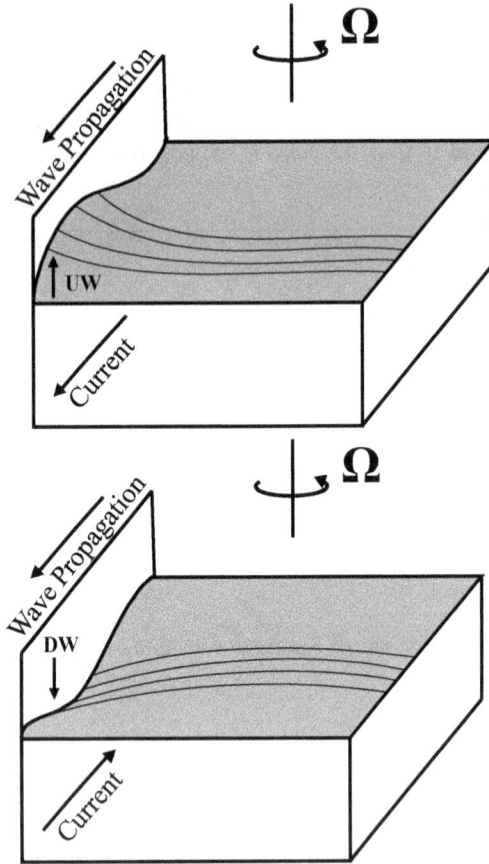

Figure 6.8: Characteristic parameters of a traveling Kelvin wave along the coast for upwelling and downwelling scenarios. Adapted from Cushman-Roisin (1994). Figure Credit: Adrienne Silver.

A great example of Kelvin waves is how winds perturb the equatorial Indian Ocean four times with changing seasons. As they propagate eastward on the equatorial region, they encounter the Eastern boundary of the Bay of Bengal (Sumatra region). Part of the equatorial Kelvin waves then turn northward (keeping the eastern boundary to their right) and go all around the Bay of Bengal at times. In fact, it is the third downwelling Kelvin wave that can be traced as the genesis of the Southward Flowing East India Coastal Current in the Bay of Bengal along the east coast of India. See Rao et al. (2010) for more details. Another example is the equatorial Kelvin waves in the Eastern pacific related to the El Nino phenomenon. We will discuss those in Section 7.3. Yet another example of the coastal Kelvin waves is found in the Brazil Current region between 22°S and 8°S (Veleda et al., 2012). Also, Illig et al. (2018) have reported on CTWs propagating for more than 20 degrees on the southeastern Pacific.

6.6   ROSSBY WAVES

Now, looking at the dispersion diagram 6.6 we can ask the following curious question.

What happens at w ≈ 0⁺, i.e. slightly above zero? – Could there be a slow evolution – not steady state; but a very slowly varying system at time periods much greater than the inertial period of "f"?

In other words, think Geostrophy to the first order. Let the long and slow waves develop on a geostrophically balanced flow field. Evolution of waves is more like perturbation on a steady-state geostrophic balance.

So, think about when $w \approx 0^+$ (slightly above zero but very close to zero).

Also, relax the $f =$ Constant, or f-plane restriction. Allow for the β-effect to work.

Thus, $f = f_0 + \beta y$, where f_0 is the Coriolis term supporting geostrophic flow over the beta-plane. And beta (β) is the ageostrophic (added on to geostrophy) component or the Coriolis parameter that supports the waves propagating with slightly positive, near-zero frequency or waves with very long time-periods compared to inertial (f_0) time-periods. These waves are called Rossby waves (after Carl Gustaff Rossby) (Rossby, 1936; Rossby & Montgomery, 1936) or planetary waves. The Rossby waves ride on geostropihic background under the influence of this planetary β-effect.

For the mid-latitude region, $f_0 = 8 \times 10^{-5}\,s^{-1}$ and $\beta_0 = 2 \times 10^{-11}m^{-1}s-1$

In terms of the motion's meridional length scale (L), this implies that

$$\beta = \frac{\beta_0 L}{f_0} << 1.$$

Now, the governing equations can be recast from equations 6.7, 6.8, and 6.9,

$$\frac{\partial u}{\partial t} - (f_0 + \beta y)v = -g\frac{\partial \eta}{\partial x} \qquad (6.66)$$

$$\frac{\partial v}{\partial t} + (f_0 + \beta y)u = -g\frac{\partial \eta}{\partial y} \qquad (6.67)$$

$$\frac{\partial \eta}{\partial t} + H\left(\frac{\partial u}{\partial x} + \frac{\partial v}{\partial y}\right) = 0 \qquad (6.68)$$

These equations have a mix of small and large order terms. The larger one (with f_0, g, and H terms) comprises the steady geostrophic balance on the constant f-plane, with f_0 being the constant. The smaller ones (the time-derivatives and the β_0 terms) come as perturbations, which, although small, can significantly govern the wave evolution.

To the first approximation, the large terms dominate the dynamics (geostrophy). Thus,

$$u \approx -\frac{g}{f_0}\frac{\partial \eta}{\partial y} \qquad (6.69)$$

$$v \approx +\frac{g}{f_0}\frac{\partial \eta}{\partial x} \qquad (6.70)$$

Using these first approximations of u and v only for the time-dependent and β_0 terms in the equations 6.66 and 6.67, we get the time-dependent velocity (noting them again as u and v, but realizing that these contain the ageostrophic components now, and different from the ones given by equations 6.69 and 6.70. These new equations which allow for both the geostrophic and the time-dependent components in u and v are:

$$-\frac{g}{f_0}\frac{\partial^2 \eta}{\partial y \partial t} - f_0 v - \frac{\beta_0 g}{f_0}y\frac{\partial \eta}{\partial x} = -g\frac{\partial \eta}{\partial x} \qquad (6.71)$$

$$-\frac{g}{f_0}\frac{\partial^2 \eta}{\partial x \partial t} + f_0 u - \frac{\beta_0 g}{f_0}y\frac{\partial \eta}{\partial y} = -g\frac{\partial \eta}{\partial y} \qquad (6.72)$$

These are easy to solve for u and v in terms of η.

$$u = \underbrace{-\frac{g}{f_0}\frac{\partial \eta}{\partial y}}_{\text{Geostrophic}} \underbrace{-\frac{g}{f_0^2}\frac{\partial^2 \eta}{\partial x \partial t} + \frac{\beta_0 g}{f_0^2}y\frac{\partial \eta}{\partial y}}_{\text{Ageostrophic}} \qquad (6.73)$$

$$v = \underbrace{+\frac{g}{f_0}\frac{\partial \eta}{\partial x}}_{\text{Geostrophic}} \underbrace{-\frac{g}{f_0^2}\frac{\partial^2 \eta}{\partial y \partial t} + \frac{\beta_0 g}{f_0^2}y\frac{\partial \eta}{\partial x}}_{\text{Ageostrophic}} \qquad (6.74)$$

Substituting the above u and v solutions in terms of η in the continuity equation 6.68 yields one single equation that describes the dynamic evolution of η.

$$\frac{\partial \eta}{\partial t} - R^2 \frac{\partial}{\partial t} \nabla^2 \eta - \beta_0 R^2 \frac{\partial \eta}{\partial x} = 0 \qquad (6.75)$$

where $\nabla^2 = \frac{\partial^2}{\partial x^2} + \frac{\partial^2}{\partial y^2}$, and $R^2 = \frac{gH}{f_0{}^2}$.

Now we can seek a wave solution for η following the generalized wave representation,

$$\eta = \eta_0 \cos(lx + my - wt)$$

Substituting this in equation 6.75, we get the following dispersion relationship for Rossby waves:

$$w + R^2 w(l^2 + m^2) + \beta_0 R^2 l = 0 \qquad (6.76)$$

or

$$w = -\beta_0 R^2 \frac{l}{1 + R^2(l^2 + m^2)} \qquad (6.77)$$

It is worth looking into the following properties of the Rossby waves.

- if $\beta_0 = 0$, then $w = 0$ – means Geostrophy
- The time-dependent terms were very small compared to the geostrophic term, i.e., $\frac{\partial u}{\partial t} \ll f_0 v$, or $U/T \ll f_0 U$, or $w \ll f_0$ – which was our starting hypothesis!
- short waves – if $L < R \longrightarrow (\frac{1}{l}, \frac{1}{l}) < R \to 1 < Rl \to 1 \ll R^2 l^2$; and from equation 6.77 $w \approx -\beta_0 R^2 \frac{l}{R^2 l^2} = -\beta_0 L$.
- Long waves – if $L > R \longrightarrow (\frac{1}{l}, \frac{1}{l}) > R \to 1 > Rl \to 1 \gg R^2 l^2$; and from equation 6.77 $w \approx -\beta_0 R^2 \frac{l}{1} = -\frac{\beta_0 R^2}{L} < \beta_0 L$
- So, in both cases for short and long waves, w is on the order of or less than $\beta_0 L$. By definition of the β-plane, $\beta_0 L \ll f_0$, i.e. the total change of f given by the Coriolis parameter variation over the length is mush smaller than the Coriolis frequency of the f-plane itself. So, the frequency of short or long Rossby waves, and w is much smaller than f_0, or it is sub-inertial.
- Finally, it is very important to note that the zonal phase speeds of Rossby waves are given by

$$C_x = \frac{w}{l} = -\beta_0 R^2 \frac{1}{1 + R^2(l^2 + m^2)}.$$

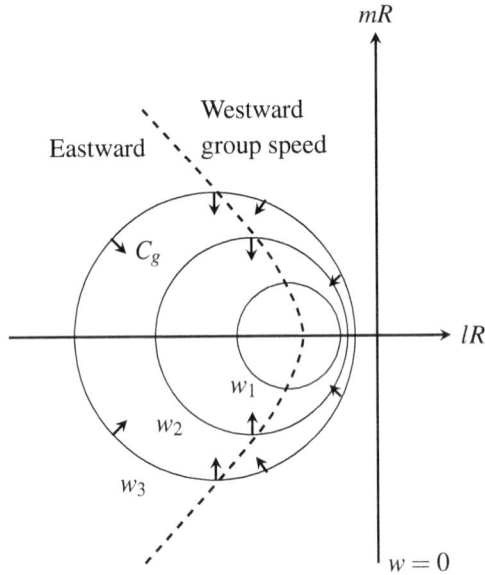

Figure 6.9: Rossby wave dispersion diagram.

This negative phase speed implies that the phase propagation of Rossby waves or planetary waves is always to the west! On the other hand, the meridional phase speed is

$$c_y = w/m = -\beta_0 R^2 \frac{l/m}{1 + R^2(l^2 + m^2)},$$

which could have any sign, meaning, the waves could travel either northward or southward. Thus, in the real ocean and in the atmosphere, the planetary waves or Rossby waves can travel only in either northwestward, westward, or southwestward direction.

In closing, let us look at the Dispersion relationship of the Rossby waves. We can rewrite the dispersion relation given in equation 6.78 as follows

$$(l + \frac{\beta_0}{2w})^2 + m^2 = (\frac{\beta_0^2}{4w^2} - \frac{1}{R^2}) \qquad (6.78)$$

This means that lines of constant frequency w are circles in the wavenumber space of (l, m). However, such circles can only exist if the RHS of equation 6.77 is greater than or equal to zero, or mathematically,

$$\frac{\beta_0^2}{4w^2} \geq \frac{1}{R^2}$$

This defines a $w_{max} = \frac{\beta_0 R}{2}$ beyond which planetary waves cannot exist.

Figure 6.9 shows the variation of w with lR and mR. A few other important points to note in this dispersion diagram are below.

- The group velocities by definition are gradients of the frequency with wavenumbers and thus normal to the w circles as shown by the arrows.
- The above naturally means that after a certain point along the periphery of a w-circle, the group of Rossby waves would travel eastward; however, their phase-velocities remain westward. The line dividing the eastward group and the westward group is a hyperbola given by $l^2 + m^2 = \frac{1}{R^2}$.
- It can be shown that the eastward propagating groups are short Rossby waves while the westward propagating groups are the long Rossby waves (lower values of l, m). The short wave group speed is much higher than the long wave group speed. This has deep implications for the generation of western boundary currents from a time-dependent perspective. The wind forcing on the subtropics creates baroclinic long and slow Rossby waves, which travel westward and, upon reflection, become fast and very short Rossby wave packets, which then become the western Boundary Current (Gill, 1982). This explanation of Western Boundary Current formation is more dynamically oriented than the steady-state models discussed in Chapter 4.
- The contours of frequency (w) are circular and reduces to a single point when the maximum $w_{max} = \frac{1}{2}\beta_0 R$ is reached. No Rossby wave can exist at a larger frequency than w_{max}, or at smaller periods than that corresponding to the time period at w_{max}.

CONCLUSION

When we allow for time-dependence in the forcing fields and response variables, the solutions are typically in the form of *waves* of multiple frequencies and wavelengths. The idea of Fourier Transform is very powerful in understanding the time-varying response of a system. The dispersion relationship characterizes different waves in their frequency-wavenumber space. We showed that Kelvin waves are generated by winds along the equator, where the Coriolis effect is negligible compared to pressure gradient and wind-stress. Similar is the situation for Kelvin waves traveling along the coast when forced by alongshore winds. In the presence of rotation, wind-stress and pressure gradient, in the limit when $\omega \approx 0^+$, or very low frequency (periods compared to inertial periods), we get Rossby waves, which always propagate westward!

FURTHER READING

Bendat, J. S., & Piersol, A. G. (2011). *Random data: Analysis and measurement procedures* (Vol. 729). John Wiley & Sons.

Cushman-Roisin, B., & Beckers, J. (2011). *Introduction to geophysical fluid dynamics: Physical and numerical aspects*. Elsevier.

French, A. P. (2001). *Vibrations and waves*, 316 pp. American Association of Physics Teachers.

Gill, A. E. (1982). *Atmosphere-ocean dynamics (international geophysics series)*. Academic Press.

Jackson, L. B. (2013). *Digital filters and signal processing: With MATLAB exercises*. Springer Science & Business Media.

Lighthill, M. J., & Lighthill, J. (2001). *Waves in fluids*. Cambridge University Press.

Newland, D. E. (2012). *An introduction to random vibrations, spectral & wavelet analysis*. Courier Corporation.

Pedlosky, J. (2013a). *Geophysical fluid dynamics*. Springer Science & Business Media.

Iguassu Falls

7 The Layering of Oceans

> And the whole is greater than the part.
>
> ———————————————
> Euclid (325 BC–265 BC)

OVERVIEW

So far in this book, we have first introduced the physical characteristics of the ocean in Chapter 1 and then described the physical setup of the governing equations in Chapter 2. Chapter 3 described the very important Geostrophic balance between pressure-gradient and Coriolis, followed by the closed-form solution for wind-driven circulation in Chapter 4 and for density-driven thermohaline circulation in Chapter 5. All of these formulations assumed steady state. In the last chapter, we introduced the idea of time-dependence and realized the existence of a series of important Geophysical waves such as Kelvin and Rossby waves. Note that all of these formulations were also done for either homogeneous ocean or in a vertically integrated system – effectively one-layer ocean.

In this chapter, we introduce the concept of layers to bring in three-dimensionality in the study of circulation, which is crucial to connect the wind-driven circulation to the abyssal flow. The very important concept of stratification or density-gradient is explained first. We use a simple 2-layer ocean to develop this layering idea in terms of density-difference or *reduced gravity*. The associated stability property of the fluid is discussed with the concept of *buoyancy frequency*. Then we apply the 2-layer idea to explore the phenomenon of El Niño and La Niña dynamically with a 2-layer model. This equatorial 2-layer model captures the essence of many related ocean-atmospheric interaction phenomena spanning the Equatorial Pacific. A brief history (from Sir Gilbert Walker to Professor J. Bjerknes) of the development of our understanding of El Niño in the ocean and Southern Oscillation in the atmosphere as a coupled phenomenon precedes the dynamical discussion. The 2-layer model highlights the propagation of both Kelvin and Rossby waves along the equator. This chapter ends with brief discussions and examples of both barotropic and baroclinic instabilities.

DOI: 10.1201/9780429347221-7

7.1   THE IDEA OF LAYERS

The ocean is a continuum. It is a fluid with a vertical extent, where the temperature and salinity vary with depth. Density increases with depth. So that is how lighter fluid stays on top of the denser fluid.

In the previous chapters, many of the ideas to facilitate the analytical considerations of the wind-driven circulation (Chapter 4) were based on the assumption of a homogeneous fluid of constant density ρ.

In reality, density, ρ is a continuous function of depth, z or $\rho(z)$.

A typical depth profile of the properties (temperature, salt, density, velocity) clearly shows two or three immediate shape characteristics. See Figure 7.1. There is a wind-driven mixed layer in the upper 50–200 meters (the Ekman Layer in Chapter 4). This is followed by a gradual but significantly steep change of temperature/salinity/density within the depth range of say 500 to 800 meters, and then there is a rather very slow decline (for temperature) or increase (for salinity and density).

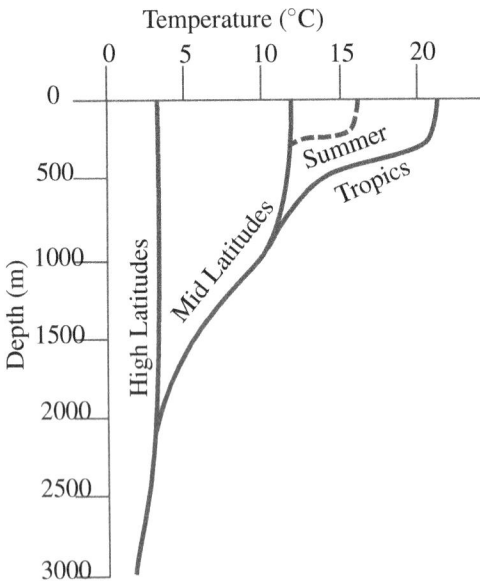

Figure 7.1: Typical temperature profiles in the open ocean. Below a relatively shallow surface layer, the ocean is uniformly cold. After Knauss (2016).

In reality, the temperature or velocity structure in the vertical is more complex than the simple one presented in Figure 7.1. An example is shown in Figure 7.2. Such a section can be thought about by representing it in multiple layers. So, the ocean can be thought of as consisting of multiple vertical layers with layer temperatures $(T_1, T_2, T_3, T_4...$ etc.$)$, with salinity $(S_1, S_2, S_3, S_4...$etc.$)$, density $(\rho_1, \rho_2, \rho_3, \rho_4...$ etc.$)$ and say velocity components $(v_1, v_2, v_3, v_4...$etc.$)$. Here the top layer is the first layer and the layer number increases as you go down the water column. Similarly, you can think of any other ocean or atmospheric variables to be represented this way. The equations of motions can then be cast for each layer and the forces on and responses of these layers can be connected in a multiple-layered system of mathematical equations.

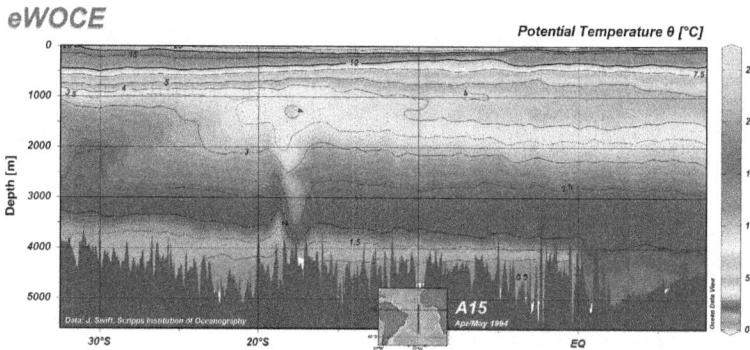

Figure 7.2: Temperature Layers across a WOCE Section P15.

Typically, temperature decreases down the water column so, $T_1 > T_2 > T_3 > T_4 > ....$ Similarly salinity increases with layers $S_1 < S_2 < S_3 < S_4 < .....$ So does density, $\rho_1 < \rho_2 < \rho_3 < \rho_4 < ....$ You can think of chlorophyll (or planktons) the same way; however, it is often the fact the chlorophyll has a subsurface maxima, so it increases down to a particular depth and then decreases again. The biology, the chemistry, and the sediments are more complex, but the layering idea helps us understand their dynamics.

The velocities also generally decrease with depth. The real ocean is more complex than this simplified picture of layers; however, if we think about the ocean being composed of infinite such layers or a very large number of layers, and construct a mathematical system of equations, then the problem of solving a force-response layered system of equations might be tractable with our computing power! This is precisely why the layering of the ocean and atmosphere is a very appealing approach for numerical modeling.

Let us explore this idea with a multi-layer ocean, as shown in Figure 7.3. A vertical profile of a parameter (u) can be decomposed into an infinite number of profiles. The first of these component profiles is the one with vertically integrated constant value or the barotropic mode. Each successive profile has an increasing number of zero-crossings. They each represent resolving additional layers for the continuous system. For example, the second profile on the right

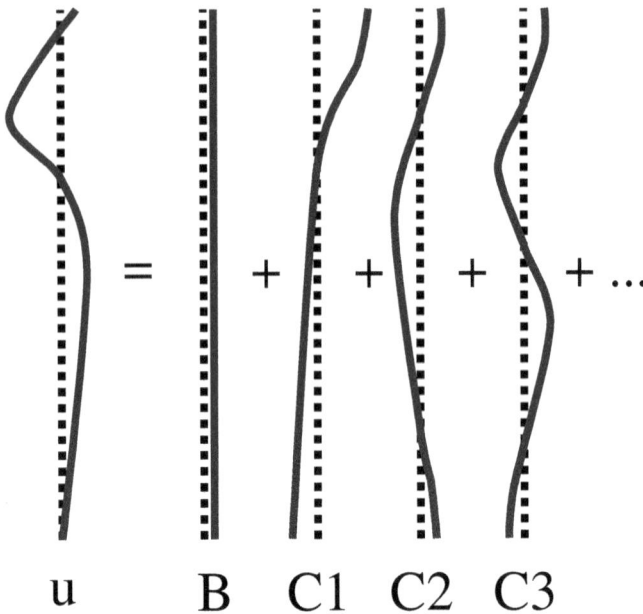

Figure 7.3: A multi-layer representation of a continuous ocean in the vertical. A vertical profile of a parameter (u) can be decomposed into multiple profiles with an increasing number of zero-crossings. Each successive mode (profile) represents resolving additional layers for the continuous system. B is Barotropic; C1, C2, and C3 are the first three baroclinic modes.

side of Figure 7.3 has one zero-crossing and represents the first baroclinic mode. It is also a representation of a "2-layer" system. The sum of the barotropic and the first baroclinic mode would then represent the total velocity profile of a 2-layer system. Similarly, the three-layer system would be represented by the barotropic and first two baroclinic modes. The fully complex velocity profile would be a summation of B+C1+C2+C3+... modes, as shown in Figure 7.3.

The simplest and often the most comprehensible way to understand the ocean circulation is to adopt a 2-layer view of the ocean. This provides a basic layering in the vertical. Once we understand the 2-layer system, it is easy to extend those ideas for an n-layer ocean or to a continuous vertical variation in a more extensive three-dimensional continuum sense. Let us look at the dynamics of a 2-layer system in detail next.

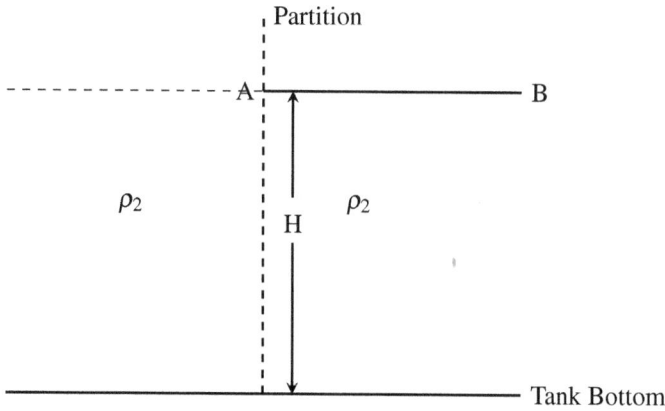

(a) Single density, Same height – No movement

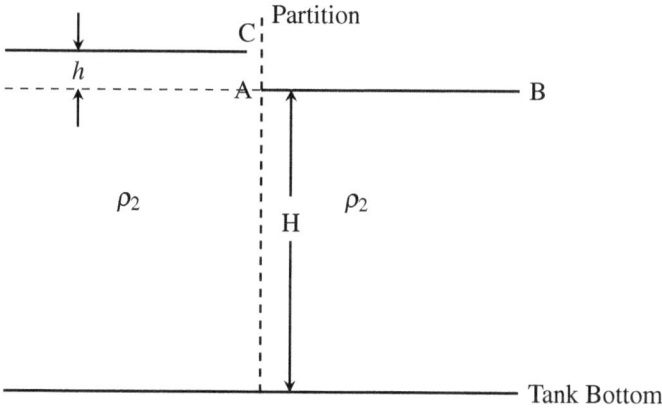

(b) Same density, different height – acceleration from left to right

Figure 7.4: Experiment 1. Same density fluid with different height across a common partition.

7.2 THE 2-LAYER OCEAN

Consider a simple laboratory experiment with three cases to understand the adjustment of two different density fluids. Think how the two different fluids will adjust or flow if we put them side by side in a tub with a removable partition first and then remove the partition suddenly.

To begin with, in the first experiment, let us put the same fluid in both the partitioned volume like in Figure 7.4. The density of the fluid is ρ_2. The upper layer of the fluid is at height H on both sides A and B. Now, if we remove the partition, there will be no movement, because the pressure on both of the free surfaces on either side of the partition is equal to the atmospheric pressure. Mathematically,

$$p_A = p_B = p_{atm} \qquad (7.1)$$

in this situation.

Now, allow yourself an additional bucket of the same density fluid and add it to the left side of the partition. This addition of water represents a height increase of h from level A to level C, as in the right panel of Figure 7.4. Now, what will happen if you remove the partition? Which way will the fluid flow?

The answer lies in comparing the pressure force at the same level on both sides, say at the level of A and B. Clearly, now the pressure at level A on the left side of the tub is the atmospheric pressure on level C plus the pressure due to the additional water between levels A and C with a height of h. Mathematically,

$$p_A = \rho_2 g h + p_C \qquad (7.2)$$

Now levels C and B are both supported by the same atmospheric pressure (one atmospheric pressure) from top (although the water columns below are of different heights); so,

$$p_C = p_B = p_{atm} \qquad (7.3)$$

And thus, the pressure difference between levels A and B is simply given by,

$$p_A - p_B = \rho_2 g h \qquad (7.4)$$

Since fluid would flow from high pressure to low pressure, once you lift the partition, the fluid from the higher elevation on the left will move to the lower elevation on the right. And the force (and the acceleration, remember $F = ma$) is proportional to the density and height difference between the levels.

Now consider the third and final experiment with two fluids of different densities (ρ_1 and ρ_2) with a step interface similar to experiment 2. This situation is illustrated in Figure 7.5. Now, how do you think the interface is going to adjust between the two layers once you remove the partition? What will be the driving force?

Why not follow the same procedure as we did for experiment 2? Let us see what the pressure difference is between the two fluids at the same level as A and B. Remember that the atmospheric pressure is now acting on the top of the first layer with density (ρ_1) at level D.

So, the pressure at A is given by the weight of the fluid of density (ρ_1) over the depth between D and C, plus the weight of the fluid of density (ρ_2) over the small thickness h between C and A (the interface difference) in addition to the atmospheric pressure on level D. Mathematically,

$$p_A = p_D + \rho_1 g H_1 + \rho_2 g h \qquad (7.5)$$

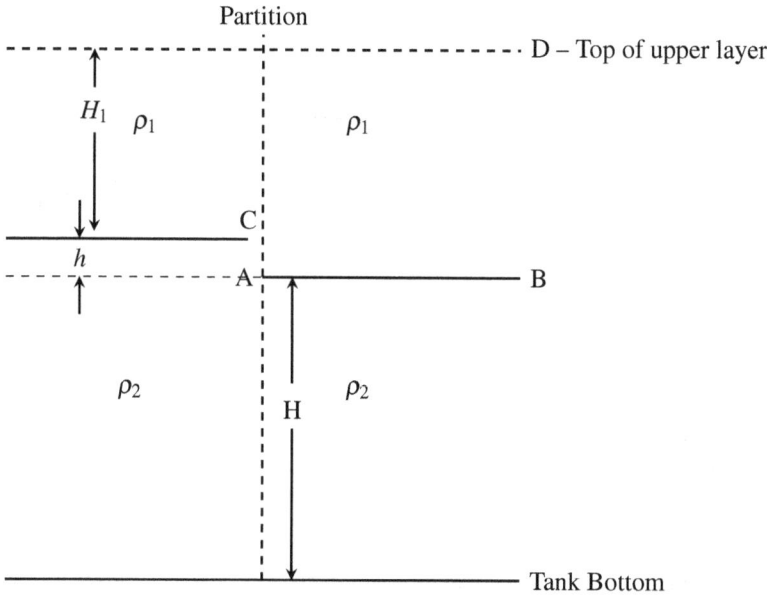

Figure 7.5: Experiment 2. Different density fluids with different thicknesses across a common partition.

In contrast, on the right side of the partition, the pressure at level B is simply the weight of the fluid of density ρ_1 over the depth $(H_1 + h)$ between D and B.

$$p_B = p_D + \rho_1 g (H_1 + h) \tag{7.6}$$

So, the pressure difference between A and B in the case of two fluids with two different densities is given by,

$$p_A - p_B = \underbrace{(p_D - p_D)}_{0} + \rho_1 g H_1 + \rho_2 g h - \rho_1 g H_1 - \rho_1 g h \tag{7.7}$$
$$= (\rho_2 - \rho_1) g h \tag{7.8}$$

Thus, the movement of the interface would be proportional to the density difference between the two layers. Compared to experiment 2, the pressure difference is reduced by a factor

$$\frac{\rho_2 - \rho_1}{\rho_2}.$$

In fact, this interface pressure difference in experiment 3 can be expressed in a similar form as in Equation 7.4, if we define a new variable, g' (pronounced "g-prime") called "reduced gravity," such that

$$g' = \frac{\rho_2 - \rho_1}{\rho_2} g \qquad (7.9)$$

Then the pressure difference in the third experiment is given by

$$p_A - p_B = \frac{\rho_2 - \rho_1}{\rho_2} \rho_2 g h \qquad (7.10)$$

$$= \rho_2 \underbrace{\frac{\rho_2 - \rho_1}{\rho_2} g}_{g'} h \qquad (7.11)$$

$$= \rho_2 g' h \qquad (7.12)$$

This idea of "reduced gravity" is the kernel of understanding how the ocean (and atmosphere) works in the vertical. A number of extrapolation of this idea is easy to think about now. Some of these are mentioned below for you to pursue in other advanced books.

- From equation 7.8, it is clear that $(\rho_2 - \rho_1)g$ is the driving force between the two layers. This is also called, aptly, "the buoyancy force" per unit volume.
- It is easy to visualize a system with multiple layers in the vertical and realize that the interface between two adjacent layers is driven by the difference of their respective densities.
- Density difference is a quantitative measure of stratification. A two-layer density difference is manifested by g'. A multi-layer problem provides a similar measure, which involves the vertical gradient of density ($\frac{\partial \rho}{\partial z}$) which is then related to N^2 or the buoyancy frequency. We will define and explain buoyancy frequency in the next subsection.
- Many interfacial phenomena such as internal tides and internal gravity waves are governed by the reduced gravity framework.
- Many real-ocean problems and processes have benefited from setting them up as a simple 2-layer system and understand the process dynamics using the g' formulation.

7.2.1 STABILITY AND STRATIFICATION (BUOYANCY FREQUENCY)

Let us now think about the continuous density variation in the vertical. The nature of a stable vertical column is by virtue of the fact that lighter fluid lies on top of the denser fluid. In other words, density ($\rho(z)$) increases with increasing depth and that is what makes the water column stable. Simply put, if

$$\frac{\partial \rho}{\partial z} < 0 \text{ as z is positive upward} \qquad (7.13)$$

then the water column is stable. The normalized rate of increase is generally taken as the measure of stability (E),

$$E = -\frac{1}{\rho} \left(\frac{\partial \rho}{\partial z} \right) \tag{7.14}$$

This slow increase of density with depth is expressed by the density gradient in the vertical or increasing dense layers of fluid as one goes down the water column. This layering of fluid is synonymous with "stratification." The fluid is divided into various "strata" (plural of "stratum" or "layer") or layers. This is stratification. So, it is the opposite of mixing! Mixing homogenizes the stratified layers. Stratification (or layering) is also opposite to the fluid's tendency to move in a vertical column under the rotational effect (potential vorticity). So, understanding the competition of these physical processes for a stratified fluid requires us to define stratification in a quantitative sense.

Now, let us consider what happens when we disturb the stability of a parcel against the ambient density gradient. Suppose we identify and isolate a small parcel in a perfectly elastic (so it can expand or shrink freely) and insulated (from heat and salt exchange) balloon. Let $\rho_0(z)$ is the density of the parcel in its undisturbed position, z. Suppose an external force is applied so that the parcel is raised to height $z + h$ without disturbing the horizontal stratification.

The upward-directed pressure force acting on the balloon at $(z + h)$ would be equal to the local volume displaced by the balloon times the local density $(\rho(z+h))$ of the fluid around it. The downward force of gravity is equal to the volume of the balloon times its original density $\rho_0(z)$ (which is higher than the local density, where the parcel is now). Thus, the balloon will be subjected to a buoyancy force given by

$$g\left[\rho(z+h) - \rho(z)\right]V$$

where, V is the volume of the parcel. The force is positive if directed upward.

Now, using Newton's law $(F = ma)$, this force must be equal to

$$\underbrace{\rho(z)V}_{\text{mass of parcel}} \times \underbrace{\frac{d^2h}{dt^2}}_{\text{acceleration}}$$

Equating these two expressions and simplifying,

$$\frac{d^2h}{dt^2} = \frac{g}{\rho(z)}\left[\rho(z+h) - \rho(z)\right] \tag{7.15}$$

Now, we can use Taylor's series expansion to express $\rho(z+h)$ in terms of the initial density $(\rho(z))$ with the ambient density gradient $(\frac{\partial \rho}{\partial z})$ as follows

$$\rho(z+h) - \rho(z) \approx \frac{\partial \rho}{\partial z} h \qquad (7.16)$$

At this time, recognize that the geophysical fluids are generally weakly stratified, which means that the density variations are significantly smaller compared to the density itself, and thus an average (or reference) density can be used to replace the density $\rho(z)$ in the denominator of equation 7.15 on the right-side. This approximation is also known as the Boussinesq approximation. Using ρ_0 as the reference density, the response of the parcel is expressed by the following equation:

$$\frac{d^2 h}{dt^2} = \frac{g}{\rho_0} \frac{\partial \rho}{\partial z} h \qquad (7.17)$$

$$\frac{d^2 h}{dt^2} = -N^2 h \qquad (7.18)$$

where the following expression defines the new variable N in terms of the static stability (E) defined earlier in equation 7.14.

$$N^2 = -\frac{g}{\rho_0} \frac{\partial \rho}{\partial z} = gE \qquad (7.19)$$

Now, let us look at some characteristics of N. What is it?

Let us see what happens to the parcel when it is displaced from its initial position at z to a height $z + h$ above it. Since it is heavier than the surrounding fluid at $z + h$, it will start sinking with a downward velocity and due to its inertia will cross its initial height (Z) and reach somewhere below. Now at some new location, it will encounter heavier fluid than itself and feel the effect of upward buoyancy being greater than its weight and will rise again. This oscillation will keep on going around the equilibrium level. So, there must be a frequency of this oscillation. Is that what N could be?

Look at equation 7.18 again and compare it with the time-dependent equations in Chapter 6. If we assume a solution for h to mimic the oscillatory behavior with the following function,

$$h = A e^{iNt}$$

then the equation 7.18 is automatically satisfied. This means that the oscillation of the parcel due to buoyancy forcing has a natural frequency of N as defined above by equation 7.19. Clearly, the unit of N^2 is s^{-2}, and so, N has a unit of s^{-1}. The period of oscillation is $T = 2\pi/N$.

N is called the "Stratification frequency" or "Buoyancy frequency" or the Brunt-Vaisala frequency to honor the two scientists, who highlighted the importance of this frequency in quantifying the behavior of stratified fluids (Emery et al., 1984).

Note that if N^2 is negative (i.e., stability E is positive due to $\frac{\partial \rho}{\partial z} > 0$), the fluid is top-heavy, resulting in instability and the solution $h = Ae^{iNt}$ grows exponentially. In case of weak disturbance and limited growth, other parts of the fluid will participate in stabilizing the system after some time. However, if the unstable situation persists, then the fluid will eventually become completely unstable vertically and result in convection. See Figure 7.6 for an illustration of this phenomenon following Knauss and Garfield (2016).

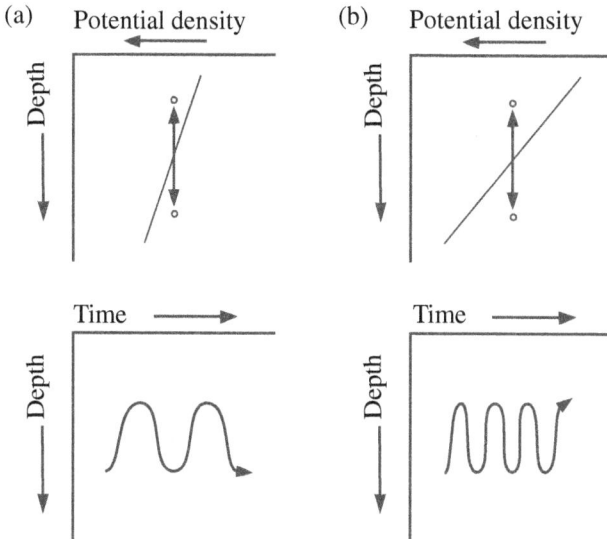

Figure 7.6: Imagine a balloon filled with seawater at rest at equilibrium depth in a frictionless fluid with gradient, as shown. When set in motion, the balloon will oscillate about its equilibrium depth with a period inversely proportional to the density gradient; the larger the gradient, the higher the frequency and the shorter the period. After Knauss and Garfield (2016).

A final point about the definition and derivation of E and N, as presented here. The assumptions of an elastic and insulated balloon can be relaxed under a more rigorous derivation by either replacing the density with the "potential density" or by considering the compressibility effect of pressure. An exact formulation for the static stability would then be given by,

$$E = -\frac{1}{\rho}\left(\frac{\partial \rho_\theta}{\partial z}\right) \text{ or} \tag{7.20}$$

$$E = -\frac{1}{\rho}\left(\frac{\partial \rho}{\partial z}\right) - \frac{g}{c^2} \tag{7.21}$$

Here, c is the velocity of the sound in seawater. In both cases ρ is the in-situ density and ρ_θ is the potential density. See Pond and Pickard, 1983 for rigorous derivation details. The calculation for N should change accordingly for more accuracy.

Let us now look at a couple of other effects of stratification on different processes: (i) vertical mixing; (ii) horizontal mixing; and (iii) energy extraction from stratified layers.

7.2.1.1 Stratification and Vertical Mixing

One thing is clear that the stratification results in the water column having different potential energy in the system due to the variation of the density gradient. The higher the stratification or the density gradient, the more the work required to mix the water or reduce the density gradient and raise its potential energy. The weaker the stratification, the more potential energy it already holds and less work is needed to mix it.

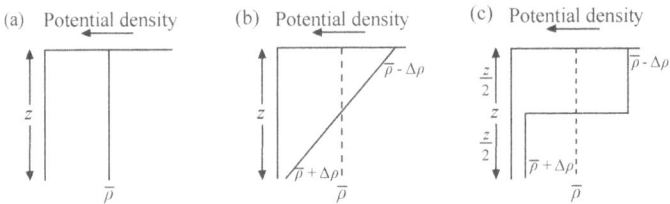

Figure 7.7: Concept of N^2 and potential energy. The potential energy associated with three columns of fluids. That which is well mixed has the largest potential energy. The potential energy for the three columns are (from left to right panels): (a) $\overline{\rho}g\frac{z^2}{2}$; (b) $\overline{\rho}g\frac{z^2}{2} - \triangle\rho g\frac{z^2}{6}$; and (c) $\overline{\rho}g\frac{z^2}{2} - \triangle\rho g\frac{z^2}{4}$. After (Knauss, 2005).

In order to understand the effect of stratification on mixing, let us consider three different cases of stratification. See Figure 7.7. The left panel shows a well-mixed water column of depth H with uniform density $\bar{\rho}$. In this case, there is no density gradient ($\frac{\partial\rho}{\partial z} = 0$), and so, $N = 0$. The potential energy of this water column is given by,

$$PE_1 = \int_{-H}^{0} \bar{\rho}gzdz = \bar{\rho}g\frac{H^2}{2} \qquad (7.22)$$

Contrast this with the middle panel, where there is a constant density gradient. The potential energy in this case for the water column given by (after similar integration),

$$PE_2 = \bar{\rho} g \frac{H^2}{2} - \triangle \rho g \frac{H^2}{6} \qquad (7.23)$$

And, then when we consider a very strong stratification, like a 2-layer system in the right panel of Figure 7.7, the potential energy is given by

$$PE_3 = \bar{\rho} g \frac{H^2}{2} - \triangle \rho g \frac{H^2}{4} \qquad (7.24)$$

So, much more energy is required to mix vertically across the thermocline for the stronger stratification cases than in the weaker stratification situations. Interestingly, strong winds at the surface can mix the upper ocean against stratification and the surface mixed-layer is created. Turbulence created by non-linearities (see Chapter 9) can also mix waters by reducing the density gradient. In the deep ocean; however, where there is weak stratification, the source of energy is limited to enable mixing.

Seasonal changes in the temperature profile within the mixed layer often depend on the input of winds and heat and buoyancy (precipitation/evaporation/runoff). See Figure 7.8 for an example of seasonal changes of the mixed layer.

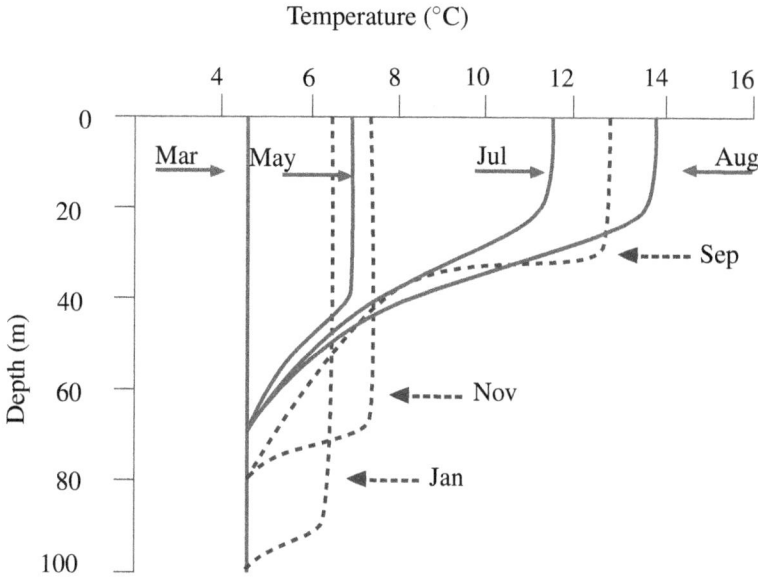

Figure 7.8: Seasonal evolution of mixed layer in the ocean.

7.2.1.2 Stratification and Horizontal Mixing

Mixing can also happen along the lines of constant density or isopycnals. It is much easier to mix along isopycnals (constant density lines or surfaces) than across isopycnals; the latter is called "diapycnal mixing." While work is required to mix to overcome stratification, much less work is required to mix along than across density surfaces. In fact, if ocean water would be an ideal fluid with no viscosity, then it would require no work to mix along constant density since there is no friction. However, the ocean is not ideal and there is lateral friction, which is generally parameterized as eddy viscosity. It has been found that the ratio of horizontal coefficients to their vertical counterparts is on the order of 10^8. This effectively means that if everything else remains the same, it is 100 million times easier to mix horizontally than it is vertically!

7.2.1.3 Energy Extraction from Stratification

When the stratified ocean has layers or interfaces which are not horizontal, there is excess potential energy in the system, which can be extracted by turbulence to convert to eddy kinetic energy (EKE). This point will be clearer once you read the section on baroclinic instability later in this chapter. In this sense, stratification becomes a source of kinetic energy. Let us just see where this energy is coming from.

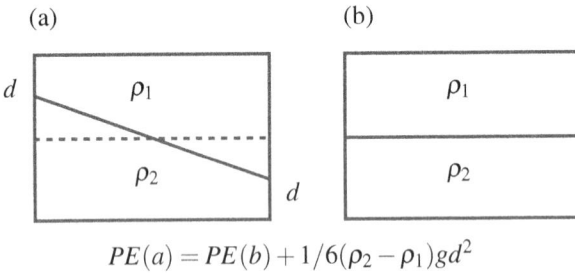

(a)   (b)

$$PE(a) = PE(b) + 1/6(\rho_2 - \rho_1)gd^2$$

Figure 7.9: In a 2-layer ocean, the one with a sloping interface has larger potential energy than the one which does not. The larger the slope, the larger the potential energy difference. After Knauss (2016).

Imagine the interface of a 2-layer ocean being slanted on one case (a) and flat (b) in the other, as shown on the left and right panels of Figure 7.9. The potential energy in the sloping interface ocean is given by

$$PE_a = PE_b + 1/6 (\rho_2 - \rho_1)\, g d^2 \qquad (7.25)$$

The excess potential energy, called the "Available Potential Energy" (APE), can be drawn upon by instabilities in the system to grow and generate eddies.

This is one of the major mechanisms for generating baroclinic instability. This process converts the APE to kinetic energy of the eddies.

7.2.1.4 Barrier Layer and Inversion Layer

As discussed earlier and shown in Figure 7.8, the upper layer of the ocean is called the mixed layer. It is often observed that the oceanic properties such as temperature, salinity, and density are uniform within this surface mixed layer. Some of the factors which influence the mixed layer are winds, thermal radiation, evaporation, precipitation, and river runoffs.

In some regions (e.g. equatorial western Pacific and India oceans), where the freshwater input in terms of rains and/or river influx is more, the upper layer becomes very fresh and the surface salinity reduces drastically. However, the layer with uniform temperature remains comparatively deeper. In such situations, the depths of the isohaline layer and that of the isothermal layer are at different levels. See Figure 7.10. Clearly, a new thin layer called the "barrier layer" forms between the upper mixed layer and the top of the thermocline(or bottom of the isothermal layer). This necessitates a re-definition of the mixed layer, barrier layer, and the isothermal layer.

The barrier layer (BL) is defined as the intermediate layer that separates the base of the mixed layer (ML) from the top of the thermocline as shown in Figure 7.10. The top of the thermocline is often defined as the starting of the steep thermocline slope at the end of the isothermal layer. This depth is called the isothermal layer depth (ILD).

Now, there are a few different ways to define the upper mixed-layer depth (MLD). One simple way is to choose the depth of the uniform density layer (up to the level where both salinity and temperature are constant). See Figure 7.10. Mathematically, this provides the barrier layer thickness (BLT) as: $BLT = ILD - MLD$. However, since observational data in the upper mixed layer is often noisy and of low-resolution (to resolve the shallow mixed layer), identifying a uniform density layer might become a challenge. Thus, the definition of the MLD has evolved and is generally based on variable density (σ_t) criteria (Sprintall & Tomczak, 1992). This process determines the depth where σ_t is equal to the sea surface σ_t plus an increment of σ_t, ($\triangle\sigma_t$) equivalent to a prescribed decrease in temperature keeping the salinity constant. This prescribed decrease in temperature varies between 0.2°C and 0.5°C (Shee et al., 2019). See Sprintall and Tomczak (1992) for a comparative study of barrier layers from climatological data in the three oceans (AO, PO, and IO). The barrier layer acts as a "barrier" for the transfer of heat, momentum, mass, and nutrients between the mixed layer and to the depths below across the thermocline.

Sometimes, the upper layer cools rather rapidly during winter in the Bay of Bengal, while the comparatively warm deeper layer (within MLD and ILD) cannot mix due to the presence of a barrier layer. In that case, the warm layer is

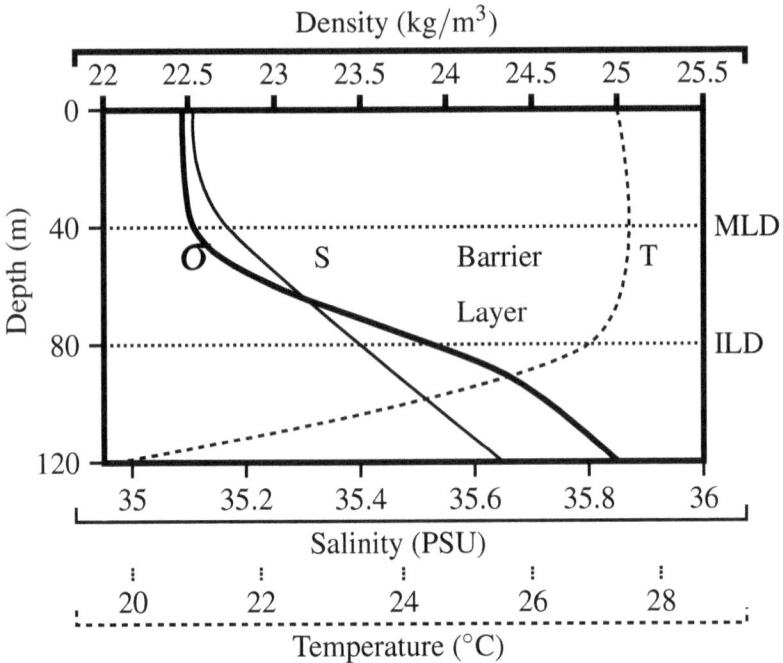

Figure 7.10: Barrier layer formation. The layer between the base of the upper mixed layer (density constant) and the top of the thermocline can be defined as the barrier layer. The MLD is often defined from density variation criteria. See text for details. Figure credit: Adrienne Silver.

trapped within the barrier layer, and a temperature inversion forms (Shee et al., 2019). An example profile is shown in Figure 7.11 in the Bay of Bengal.

7.3   EL NIÑO AND LA NIÑA

A very unique example of a 2-layer approach is its application to the understanding of the circulation in the Equatorial Pacific. We are discussing a very well-known phenomenon called El Niño. What is El Niño? You may have been thinking what's the connection to Spanish since, in its originality, the Spanish expression means "a boy" and its counter process "La Niña" means "a girl." Well, it is an oceanographic phenomenon that happens every 3–8 years and it was known to the Peruvian fishers for over centuries.

What happens during an El Niño year? The fish ecosystem is disrupted because of colder water due to upwelling along the equatorial eastern Pacific coast of South America. The Peruvian coastal fishery (which spans a large

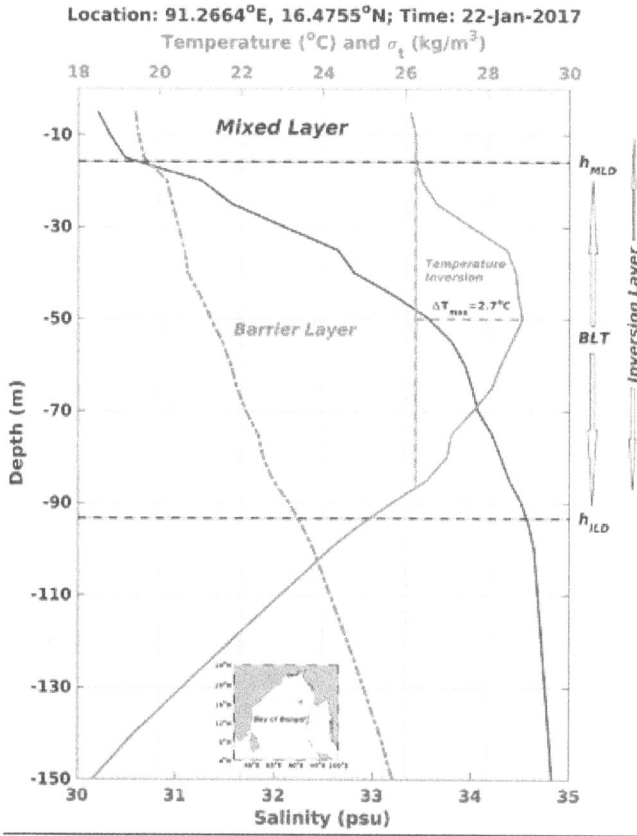

Figure 7.11: Example of an inversion layer in the Bay of Bengal. Temperature in °C (solid red), σ-density in kg/m^3 (dotted red), and salinity in psu (blue line) from Argo (WMO 5904302) at 91.3°E, 16.5°N on 22 January 2017. Figure credit: Sourav Sil.

portion in the equatorial eastern Pacific) suffers during an El Niño year because of lack of fish in these waters every so many years. They would relate this absence in otherwise generally colder, nutrient-rich, and fish-plenty water to a much warmer, nutrient-poor warm December name it as "a Christmas boy" or "El Niño" year. Following the year of the El Niño, there is a boom year of excess fish harvest with high-nutrient water, which they called 'the girl' or "La Niña."

7.3.1 A BRIEF HISTORY

This oceanic phenomenon was known to early oceanographers, and they could relate the El Niño year to the cessation of upwelling, which would otherwise bring colder, nutrient-rich waters from greater depths. However, the reason of the upwelling or the cessation of it in a year and then followed by an excessive return was largely unknown, even in the early decades of the second half of the twentieth century.

Meteorologists had also been interested in Equatorial Pacific and its atmospheric circulation for a long time. Because of its vastness the Pacific was thought to control the weather in all the surrounding countries and even in the countries around the Indian Ocean. Remember that the surface winds on the equator converge from north and south to flow from east to west (Figure 1.9) due to the earth's rotation. As the surface air comes from the north in the northern hemisphere near the equator, it veers to the right due to the Coriolis effect and turns westward. Similarly, the equator-ward surface air in the southern hemisphere turns to its left, again westward. These are called the trade winds, which the European explorers and traders used to sail across the globe in the fifteenth to twentieth century.

In 1896, there was a massive famine in the Indian subcontinent, which was mostly under British occupation at that time. Interestingly, it was Mahatma Gandhi who wrote a letter to the Queen of England reiterating the possibility of the famine being caused by the lack of monsoon rain in successive years. Gandhi requested help in seeking a better understanding of the Indian Monsoon. The Queen passed on the message to the House of Lords and they promised a more systematic investigation. Back in 1875, the Indian Meteorological Department (IMD, under the Royal Meteorological Department) was instituted for bringing multiple weather centers to work for an organized effort to understand monsoon or the lack of it. This effort was led by H. F. Blanford. In 1889, Sir John Eliot was appointed as the first Director-General of the IMD in Kolkata (then Calcutta), India. Calcutta was the capital of India till 1905, when the capital was moved to Delhi.

A young mathematician, Gilbert Walker, was sent as the Director-General of the IMD in 1903, to follow Sir John Eliot. Walker was very interested in studying the monsoon and started investigating the cause of monsoon from a force-response perspective. He wanted to find out about the distribution of the sea level pressure around the globe, and he believed that the whole atmospheric system is inter-connected with their dominating cells of action. By writing letters and collecting data from all of the existing meteorological stations (many under British and American and other European authorities) and Royal Navy ships, while sitting in his office in Calcutta, India, he drew up maps and tables of the Sea Level Pressures around the globe and started seeing patterns of oscillations! He discovered that the atmospheric systems have these centers of pressure cells, he named them and described their oscillating behaviors and related them to the weather patterns around various oceans using correlations. These include the Southern Oscillation in the Indo-Pacific region, the North Atlantic Oscillation in the Atlantic, and the North Pacific Oscillation in the North Pacific. His seminal works are published in multiple papers between 1904 and 1924 in different volumes of the IMD bulletin, e.g., (G. T. Walker, 1910, 1924). A summary was published by S. G. T. Walker and Bliss, 1928 by the Royal Meteorological Society and many of Walker's later works and their impact on our understanding of ENSO and other teleconnections are found in the papers by Wolter and Timlin (2011), Wallace and Gutzler (1981) and in the article by Stephenson et al. (2003). As an aside, it is the same Sir Gilbert Walker whose name is penned with the famous "Yule-Walker Equation" in time-series analysis, specifically for autoregressive (AR) modeling. Walker used to think that most of the atmospheric phenomena could be explained by some low-order autoregressive model, at least for their dominant modes of variability. He always wanted to investigate a natural process from a force-response system perspective. More on Walker's life and contribution in various branches of science can be found in the articles by Katz (2002).

Another aside about Sir Gilbert Walker. While looking at the patterns of sea level pressure distributions around the globe, Walker discovered the North Atlantic Oscillation (NAO) – a seesaw pattern for the variations of sea level pressure difference between the Azores High (the subtropical high on the Atlantic) and the Icelandic Low (the subpolar low on the Atlantic). This pattern is what we call the North Atlantic Oscillation (NAO). Walker connected the phases of the NAO with the storms and weather in the Americas and Europe as well as with the strength of the Gulf Stream. More on those in a later section.

Looking back at the pacific, Walker first discovered the atmospheric pressure fluctuation in an oscillatory manner (seesaw) over the tropical Indo-Pacific region, which he termed the "Southern Oscillation." He noticed that a high sea level pressure on the eastern pacific (Tahiti) would coincide with a low on the western Pacific (Darwin, Australia) in a typical year. The trade winds would flow from the east to the west along the equator. The sinking air on the eastern side is dry, which, upon traveling along the warm equator westward, becomes light and moist and rise on the western Pacific. This rising air, reaching higher altitudes, would become colder and heavier and precipitate heavily on the western Pacific countries (Indonesia and parts of the Indian subcontinent). The higher-level winds will flow eastward, becomes dry, reaching the eastern Pacific, and sinks due to the high pressure. He also discovered that at particular times (every so many years) this high-low pattern of the sea level pressure changes completely. The pressure on Tahiti becomes lower than the pressure at Darwin, Australia – which leads to reversing winds, shutting down the convection on the west causing droughts in parts of Indonesia and the Indian subcontinent and more rain on the central and eastern Pacific. This atmospheric circulation pattern is now known as the "Walker Circulation" and it is a common feature of the equatorial atmosphere over all the three major oceans. While the Oceanographers knew about El Niño in the Eastern pacific, and meteorologists understood that phases of the Southern Oscillation are related to the east-west movement of the Walker Cell and resulting precipitation patterns; the inter-dependence or inter-connectivity between the ocean and the atmosphere was not established till the 1960s. In the early years of the second half of the twentieth century, we were beginning to realize that the ocean and the atmosphere interact with each other in a significant way and that many of the phenomena that we observe in the atmosphere are affected by the underlying sea surface temperature and its evolution, like cyclones getting energy and intensifying from the warm currents, while the upper ocean is primarily driven by the winds above it (Ekman, Sverdrup, Stommel, Munk, Robinson).

It was such inter-disciplinary thinking that led to Dr. Jacob Bjerknes of UCLA, who was the son of Vilhelm Bjerknes, Norwegian Metereologist, one of the pioneers of modern numerical weather forecasting (see Chapter 14), (J. Bjerknes, 1969) to propose a combined hypothesis for an ocean-atmosphere interaction process on the Equatorial Pacific. He realized that the "El Niño" phenomenon observed in the ocean and the "Southern Oscillation" observed in the atmosphere are connected to each other in a larger ocean-atmosphere interaction phenomenon. This is what we call today the "El Niño – Southern Oscillation" or "ENSO." Jacob Bjerknes was a support meteorologist when Roald Amundsen made the first crossing of the Arctic.

Bjerknes noticed that the typical (or normal) state of the sea surface temper-atures (SSTs) at the eastern pacific is to be very cold, much colder to be near the equator. Since the western pacific is much warmer, a general temperature gradient has to exist along the equator. He maintained that the sea surface tem-perature gradient (cold near Peru and warm near the western end) is necessary for the atmospheric pressure gradient to exist that drives the Walker circulation.

Bjerknes also hypothesized that, in some odd years, a warming of the eastern tropical Pacific weakens the Walker circulation and causes the convective zone of heavy rainfall to move eastward, from the western side to the central and eastern tropical Pacific. This would be the other phase of the Southern Oscillation and the El Niño condition will prevail.

The problem is this. Think about it. From a meteorological perspective, SST variations cause the Southern Oscillations and its two phases of the seesaw pattern and affect the Walker Circulation. But from the oceanographic point of view, the SST changes are caused by the surface wind fluctuations associated with the Southern Oscillation. So, it's circular!

For example, relaxation of trade winds (which carry the warm water westward) causes the warming of central and eastern pacific, which reduces the east-west pressure gradient and relaxes the wind further, which causes further warming, which causes further relaxation – which leads to El Niño!

Bjerknes proposed "a never-ending succession of alternating trends by air-sea interaction in the equatorial belt" – as a possible perturbation mechanism for El Niño – but he was uncertain about the mechanism that causes a turnaround from its warm to cold phase. In summary, we know how it (El Niño or La Niña) happens to evolve; we just don't know (yet) how it gets started – is it in the winds or is it in the temperature? In these situations, models are very helpful as they can predict the final outcome given an initial perturbation which is observed by a set of instruments. A simple 2-layer model of the ocean and atmosphere were able to predict the evolution of El Niño and La Niña. We will discuss some of these attributes here.

A final remark about upwelling associated with the thermocline movement along the equator: while the surface winds are being driven westward along the equator by the zonal SST gradient, they act to create the cold upwelling ocean water in the east. The cold of the eastern equatorial Pacific waters is explained by the horizontal advection of westward currents along the equatorial Pacific, upwelling along the equator, and upward thermocline displacement.

Bjerknes associated the feedback loop of the oceanic and atmospheric circulation over the tropical Pacific as a "chain reaction," noting that "an intensifying Walker Circulation also provides for an increase of east-west temperature contrast that is the cause of the Walker Circulation in the first place." Bjerknes also found that the interaction could operate in the opposite: a decrease in the equatorial easterlies diminishes the supply of upwelling cold water and the lessened east-west temperature gradient causes the Walker Circulation to slow down. He thus provided an explanation for the association of the low phase of the Southern Oscillation with El Niño as well as the association of the higher phase with normal cold state of the eastern Pacific. A very high phase is associated with an extremely cold phase of the eastern Pacific, La Niña.

7.3.2 THE PHYSICAL SETUP

The ocean-atmosphere interaction on the equatorial Pacific can be best described if we consider, compare, and contrast the three different conditions: (i) the Normal condition; (ii) the El Niño condition, and (iii) the La Niña condition.

It is also instructive to look at these conditions in terms of three oceanographic variables and three atmospheric variables. The three most important atmospheric variables are: (i) the trade winds blowing from east to west; (ii) the sea level pressure; and (iii) the precipitation. The three most relevant oceanographic parameters/processes are (i) the SST (also SSH) across the Pacific, (ii) the thermocline depth variation from east to west, and (iii) the upwelling on the eastern side of the Pacific. Additionally, let us think about what happens to the nutrient availability due to more or less upwelling, affecting in turn the fish population with abundance or the lack thereof.

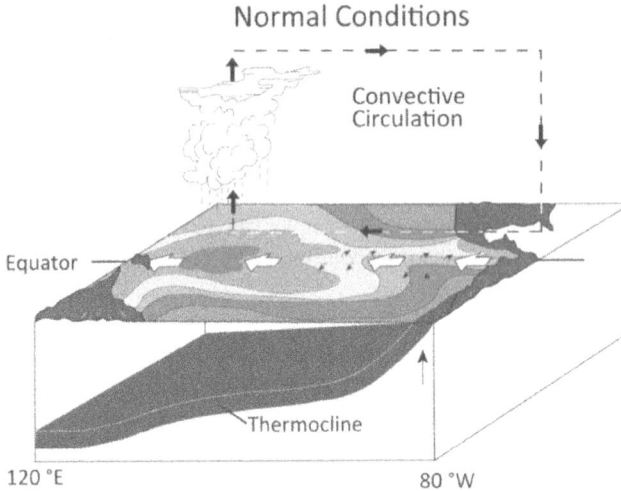

Figure 7.12: Normal Condition on the Equatorial Pacific. See the winds (white arrows) blowing from west to east, piling up warm water in the western Pacific (red blob). The thermocline (blue subsurface curtain) rises up to the east, bringing nutrients and colder waters to Eastern Pacific. The warm water on the west enhances convection on the western side Pacific on the western limb of the Walker Cell. Image credit: pmel.noaa.gov.

In the *neutral* condition in the tropical Pacific,

- The sea level pressure is high on the east and low on the west, creating trade winds flow from east to west.
- The westward winds move the warm water towards west to create what is called the "warm pool' in the western Pacific and leaving a "cold tongue" in the eastern pacific.
- Contrast in SSTs between the western and eastern tropical Pacific is about 8 Celsius degrees or 14.4 Fahrenheit degrees.
- Over the western tropical Pacific, warm surface waters heat the overlying air, strengthening convection currents that produce heavy rainfall.
- In response to the SLP variation (high on the east to low on the west), the thermocline dips in the west and rises on the east.
- In the eastern pacific, under the cold tongue, deeper nutrient-rich water upwells (due to continuity) on the raised thermocline bringing food

and habitable temperature for fish (anchovies) to be exploited by the Peruvian fishermen.

El Niño Conditions

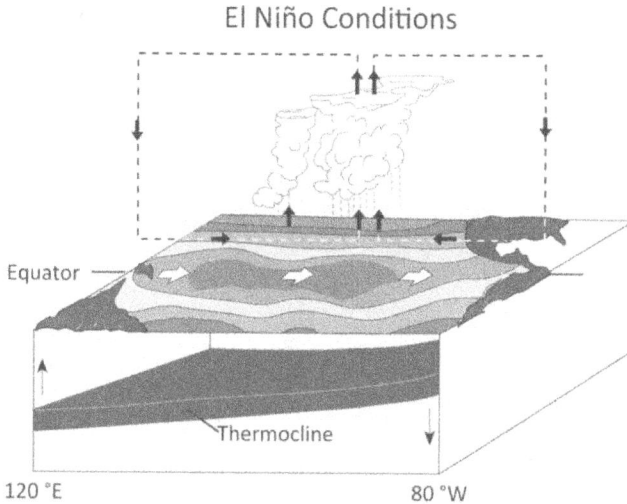

Figure 7.13: El Niño condition on the Equatorial Pacific. Note the relaxing trade winds, flattening of the thermocline, and Eastward movement of the Walker cell bringing precipitation to Central and eastern pacific. Image credit: pmel.noaa.gov.

In the *El Niño (warm phase)* condition in the tropical Pacific

- Air pressure gradient across the tropical Pacific weakens.
- Relaxing trade winds weaken the westward flow of the equatorial currents, even reversing direction.
- The thick layer of warm surface water drifts eastward, until deflected toward the north and south by the continental landmasses.
- In the western Pacific the trade winds slacken, SSTs drop, sea level falls and the thermocline rise.
- In the eastern tropical Pacific, SSTs rise, sea level climbs and the thermocline deepens.

In the La Niña (Cold phase) conditions in the tropical Pacific, which is opposite to the El Niño phase, are characterized by,

- Periods of unusually strong trade winds
- Colder than usual surface waters over the central and eastern tropical Pacific, with exceptionally vigorous upwelling

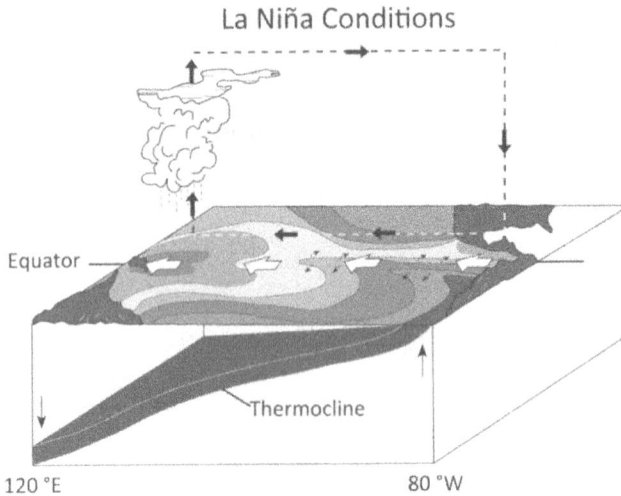

Figure 7.14: La Niña condition on the Equatorial Pacific. Note the enhanced trade winds, sharper thermocline rise to the east and westward movement of the Walker cell, bringing very dry (or drought-like) conditions to the central and eastern Pacific regions and excessive rain (flood-like) over the western Pacific island countries. Image credit: pmel.noaa.gov.

- Somewhat warmer than usual surface waters over the western tropical Pacific
- SST anomalies are essentially opposite of those observed during El Niño
- La Niña tends to persist for 9 to 12 months
- Some episodes may persist for as long as two years

Finally, realize that because of the spread of the Pacific, there is a series of linkage between changes in atmospheric circulation occurring in widely separated regions of the globe, often over thousands of kilometers. These are often referred to as natural telecommunications between weather systems.

7.3.3 A 2-LAYER MODEL OF THE EQUATORIAL PACIFIC

Now that we understand the physical setup of ocean-atmosphere interaction on the equatorial pacific, we can think about a simple 2-layer dynamical model where the primary variables can be those specific ones that we know are probably the most important.

It is clear that the wind stress and thermocline depth are definitely two key characteristics of the Neutral-El Niño-La Niña conditions, as shown in

Figures 7.12, 7.13, and 7.14. The 2-layer equations of motions can now be sought with the interface height η above the mean thermocline depth of H with a rigid lid at the surface (no vertical velocity or pressure variation at the surface). The horizontal velocities u and v are for the upper layer, the velocities in the lower layer are zero. The balance is due to reduced gravity g'. See Figure 7.15 for the mathematical model set up.

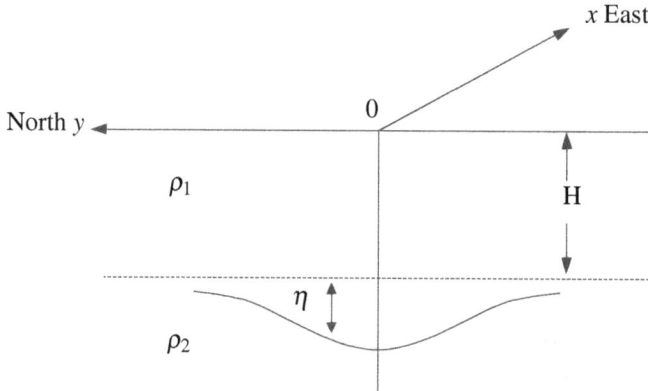

Figure 7.15: The 2-layer model setup on the Equatorial beta-plane.

The linear hydrostatic motion in the upper layer is driven by the zonal wind stress (trade winds), τ_x that acts as a body force uniformly distributed over the upper layer mean depth H. The motion is associated with the displacement η of the interface and can be described by the following momentum equations:

$$\frac{\partial u}{\partial t} - fv + g'\frac{\partial \eta}{\partial x} = \frac{\tau_x}{H} \tag{7.26}$$

$$\frac{\partial v}{\partial t} + fu + g'\frac{\partial \eta}{\partial y} = \frac{\tau_y}{H} \tag{7.27}$$

These two horizontal equations clearly will lead to a wave solution of the form

$$g'\frac{\partial \eta}{\partial t} + c^2(\frac{\partial u}{\partial x} + \frac{\partial v}{\partial y}) = 0 \tag{7.28}$$

Since $y = 0$ at the equator, the rotational effect is entered through the equatorial beta plane approximation, or $f = \beta y$, where $\beta = \frac{\partial f}{\partial y} = \frac{2\Omega \cos \phi}{R}|_{\phi=0} = \frac{2\Omega}{R}$, and R is the length scale over which the rotational effects are comparable to the inertial effects or the radius of deformation.

Clearly, $c^2 = g'H$ and c is the Kelvin wave speed riding on the thermocline interface. The speed c could be thought of as that of an equivalent surface gravity wave with an equivalent depth of h where $c = \sqrt{gh}$.

Using a typical density variation for the 2-layer system, $g' = \frac{\rho_2 - \rho_1}{\rho_2} g$ and $\frac{\rho_2 - \rho_1}{\rho_2} = 0.002$, for a typical thermocline depth ($H = 100m$), the equivalent depth h is only 20 cm, and $c = 1.4 \text{ ms}^{-1}$.

Now the set of equations 7.26, 7.27, and 7.28 are time-dependent equations. To seek solutions, the goal would be to get one equation in v, eliminating u and η.

To do this, one needs to first develop the vorticity balance by cross-differentiating equations 7.26 and 7.27 and subtracting one from the other ($\frac{\partial}{\partial x}[7.27] - \frac{\partial}{\partial y}[7.26]$) to get the following equation

$$\frac{\partial^2 v}{\partial x \partial t} - \frac{\partial^2 u}{\partial y \partial t} + f\left(\frac{\partial u}{\partial x} + \frac{\partial v}{\partial y}\right) + \beta v = \frac{1}{H}\left(\frac{\partial \tau_y}{\partial x} - \frac{\partial \tau_x}{\partial y}\right) \qquad (7.29)$$

Note the difference between this equation 7.29 with those vorticity equations developed in Chapter 4 for steady-state solutions. The time-dependent terms are combined here with spatial derivatives of the velocity fields. In order to decouple u and v, we need to go further down to another level of differentiation and subtraction and combine with equations 7.26 and 7.27 in some manner.

This is done by operating on 7.26 and 7.27 as $\frac{f}{c^2} \times \frac{\partial}{\partial t}(7.26) - \frac{\partial^2}{\partial t^2}(7.27)$, which eliminates u and leaves one equation in v and η as

$$-\frac{v_{ttt}}{c^2} - \frac{f^2}{c^2}v_t + \frac{g'f}{c^2}\frac{\partial^2 \eta}{\partial x \partial t} - \frac{g'}{c^2}\frac{\partial^3 \eta}{\partial t^2 \partial y} = \frac{f^2}{c^2}\left(\frac{\tau_x}{H}\right)_t - \frac{1}{c^2}\frac{\partial^2}{\partial t^2}\left(\frac{\tau_y}{H}\right) \qquad (7.30)$$

Here the subscript notation for derivatives is used for ease of understanding (e.g., $v_{ttt} = \frac{\partial^3 v}{\partial t^3}$).

A little more algebraic manipulation as in step 1: rewrite last 2 terms of the LHS of equation 7.29 using equation 7.28; and in step 2: do $\frac{\partial}{\partial x}$ of (7.29) and use it in step 1 to eliminate the terms containing u, can yield the following final equation in one variable v,

$$(v_{xx} + v_{yy})_t + \beta v_x - \frac{v_{ttt}}{c^2} - \frac{f^2}{c^2}v_t = F \qquad (7.31)$$

where

$$F = \frac{f}{c^2}\left(\frac{\tau_x}{H}\right)_t - \frac{1}{c^2}\left(\frac{\tau_y}{H}\right)_{tt} + \frac{1}{H}\frac{\partial}{\partial x}\left(\frac{\partial \tau_y}{\partial x} - \frac{\partial \tau_x}{\partial y}\right) \qquad (7.32)$$

Look at the terms of this equation.

In regions away from the equator, and that had a latitudinal extent of L, and the time-scale of interest is much greater than the local inertial period, or,

$fL/c \approx 1$, let us use the streamfunction formulation $\psi_x = v$ and $-\psi_y = u$. One can then write the equation 7.32 as

$$\left(\psi_{xx} + \psi_{yy} - \frac{f^2}{c^2}\psi\right)_t + \beta\psi_x = curl_z\,\tau \tag{7.33}$$

Note that in the above equation, as soon as you assume steady-state, you recover back the sverdrup balance of Chapter 4,

$$\beta v = curl_z\tau \tag{7.34}$$

Now, if the curl of the wind stress is zero, then $u = v = 0$; and $g'\eta_x = \frac{\tau_x}{H}$ or it's a steady zonal wind-stress creating a sea surface elevation. This just means that a steady uniform zonal wind maintains a pressure gradient that does not drive any current. How? Because the Coriolis force is non-existent on the equator, so it cannot support any flow.

7.3.4 THE EQUATORIAL JET

Consider the motion when a sudden onset of spatially uniform zonal winds happens, which then remain steady after the initial pulse. Initially, the flow in the interior of the ocean basin, far from the coasts (eastern and western boundaries of the equatorial ocean) is independent of the longitudes. Zonal variations become dominant and important. So, we can assume, $\frac{\partial}{\partial x} = 0$ for u, v, and η in equation 7.32, which results in a simplified equation

$$v_{tt} + f^2 v - c^2 v_{yy} = -f\frac{\tau_x}{H} \tag{7.35}$$

At large distance L from the equator, and after time larger than the local inertial period of $2\pi/(\beta L)$, the second term would dominate the LHS of equation 7.35 and the second derivatives with respect to time and space would be small. This leads to the familiar Ekman balance equation (4.28) at a distance, or

$$v = -\frac{\tau_x}{H} \tag{7.36}$$

So, for a sudden eastward zonal wind pulse on the equator, the meridional velocity would be equatorward in both the northern and southern hemispheres, thus creating a convergence zone on the equator! This convergence zone would demand a downwelling on the equator due to continuity. This Ekman drift converges on the equator, increases with decreasing latitude so that downwelling must be very intense near the equator.

Note that not only the vertical component is strong, the zonal velocity or u is sufficiently strong given by

$$u_t = \frac{\tau_x}{H} \quad \text{at} \quad y = 0$$

since $\frac{\partial}{\partial x} = 0$ (no zonal gradient).

Thus a distinctive equatorial zone can exist for a sudden pulse of eastward wind stress. This is what happens during the onset of El Niño. Let us look at the time to setup this equatorial zone, as well as the width of the zonal jet.

We use a technique called scale analysis to estimate the time to setup and the distance of influence. We compare the time-independent terms on the LHS of equation 7.35 to estimate the spatial scale (λ). These are the second and third terms. This yields,

$$f^2 v = c^2 v_{yy}$$
$$\beta^2 \lambda^2 \cancel{v} = c^2 \frac{\cancel{v}}{\lambda^2}$$
$$\lambda^2 = c/\beta$$
$$\lambda = \sqrt{c/\beta} \tag{7.37}$$

For a typical value of $c = 1.4$ ms^{-1}, and $\beta = 2\Omega/R$, we obtain a $\lambda \approx 250$ km. This is the equatorial radius of deformation.

Similarly, the time-scale T_0, for the attainment of the steady zonal jet can be found by comparing the equivalence of the spatially independent terms with each other or the first two terms. This yields,

$$\frac{v}{T_0^2} = (\beta\lambda)^2 v$$
$$T_0 = (\beta\lambda)^{-1}$$
$$T_0 = (\beta c)^{-\frac{1}{2}} \tag{7.38}$$

using $\lambda = \sqrt{c/\beta}$ from equation 7.37. Applying the typical values of β and c as above, we get $T_0 = 1.5$ days!

Shortly after the wind starts the eastward pulse ($0 < t << T_0$), the first term is much greater than the second term. As time t approaches $T_0 = 1.5$ days, a distinctive eastward zonal jet is formed. See Figure 7.16 for the solution of equation 7.30 for various parameters of the 2-layer system which was constructed by D. W. Moore and Philander (1977).

The meridional Ekman drift is steady and is maintained by a steady deepening of thermocline near the equator. The associated increase in latitudinal density gradient is maintained by the geostrophic balance for the 2-layer system:

$$f u + g' \eta_y = 0 \tag{7.39}$$

This balance continues in the absence of any additional forcing.

To destroy the equatorial jet generated by eastward wind pulses, the winds must blow westward for the same amount of time T_0.

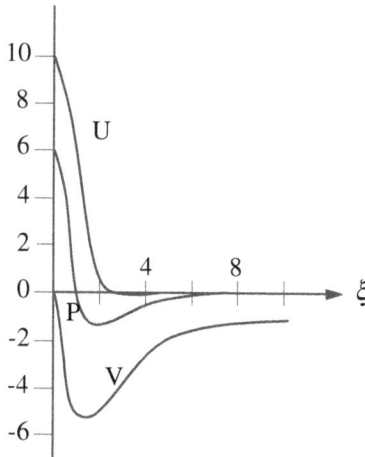

Figure 7.16: The latitudinal structure of the accelerating equatorial jet (U), of the steady meridional flow (V), and of the thermocline displacement (P) in response to spatially uniform eastward winds. The unit for latitude (ζ) is the equatorial radius of deformation. After Moore and Philander (1977).

7.3.5 EQUATORIAL KELVIN AND ROSSBY WAVES

Let us look at the understanding developed so far for the equatorial Pacific from a wave propagation perspective along the equator. Note that the equatorial Kelvin Wave (discussed in Section 6.5) plays a key role and its elevation (η) can be now assumed to be from the mean depth of thermocline H.

The weakening of the eastward winds in the central Pacific region initiates two waves at the thermocline level: one traveling westward, another traveling eastward. The one traveling eastward is a downwelling Kelvin Wave (with speed of about 1.4 ms^{-1}) and thus takes about 2–3 months to cross the Pacific. The other is the long baroclinic (2-layer ocean produces one barotropic Rossby wave at the surface and another long wave which is very slow) Rossby wave riding on the thermocline. This Rossby wave takes about 7 months to cross the Pacific. The Kelvin wave travels east and warms up the ocean to the east. The thermocline starts to deepen on the east from its normal rise to the east. Contrast the thermocline slopes between Figures 7.12 and 7.13. El Niño happens.

A pictorial depiction from the original 2-layer model of S. Philander et al. (1984) is shown in Figure 7.17.

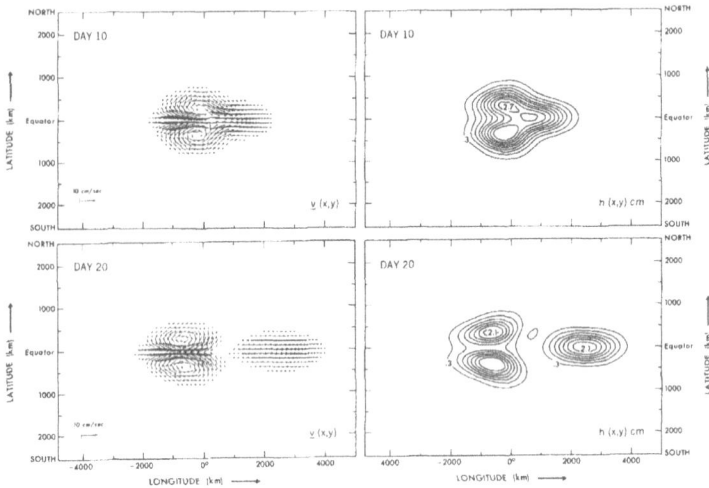

Figure 7.17: The dispersion of an initially hell-shaped thermocline displacement into an eastward-propagating Kelvin wave and westward-propagating Rossby waves. The left-hand panels show the horizontal currents and the right-hand panels the thermocline displacements. Republished with permission of Elsevier Science; Academic Press, from "El Niño La Niña, and the Southern Oscillation" George Philander (1989); permission conveyed through Copyright Clearance Center, Inc.

During the same time, the downwelling long Rossby waves propagating westward is slowly reaching the west, and upon reaching the western boundary of the equatorial Pacific return to the west as upwelling Kelvin wave (The reflection of waves is left to more advanced Texts). These reflected Kelvin Waves cool the eastern Pacific and bring an end to the El Niño.

The idea of predicting El Niño has been not new and has been there from the time of Sir Gilbert Walker. The first attempt to forecast an El Niño was made in 1974 by Bill Quinn of Oregon State University and Klaus Wyrtki of the University of Hawaii (McPhaden et al., 2015). Unfortunately, their prediction did not materialize as the westerly wind bursts, which they thought would initiate the Kelvin waves, did not do so. Ten years after that, with the success of the 2-layer model in predicting the El Niño of 1986–87 by Mark Cane (Cane et al., 1986), it became apparent that an observational network on the Pacific would lead us to detect the onset of El Niño and then the Peruvian fisheries could be warned about the incoming El Niño 3–6 months in advance. Such a system has been in existence from the early 1990s and has been known as the TAO (Tropical Atmosphere Ocean)/TRITON system. It was built over the 10-year period between 1985 and 1994. It regularly monitors an array of moorings spread over the $\pm 10°$ latitude from the equator. It is worth taking a look at the website https://www.pmel.noaa.gov/gtmba/pmel-theme/pacific-ocean-tao for today's condition of the Pacific Ocean. Look at the following maps available for today and 3–6 months back to see if you can detect the El Niño that might be happening now!

- The TAO/TRITON region (140°E–105°W), (10°S–10°N)
- SST, SSH, and wind maps
- Temperature-depth (2°S to 2°N, 500 m deep) map
- Look at the thermocline (depth of the 20°C isotherm) variation
- Look at the cool 3D animations (SST, winds, currents, equatorial temperature with depth, and 20°C isotherm) to see if you can see the El Niño-La Niña over the recent years!
- Try to see the downwelling and upwelling Kelvin waves.

7.4   THE INDIAN NIÑO – INDIAN OCEAN DIPOLE

There is a similar phenomenon with east-west temperature asymmetry and sea level pressure difference oscillation in the tropical Indian ocean. This basin-scale climatic variability is known as the 'Indian Ocean Dipole' (IOD) or the "Indian Niño." When the SST in the western IO is high and that in the eastern IO is low, the IOD is said to be in its "positive phase." This setup brings more rain in the western IO and its surrounding countries in the middle-east and Africa. This is similar to the La Niña situation in the Pacific (warmer-wetter west and cooler-drier east). During this positive IOD phase, the eastern IO and its surrounding countries, such as Australia's west coast and Indonesia, becomes drier with less rain. Opposite happens during the 'negative phase' of IOD, when the western IO gets less rain and the Australian side gets more rain. This setup of the negative IOD is similar to El Niño) setup in the pacific.

Now, think about the Indian Ocean and Pacific Ocean as two oceans connected in between through the Indonesian throughflow with the islands and Australia in between the oceans. Clearly, Australia and Indonesia would get a lot of rain if the negative IOD coincides with a La Niña event. Similarly, there will be extremely dry conditions if an El Niño in the pacific ocean coincides with the positive phase of IOD in the Indian ocean. Such coincidences have generally been observed Saji et al. (2005). However, the connection between ENSO and IOD is a topic of current research and is complicated by the surrounding landmasses, the Indonesian throughflow, the atmospheric teleconnections, heat content increase due to climate change, and many other factors.

7.5   INSTABILITY

Now that we have some idea of how the ocean and atmosphere work with wind-driving, thermohaline circulation, time-dependence, and in a layered vertical structure (with 2-layer as a foundation exercise), let us ask the question: How do the waves grow once you perturb the system?

In other words, now that we know how the ocean looks in a steady-state (Chapters 4 and 5), how the waves are the solutions for a set of time-dependent equations of motion, and we know how the ocean could be thought in terms of multiple layers in the vertical, how do we put them together to understand the ever-evolving, dynamic ocean with their eddies and currents and the atmosphere with its cyclones, hurricanes, typhoons and jet stream and trade winds? One way to think of a time-dependent evolution is to think about the ocean or the atmosphere as a system which allows for perturbations to grow and develop in time while being supported in a background (mean) steady-state flow on the rotating earth (Geostrophy). Mathematically, the total velocity would then be comprised of a mean velocity and a perturbation velocity. The mean velocity could be in simple geostrophic balance and the perturbation velocities could evolve in time.

$$u(x,y,z,t) = \bar{u}(x,y,z) + u'(x,y,z,t)$$

The idea is to allow for variations of the mean flow ($\bar{u}$) to support the growth of the perturbations (u'). The structure of the mean flow is generally represented by the shears of the flow $-(\frac{\partial u}{\partial x}, \frac{\partial u}{\partial y})$ in the horizontal and ($\frac{\partial u}{\partial z}$) in the vertical. For convenience, we can assume zonal flow for the mean, and then the horizontal shear is only in the meridional direction ($\frac{\partial u}{\partial y} \neq 0$).

The process through which the perturbations grow and develop into other entities other than the mean flow (e.g., eddies, filaments, anomalous pools etc.) is called "instability." You can imagine two extreme situations. In one extreme, when the mean flow has only horizontal shear ($\frac{\partial u}{\partial y}$) and no vertical shear ($\frac{\partial u}{\partial z} = 0$), the perturbation will grow with *barotropic* instability under certain conditions. In the other extreme, when the mean flow has purely vertical shear ($\frac{\partial u}{\partial z}$) and no horizontal shear ($\frac{\partial u}{\partial y} = 0$), the perturbation will grow with *baroclinic* instability under certain conditions.

7.5.1   BAROTROPIC INSTABILITY

Large currents with strong velocities carry a lot of kinetic energy. It is thus possible that, if perturbed, these currents can support those perturbations as waves riding on the basic state of the current. Eventually, in time, these waves (instabilities) can grow in time, extract energy from the mean current and detach from the current as eddies.

In other words, the currents have horizontal shear $u(y)$ and can support waves as perturbations which extract energy from the mean flow's kinetic energy. Let us look at this growth problem (waves in a shear flow) in detail.

The equations of motions are with time-dependence (we want to allow for waves to ride on the basic state), Coriolis, and pressure gradient in an inviscid (no-friction) fluid. Clearly, this can be given by (modified from 6.1 and 6.2):

$$\frac{\partial u}{\partial t} + u\frac{\partial u}{\partial x} + v\frac{\partial u}{\partial y} + w\frac{\partial u}{\partial z} = +fv - \frac{1}{\rho}\frac{\partial p}{\partial x} + F_x^{0} \qquad (7.40)$$

$$\frac{\partial v}{\partial t} + u\frac{\partial v}{\partial x} + v\frac{\partial v}{\partial y} + w\frac{\partial v}{\partial z} = -fu - \frac{1}{\rho}\frac{\partial p}{\partial y} + F_y \qquad (7.41)$$

Add the continuity to these horizontal momentum equations,

$$\frac{\partial u}{\partial x} + \frac{\partial v}{\partial y} + \frac{\partial w}{\partial z} = 0 \qquad (7.42)$$

For the basic state, we assume no meridional velocity and a beta-plane. $u = \bar{u}(y)$, $v = 0$, and $f = f_0 + \beta y$

This is an exact solution of the above equations as long as the pressure, $p = \bar{p}(y)$ satisfies the following equation

$$(f_0 + \beta y)\bar{u}(y) == \frac{1}{\rho_0}\frac{\partial \bar{p}}{\partial y} \qquad (7.43)$$

Now, add a slight perturbation, which can represent an arbitrary wave of weak amplitude.

So, the variables can now be expressed as follows:

$$u = \bar{u}(y) + u'(x,y,t) \qquad (7.44)$$
$$v = v'(x,y,t) \qquad (7.45)$$
$$p = \bar{p}(y) + p'(x,y,t) \qquad (7.46)$$

Here, the perturbations are much smaller than the mean variables, or

$$u', v' << \bar{u}; \ p' << \bar{p}.$$

Substituting these variables in equations 7.40 and 7.41, we get

$$\frac{\partial u'}{\partial t} + \bar{u}\frac{\partial u'}{\partial x} + v'\frac{\partial \bar{u}}{\partial y} - (f_0 + \beta y)v' = -\frac{1}{\rho}\frac{\partial p'}{\partial x} \qquad (7.47)$$

$$\frac{\partial v'}{\partial t} + \bar{u}\frac{\partial v'}{\partial x} + (f_0 + \beta y)u' = -\frac{1}{\rho}\frac{\partial p'}{\partial y} \qquad (7.48)$$

The continuity equation for the perturbation wave variables satisfy the following equation

$$\frac{\partial u'}{\partial x} + \frac{\partial v'}{\partial y} = 0 \qquad (7.49)$$

which lets us define a perturbation streamfunction, ψ as follows

$$u' = -\frac{\partial \psi}{\partial y}; \; v' = +\frac{\partial \psi}{\partial x} \qquad (7.50)$$

It is useful to remember that this definition of the streamfunction *psi* assumes higher streamfunction values to the right (increasing x) of a particular streamline.

Cross-differentiating equations 7.47 and 7.48 and using ψ, we get one equation in ψ,

$$\left(\frac{\partial}{\partial t} + \bar{u}\frac{\partial}{\partial x}\right)\nabla^2\psi + \left(\beta_0 - \frac{d^2\bar{u}}{dy^2}\right)\frac{\partial \psi}{\partial x} = 0 \qquad (7.51)$$

Since ψ is the perturbation wave, which is being supported by the basic state $\bar{u}(y)$, we can assume that it only grows in the x or alongstream direction – this is often perceived as the meander growth in a dynamical system like the Gulf Stream, the Kuroshio, the Brazil Current and so forth. Note that the basic state is in geostrophic balance, while the waves are time-dependent. This is why these equations are also called 'quasi-geostrophic equations.' equations.

So, we can now write ψ in a wavelike form as

$$\psi(x,y,t) = \phi(y) \, \exp(i[lx - wt]) \qquad (7.52)$$

Substituting this in the equation 7.51 we find that the distribution of $\phi(y)$ has to satisfy,

$$\frac{d^2\phi}{dy^2} - l^2\phi + \frac{\beta_0 - \frac{d^2\bar{u}}{dy^2}}{\bar{u}(y) - c}\phi = 0 \qquad (7.53)$$

where $c = w/l$ has been used to bring in the phase speed of the perturbation wave. This equation 7.53 is known as the Rayleigh equation and is the kernel of the Barotropic instability criterion.

Remember that the expression $\exp(i\theta) = \sin\theta + i\cos\theta$ is used to invoke complex algebra operations for ease of mathematical solutions using both sines and cosines.

Multiplying the Rayleigh equation 7.53 by the conjugate of ϕ i.e., ϕ^* and integrating over the domain of the mean current (basic state), one arrives at a condition that the following relationship must hold.

$$\int_{o}^{L} (\beta_0 - \frac{d^2\bar{u}}{dy^2}) \frac{|\phi|^2}{|\bar{u}-c|^2} dy = 0 \qquad (7.54)$$

This means that the term within brackets inside the integral (which has a negative sign within it) must change sign within the domain across the mean current within the mean current somewhere!

Let us look at this term carefully.

$$\left(\beta_0 - \frac{d^2\bar{u}}{dy^2}\right) = \frac{d}{dy}\left(\underbrace{f_0 + \beta y}_{\text{Planetary Vorticity}} + \underbrace{-\frac{d\bar{u}}{dy}}_{\text{relative vorticity}}\right)$$

$$= \frac{dq}{dy} \qquad (7.55)$$

where

$$q = f_0 + \beta y - \frac{d\bar{u}}{dy} \qquad (7.56)$$

which is the total vorticity. For the barotropic flow it is the potential vorticity.

This leads to the condition for barotropic instability or the perturbation waves to exist and grow on the basic mean current ($\bar{u}(y)$), we must have the gradient of the total vorticity q change sign within the domain, or $\frac{\partial q}{\partial y} = 0$, somewhere across the current.

So, effectively, the total vorticity of the basic zonal flow must reach an extremum ($\bar{u}(y)$ would have to be maximum or minimum) within the domain to support and help grow barotropic instabilities.

For most of the zonal jets, e.g., Gulf Stream, Kuroshio, Brazil Current, after leaving their respecting western boundaries (when they meander zonally and generally have a maximum velocity along their axes) they can support evolving instabilities and form eddies.

7.5.2 BAROCLINIC INSTABILITY

The ocean is a huge storage of potential energy due to its thermal structure. For example, the tilted thermocline structure across the fronts, when perturbed, allows for the conversion of some of its potential energy (called the "available potential energy") to kinetic energy and generates eddies or other multiscale features. This energy transfer process is, in effect due to the generation of instabilities in the system and their growth (much like what happened for Barotropic instability). Since the conversion of potential energy to kinetic energy or for that matter, to eddy kinetic energy requires the presence of the vertical gradient of temperature ($\frac{\partial T}{\partial z}$ and $\frac{\partial u}{\partial z}$), the vertical shear of velocity, this process is known as "baroclinic instability." Note that without the vertical shear, the fluid will fall back to its barotropic nature (no-depth variation).

The "pure" baroclinic instability can be derived by following the barotropic instability procedure assuming this time that the basic flow is exclusively z-dependent, i.e., $\bar{u}(z)$. The perturbations (u' and v') can then be functions of (x, z, t). The key to understanding baroclinic instability is the fact that a north-south temperature gradient has to exist for the basic state to allow for vertical shear of the zonal geostrophic flow on a rotating earth.

Without getting into the full mathematical exposition of deriving the Baroclinic instability in a fully three-dimensional ocean (which is most elegantly done by multiple books including (Cushman-Roisin & Beckers, 2011; Gill, 1982; Pedlosky, 2013a); we present here the final equation and heuristically discuss the terms for satisfying the conditions of baroclinic instability. Some of the original work on this topic was done by Jule Charney and Melvin Stern, among others. See Charney and Stern (1962) and Eliassen (1983) and references therein.

Let us now consider a more general basic flow which can allow for both barotropic and baroclinic instability, i.e. the basic flow is zonal and has shear in both vertical (z) and horizontal (y) directions. Following the development of the barotropic instability, in this case, the basic zonal flow $u(y, z)$ is also associated with a stream function $\psi(y, z)$ and potential vorticity $q(y, z)$, which are related by,

$$q(y,z) = \frac{\partial^2 \psi}{\partial y^2} + \frac{f^2}{N^2}\frac{\partial^2 \psi}{\partial z^2} + \beta y \qquad (7.57)$$

using the linearized perturbation equations for small amplitude perturbations $q'(y, z)$, and allowing the stream function perturbations (ψ') to grow in the x direction in time (t), this perturbation stream function can be expressed as

$$\psi'(x,y,z,t) = Re[a(y,z)\,exp[il(x - ct)]] \qquad (7.58)$$

where a is the amplitude of the perturbation.

After a series of algebraic manipulation (see Cushman-Roisin and Beckers, 2011, Chapter 16 for details), one can arrive at the following mathematical condition (similar to the barotropic instability development),

$$\int_0^H \int_o^L \frac{\partial \bar{q}}{\partial y}dydz + \int_o^L \frac{\partial \bar{u}}{\partial z}\Big|_0^H dy = 0 \qquad (7.59)$$

The necessary condition for baroclinic instability is then that at least any one of the following three conditions need to be satisfied.

- $\frac{\partial \bar{q}}{\partial y}$ has to change sign across the domain;

- the sign of $\frac{\partial \bar{q}}{\partial y}$ would be opposite of that of $\frac{\partial \bar{u}}{\partial z}$ at the top;
- the sign of $\frac{\partial \bar{q}}{\partial y}$ would be the same as that of $\frac{\partial \bar{u}}{\partial z}$ at the bottom.

Note that the last two conditions are actually a comparison of the horizontal shear of potential vorticity (or the second derivative of the zonal velocity) to the vertical shear of the velocity. Thus baroclinic instability is directly related to the stratification (N^2) and vertical shear of the velocity ($\frac{\partial u}{\partial z}$); while the barotropic instability is related to the horizontal gradient of potential vorticity ($\frac{\partial q}{\partial y}$).

More details on baroclinic instability can be found in more advanced text books such as Cushman-Roisin and Beckers (2011) (Chapter 16), Gill (1982) (Chapter 13), and Pedlosky (2013a) (Chapter 7). A fascinating advanced description and review of the development of the idea and the history of baroclinic instability is given by Lindzen et al. (1990), which is a tribute to the original theoretical work on the baroclinic instability problem by Jules J. Charney (see Charney, 1990 and references therein).

7.6 EXAMPLES OF BAROTROPIC AND BAROCLINIC INSTABILITIES

A final note on Barotropic/Baroclinic instabilities on how the current's energy is used to develop the instability growth is discussed here. Both instabilities lead to the basic current converting energy to kinetic energy to grow the perturbations into eddies. In a baroclinic instability-dominated process, the current's available potential energy (APE) gets converted to the eddies Eddy Kinetic Energy (EKE). In a barotropic instability process, the current's mean kinetic energy (MKE) in the basic state gets converted to the eddies Eddy Kinetic Energy (EKE). Such energy conversion pathways can be detected during eddy formation using high-resolution ocean models.

A few examples of application of barotropic and baroclinic instabilities are in order.

- An example of the BTI (Rayleigh-Kuo) for the Gulf Stream. Let us consider a typical analytical distribution of the zonal velocity across the Gulf Stream. This is given by a Gaussian distribution for a symmetric profile as follows:

$$u(y) = u_0 \, exp(y^2/d^2) \qquad (7.60)$$

where, u_0 is the axis velocity and d is the half-width of a symmetric jet along the cross-stream direction, y. The potential vorticity is simply $\frac{\partial q}{\partial y}$ and is shown in Figure 7.18. Clearly, the potential vorticity changes sign across the flow, and thus, the stream is barotropically unstable.
- An example of the BCI (Charney-Stern type) for the Brazil Current is shown in Figure 7.19. The panel on the left shows the gradient of

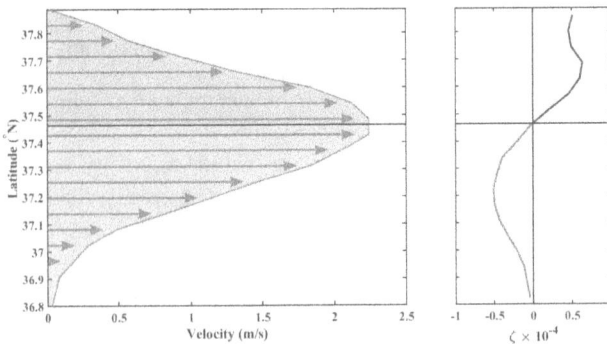

Figure 7.18: (Left): Synoptic velocity structure across the Gulf Stream along a satellite track (7 km resolution) for November 2002. The portion of the track used has a mean longitude of 70.8 W. (Right): relative vorticity profile across the stream. Figure credit: Adrienne Silver.

the potential vorticity based on the stretching term. It was developed by Professor Ilson Silveira of the University of Sao Paulo using one mooring line available within the instability region circled in pink on the right panel. The change of sign of the potential vorticity at around 300 m depth triggers the baroclinic instability within the Brazil Current to shed the unstable feature off of the meander within days, which then grows into an eddy-like feature within 2–3 weeks (right panels).

• Recent years have seen a surge in the installation of high-frequency radars in the coastal regions all across the world oceans. Naturally, observations of new coastal eddies from such high frequency (order of hours) and high resolution (order of 6–10 km) data sets are appealing to researchers to discover and investigate a new process for eddy genesis. However, note that HF Radar provides only surface velocity information and thus the vertical shear of velocity (required for a full baroclinic instability analysis) is still unavailable. One can of course do a full barotropic instability analysis from HFR data. To do a full baroclinic instability analysis, one has to supplement the HFR data with high-resolution data-assimilative numerical models, which can provide the full three-dimensional t, s, u, v structure.

• A very interesting and continuing problem in physical oceanography is that of understanding the eddy formation process from strong currents (with comparable horizontal shear and vertical shear for the velocity) such as the Gulf Stream, the Kuroshio, the Brazil Current, the WBC, and EICC in the Bay of Bengal, the Agulhas, the EAC, etc.

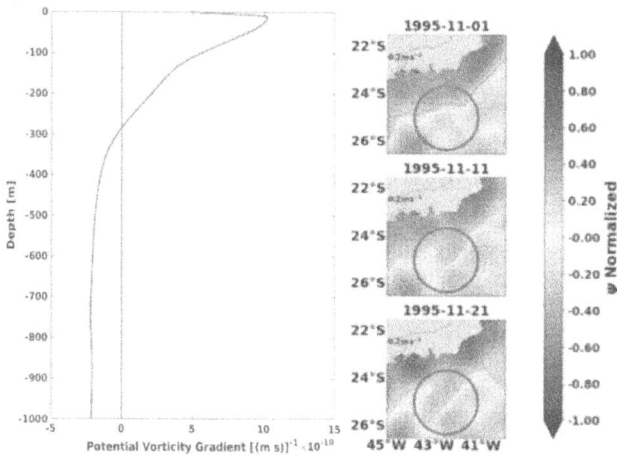

Figure 7.19: Example of Baroclinic instability in the Brazil Current. Figure credit: Professor Dr. Ilson Silveira.

- For example, for the Gulf Stream system, conventional wisdom suggests that both barotropic and baroclinic instability are important for ring formation processes. Numerical models, targeted shipboard experiments, ARGO floats, and satellite data could be combined to quantify and understand such complex problem of genesis of cyclonic versus anticyclonic rings, their seasonality (dependence on stratification and transport), and their regional formation preference, which are topics of current research.

CONCLUSION

The concept of layers was introduced to understand the three-dimensional ocean circulation. It will also help in developing numerical models in the next few chapters. The very important concepts of stratification, buoyancy frequency, and reduced gravity are key to understanding circulation and dynamics of different phenomena including instabilities. These were discussed using a simple 2-layer framework in detail. An equatorial 2-layer model was discussed in detail for the Pacific, which outlines several aspects of the ocean-atmospheric interaction phenomenon called ENSO (El Niño Southern Oscillation). A brief and interesting history of our understanding of this coupled ocean-atmosphere interaction phenomenon should guide us to look for such connectivity in other regions in the world. The results of the 2-layer model regarding the propagation of both Kelvin and Rossby waves along the equator can be extended to multi-layer system. The final brief discussions and examples of both barotropic and baroclinic instabilities will be more meaningful (and should be reread) once you are exposed to numerical models with multiple layers in the next few chapters.

FURTHER READING

Bjerknes, J. (1969). Atmospheric teleconnections from the equatorial pacific. *Monthly Weather Review*, *97*(3), 163–172.

Cushman-Roisin, B., & Beckers, J.-M. (2011). *Introduction to geophysical fluid dynamics: Physical and Numerical Aspects*. Academic Press.

Gill, A. E. (1982). *Atmosphere-ocean dynamics (international geophysics series)*. Academic Press.

Katz, R. W. (2002). Sir Gilbert Walker and a connection between El Nino and statistics. *Statistical Science*, *17*(1), 97–112.

Marshall, J., & Plumb, R. A. (2016). *Atmosphere, ocean and climate dynamics: An introductory text*. Academic Press.

Pedlosky, J. (2013a). *Geophysical fluid dynamics*. Springer Science & Business Media.

Philander, S. G. (1989). El Niño, La Niña, and the southern oscillation. *International geophysics series*, 308 pp. Academic Press.

Stern, M. E. (1975). Ocean circulation physics, 246 pp. Academic Press, London.

Walker, S. G. T., & Bliss, E. (1928). *World weather, III*. Edward Stanford.

Iemanjá - Goddess of the Sea

8 Introduction to Modeling

> An equation means nothing to me
> unless it expresses a thought of
> God.
>
> _____
>
> Srinivas Ramanujan (1887–1920)*

OVERVIEW

In the last seven chapters, we have discussed various aspects of oceanic circulation. We started with defining the response variables and forcing function of the ocean, discussed how to set up the governing equations of motion, and then dived into different balances in the steady-state. We then investigated the time-dependence and layered three-dimensional aspects of circulation. We also had a glimpse of a 2-layer model and how it captures the essential behavior of equatorial dynamics.

In this chapter we first present the purpose and general approach of modeling the equations with a brief history of early ocean models. Then we discuss the general approach for the numerical methodology using one of the earliest version of the GFDL modular ocean model (MOM). We describe the finite difference method in some detail for a better understanding of the numerical algorithms and their adaptation to the continuous equations of motion. A computer program flowchart completes this chapter with a first exposure to numerical modeling.

8.1 CONTEXT OF MODELING

Now, it's time to dive into modeling ocean circulation. As we start modeling, four questions arise: (i) What is a model? (ii) What is the purpose of modeling? (iii) What are we modeling? and (iv) How do we model?

So far, we have discussed how the ocean circulates under different forcing conditions and we have learned how to describe the responses in terms of the various parameters of the oceanic system, the homogeneous ocean's response, the two-layer response, the steady-state response, the time-dependent response and how to analyze the responses in a layered system framework. We have learned that there are a prognostic set of equations that governs the behavior of the oceanic system.

_____

*Credit/Source: **A Passion for Mathematics**. Numbers, Puzzles, Madness, Religion, and the Quest for Reality. **Clifford A. Pickover**. John Wiley & Sons, **2005**. (with Author's permission)

DOI: 10.1201/9780429347221-8

We understand that the primary reasons behind the movement of the oceanic water-masses are: (i) the gravitational pull exerted by the sun and the moon (tides) – we have not discussed tidal modeling in this book and leave the readers to references in Further Reading; (ii) the difference in atmospheric pressures at sea levels (the pressure gradient forces); (iii) the wind stress over the sea surface (curl of the wind stress); (iv) convection resulting from atmosphere cooling and evaporation (thermohaline circulation); and (v) other forces such as river outflow, precipitation, and extreme events (e.g., tsunami, storm surges, marine heat waves).

8.1.1 WHAT IS A MODEL?

A model represents such force-response relationships in a mathematical framework. In this sense, all of the wind-driven circulation theories are also models – with analytical expressions as their solutions – and thus can be classified as "analytical models." Now, recollect the force-response system that we developed in Chapter 2. At the end of Chapter 2, we obtained seven equations in seven primitive variables. These are non-linear equations and can describe the system's behavior in time; however, they cannot be solved analytically. Fortunately, they can be solved if we discretize the continuous forms, i.e. write the temporal and spatial gradients using their discrete versions. This would then allow us to represent the variables at one time-step to those (and their gradients) at the previous few time-steps. Then we can use the power of the computing machines to calculate all variables at every time step and move forward. In other words, we seek a computational solution of the set of partial differential equations describing ocean circulation. This is the idea of "Numerical Modeling." In short, numerical models are mathematical models that use the time-stepping procedure on a computational platform to obtain the system's behavior (governed by a set of dynamical equations) over time. The solution is a set of response fields which has evolved over time.

8.1.2 PURPOSE OF MODELING

Modeling is usually done with a specific purpose in hand. Let us consider three examples. First, we might want to reproduce past history realistically in order to understand certain processes and/or climatic events. These processes or events might be related to multiscale or scale-to-scale transfer of matter or energy. A second example is that we might want to predict the future state of the ocean-atmosphere-geosphere-biosphere system or our climate system 100 or 200 or 1000 years from now so that our future generations can live better. Since there has been an observed trend of warming of the planet as a whole during the last 200 years, there is a need to quantitatively predict the future state of the climate.

A third and practical example is that we want to have some models that we can use to benefit the human society in the short term by providing scientific information regarding the ecosystem, the weather, the storms. We want to use the model for predicting the currents or eddies for naval or industrial (oil and gas) applications, for search and rescue, oil spill, storm surge, or for river flooding. A single model might not be able to solve all these different issues. Usually, a model experiment focuses on a selected number of issues (this idea will become clearer as we go through the next few chapters). Perhaps you want to set up a model to investigate a particular phenomenon in a dynamical sense. In another case, you might want to set up a model to investigate a chlorophyll bloom or a propagation of an oil spill, or a propagation of a Tsunami, or help investigate the movement of plastic. You might also want to find out why and how the great white shark from Florida landed up on the beach in Maine, or how the fish migrates in response to climate change, or how does the upwelling change if the wind changes its direction around a cape.

8.1.3   WHAT DO WE MODEL?

That brings us to the third question. What do we model? The purpose of the modeling should help us select a model or choose a set of models. Let us consider a few examples to illustrate this.

Say you have decided to investigate the interaction of ocean and atmosphere, for example, the ENSO (El Niño Southern Oscillation – see Chapter 7) cycle or the Indian Monsoon. So, we need to employ an ocean-atmosphere coupled model. Some might want to predict the ecosystem in the Gulf of Maine for fishery needs or the abyssal circulation over the deep oil and gas fields in the equatorial margin or in the Santos Basin. So we need a model that can evolve the behavior of the physics, the chemistry, or the biology or be interdisciplinary. We also need these interdisciplinary models to span the ocean, the atmosphere, the land surfaces, the icecaps – in a coupled mode. So, we need coupled interdisciplinary models. See Chapters 13 and 14 for description of some of these models.

8.1.4   HOW DO YOU MODEL?

The art and science of prediction of natural phenomenon is not new. Meteorologists have tried to forecast the weather for a long time. For example, the idea of monsoon prediction is as old as the age of civilization. We want to know how the temperature and precipitation will be tomorrow, this month, next month, next season, and so forth all the time. But those predictions were based on observations, hints, hunches, and yes, statistics. The idea of setting up a mathematical model to forecast the weather is relatively new, started in the late nineteenth century (the 1880s).

The idea is to set up a typical three-dimensional domain with horizontal and vertical grids ($\triangle x, \triangle y, \triangle z$) in space and use the mathematical equations

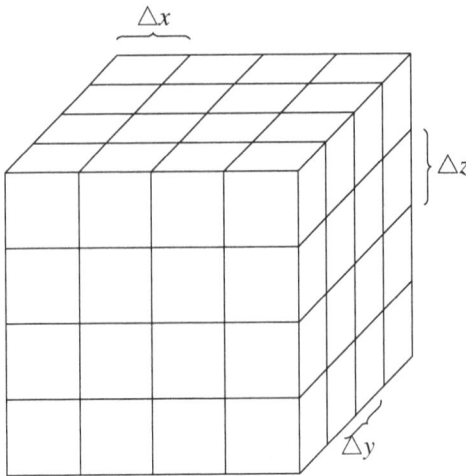

Figure 8.1: A three-dimensional grid for the ocean. The resolution or the grid size is shown as $\triangle x$, $\triangle y$, and $\triangle z$ in the three directions.

to describe the evolution of variables of interest that are dependent upon some known forcing fields. See Figure 8.1. Then use the mathematical model to evolve in time with time-stepping of ($\triangle t$).

Note that the early numerical ocean model development started following the footsteps of weather forecasting, and it used to be fashionable to say that "Weather forecasting is 30 years ahead of ocean forecasting." With the advances in computing power and algorithmic developments with more efficient data assimilation techniques, the gap between the two is reducing at a fast pace. Professor Allan Robinson (a pioneer in regional ocean forecasting and data assimilation from Harvard University) once said, "In the twentieth century, scientific innovations have brought forth technology and computers, the twenty-first century will see how technology drives scientific innovation." We are seeing things that we could not have seen or imagined before computer modeling and simulations. New results of non-linear simulations bring forth new ideas to science to theorize and study with the power of technology. It's a new way of doing science by mixing the power of technology and the approach of numerical modeling, simulations, and prediction. Let us explore some of these ideas in this chapter and the rest of the book.

8.2 GENERAL APPROACH TO MODELING

Model is a tool. It solves the primitive equations of the seven primitive variables in a numerical space-time domain. It is a mathematical construct of the equations of motion that we know and learned in earlier chapters. The model will help you better understand the system and the processes that govern the dynamics of the system.

To solve the primitive equations, we need the following: (i) a three-dimensional model ocean with appropriate grid-resolution ($\triangle x, \triangle y, \triangle z$); (ii) initial conditions on the model grid; (iii) boundary conditions at the model boundaries with forcing fields, and (iv) parameters for viscosity and diffusivity. Let us explore these four aspects below in a little more detail.

8.2.1 DOMAIN AND GRID-RESOLUTION

As soon as you bring in this idea of grid-spacing ($\triangle x, \triangle y, \triangle z$), and time-step $\triangle t$, it becomes apparent that a numerical modeling set up can only resolve (i) features of circulation whose scales are more than the *resolution* of the model and (ii) phenomena, whose time-evolution can be resolved by waves whose speeds are less than $\frac{\triangle x}{\triangle t}$. Grid size is called "resolution." The smaller the grid-size ($\triangle x, \triangle y, \triangle z$), the higher the resolution and vice-versa. Any phenomenon that requires a description within one grid-spacing and one time-step cannot be resolved by the numerical resolution of the model. Consider a biological model – it needs a spatial scale of 1 mm to resolve many biological activities. Consider a biophysical scale – we can probably live with 1 m to 10 km for distribution of larvae and fish assemblage and plankton distribution in patches. Consider a physical feature like a front or an eddy or an upwelling or river plume – we can resolve those with 1 m to 10 to even 30 km for large-scale features like ENSO, its warm pool or cold tongue.

This brings us to select the size of the domains. We can now think about global-scale models, which cover the whole Earth at a 1-degree or about 100 km resolution ($360 \times 180 \times 30$ grid points for the ocean model only). They won't be able to resolve the mesoscale eddies in the oceans, but they can be useful to generate the large-scale features of water-mass distribution of the ocean, the slow thermohaline circulation, and interact with the global atmosphere and ecosystem for large-scale variations. Now you can use these global-scale models to study climate-scale problems over 100 years or even 1000 years with today's computers. Next are basin-scale models (North Atlantic, North Pacific, South Atlantic, Indian Ocean, etc.) at say 30–50 km resolution Typical domain size of $120 \times 180 \times 30$). They used to be known as eddy-permitting models. Typical application of such models was in Equatorial regions, for ENSO, for Monsoon to study the seasonality of Indian Ocean and so forth. And then there are regional models for forecasting and process studies with 5–10 km resolution.

8.2.2 INITIAL CONDITION

The three-dimensional domain requires the specification of seven primitive variables at every grid point at the initial time, $t = 0$. This is called the initial condition for the model. For basin-scale simulations, generally, $\vec{V} = 0; t, s, \rho$ at every (x, y, z) point can be specified from climatologies such as (WOA18) (Locarnini et al., 2018). The model will evolve and create velocities in what is called a "spin-up period." The spin-up period is the time the model takes to reach a dynamical equilibrium under the applied forcing when the domain-averaged kinetic energy and other parameters vary within a reasonable seasonal range and do not increase or decrease monotonically. There are various other ways to initialize from data (observations). Chapter 11 will discuss more details. Chapter 12 will discuss initialization for prediction on short time-scales (days to weeks), known as the oceanic "synoptic scale."

8.2.3 BOUNDARY CONDITIONS AND FORCING

As the discrete model equations are solved in each time step, the variables are evolved in time in a time-integration mode. This process generally requires the specification of surface forcing fields, e.g. the wind-stress fields, the heat flux, evaporation, precipitation, and river runoff. Additionally, tidal forcing and any other additional appropriate fluxes are prescribed around the lateral boundaries.

Sometimes, for maintaining the surface temperature and salinity fields, their climatological values are imposed with a relaxation time-scale. Furthermore, temperature, salinity, and velocity values are also needed at open boundaries for a regional model. In short, different model configurations demand different modes of surface and lateral boundary conditions and forcing fields for process-appropriate simulations.

In present days, the wind stress and heat flux fields field is often prescribed as climatology – monthly averages of zonal and meridional stress and heat flux component values over the domain from a reanalysis available from multiple sources (e.g., JRA55, ECMWF, SCATSAT – see Chapter 11). Some example simulations are presented later with details on boundary conditions and forcing fields in both Chapters 11 and 12.

8.2.4 PARAMETERS

Additionally, sufficient friction is generally required to maintain stable model simulations. Remember the coefficients of viscosity (A_H and A_z) in the momentum and heat/salt equations. They depend on the gradients of velocity and temperature/salt. These effects are produced in some turbulence-like fashion by motions on scales too small to resolve by the grid. As it is, as discussed in chapter 9.1.1, turbulence is a yet-unresolved physical process which is generally parameterized to the best of our understanding. These are generally called the "sub-grid scale" effects and are included in the model in different ways (depending on which model – one other reason why there are so many choices for models). Some of the early and still among popular ways are:

- A_z and A_H are parameterized
- they are chosen as constant for a region during the whole simulation;
- a turbulent kinetic energy model $k - \varepsilon$ (see Section 9.1.1).
- variable eddy viscosity being proportional to the rms of the rate of strain (O'Brien, 1971).

8.3   AN EARLY OCEAN MODEL

Let us now briefly look at some of the important characteristics of one very early numerical model, Cox's model of the Indian Ocean. Michael Cox (1941– 1989) was the pioneer of modern ocean modeling to study ocean circulation. He simulated the response of the Indian Ocean to the monsoonal changes in the winds on a seasonal scale. His simulation showed that the reversal of the Somali Current was actually due to local monsoon-driven coastal upwelling and not due to propagating Rossby waves along the equatorial region (waveguide). This was a clear vindication that numerical models are an essential tool (in combination with observations of temperature, salinity, velocities, and winds) for understanding and even discovering new processes and their underlying dynamics.

Some of the important elements of the Cox (1970) model are listed below.

- Wind driving is seasonal; 3-month average Hellerman-Rosenstein (1983) climatological wind stress (Hellerman & Rosenstein, 1983); More accurate high-resolution satellite winds and reanalysis fields are available now.
- Thermohaline driving was included by imposing surface temperature and salinity; More accurate temperature and salinity data and climatology fields are used now with high vertical resolution and even bottom boundary layers to resolve abyssal flows.
- A "rigid lid" approximation with $w = 0$ at the surface was imposed. This is still followed in some of the models where the effect of real variation of surface elevation appears through an "equivalent pressure distribution."
- Initialization was done from a state of rest, with temperature and salinity to be horizontally uniform with a vertical distribution mimicking observational data. Model initialization has improved with increased availability of data, real-time information gathering capability, and new techniques of data assimilation. (See Chapter 11.)
- Friction and diffusion (see Chapter 4 to recall these values)

- A_z – eddy viscosity for vertical friction – $10^{-2}\text{m}^2\text{s}^{-1}$
- A_H – eddy viscosity for horizontal friction – $2 \times 10^5\text{m}^2\text{s}^{-1}$ to $5 \times 10^3\text{m}^2\text{s}^{-1}$ as different grids were used.
- K_z – vertical eddy diffusivity in the heat/salt equations – $10^{-4}\text{m}^2\text{s}^{-1}$
- K_H – Horizontal eddy diffusivity in the heat/salt equations – 10^4 to $5 \times 10^3\text{m}^2\text{s}^{-1}$

Note that the above general approach is somewhat similar to even today's most modern model setups with much higher resolution and more advanced numerical techniques and algorithms for efficient treatment of non-linearity, boundary conditions, and integral constraints. And of course new power of computers has resulted in a more robust and enhanced approach to real-time, probabilistic, interdisciplinary, interconnected, nested-resolution modeling set up in a multiscale data-assimilative framework. Our ability to model the ocean for different processes and different purposes in the last six decades from the time of Cox has grown tremendously so that we can now predict the ocean with some confidence on a regular basis for 48–72 hours using the global network of a system of models and observational platforms. See Chapters 11 and 12 for some examples.

8.4   NUMERICAL METHODS

Since we are dealing with spatial and temporal derivatives, it is important to understand how best to compute them numerically on a grid and what the grid might look like. Most of the numerical models use the finite difference method (FDM), while a number of recent models have adopted finite element and finite volume methods as computer power is becoming more inexpensive.

It is, however, instructive and relatively easy to understand the development of models using the FDM and then apply similar logic while changing the numerical method for better accuracy in search for better simulations and predictions.

8.4.1   FINITE DIFFERENCE

In the FDM, the gradient or the derivative of a function $u(x)$ at a point x is given in terms of its values at adjacent points $(x+h)$ and/or $(x-h)$. See Figure 8.2. In the following, we will replace the grid-cell width "h" in this figure by $\triangle x$ to conform to our Cartesian coordinate convention. The first-order derivative of the function, $u'(x)$ can be written as the value of the slope of the tangent to the curve $(u(x))$ at x. This can be computed in three different ways; mathematically,

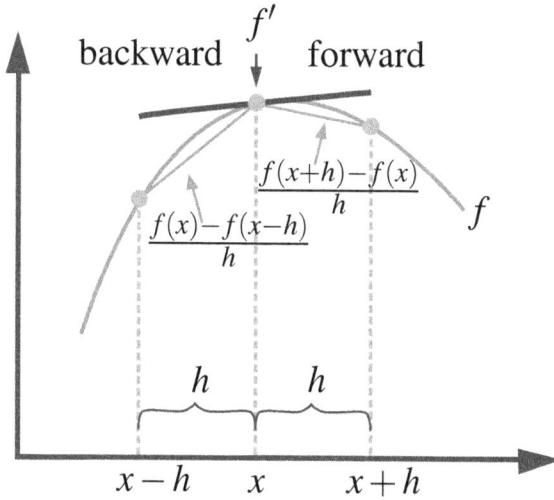

Figure 8.2: Gradient calculation for a Finite Difference Method using a backward or forward difference scheme. $h = \triangle x$.

$$\frac{\partial u}{\partial x} = \frac{u_{j+1} - u_j}{\triangle x} \qquad \text{Forward difference} \qquad (8.1)$$

$$\frac{\partial u}{\partial x} = \frac{u_j - u_{j-1}}{\triangle x} \qquad \text{Backward difference} \qquad (8.2)$$

$$\frac{\partial u}{\partial x} = \frac{u_{j+1} - u_{j-1}}{2\triangle x} \qquad \text{Central difference} \qquad (8.3)$$

So, in the forward difference, we are assuming that the derivative is equal to the slope as we move forward to the next grid point; in the backward difference, we are assuming that the derivative is equal to the slope between the current grid point from the point before. In the central difference, one utilizes the information from both sides of the current location.

Let us look into the details of constructing these rather simple-looking, easy-to-grasp equations. We know that Taylor series expansion of the continuous function at any point can be given in terms of its derivatives at that point as follows.

$$u_{j+1} - u_j = \frac{\partial u}{\partial x} + \frac{1}{2}\frac{\partial^2 u}{\partial x^2}(\triangle x)^2 + \frac{1}{6}\frac{\partial^3 u}{\partial x^3}(\triangle x)^3 + \frac{1}{24}\frac{\partial^4 u}{\partial x^4}(\triangle x)^4 + O(\triangle x)^5 \quad (8.4)$$

$$u_j - u_{j-1} = \frac{\partial u}{\partial x} - \frac{1}{2}\frac{\partial^2 u}{\partial x^2}(\triangle x)^2 + \frac{1}{6}\frac{\partial^3 u}{\partial x^3}(\triangle x)^3 - \frac{1}{24}\frac{\partial^4 u}{\partial x^4}(\triangle x)^4 + O(\triangle x)^5 \quad (8.5)$$

Note that the denominators (1/2, 1/6, 1/24, etc.) are the factorials of the order of $\triangle x$ in that term. The error of truncation due to discretization is shown by the order after the last explicit term. As more terms are included, the error becomes less.

Clearly, the forward and backward differences in equations 8.1 and 8.2 are first-order approximations of equations 8.4 and 8.5 with error of order $(\triangle x)^2$.

Adding equations 8.4 and 8.5, and dividing by 2, the central difference formula in equation 8.3 can be retrievable from the following.

$$\frac{u_{j+1} - u_{j-1}}{2\triangle x} = \frac{\partial u}{\partial x} + \underbrace{\frac{1}{6}\frac{\partial^3 u}{\partial x^3}(\triangle x)^2 + O(\triangle x)^5}_{\text{2nd order accurate}} \quad (8.6)$$

Clearly, the central difference is more accurate than either the forward or the backward difference schemes. Extending the span of influence to two adjacent grid points on either side, one can get

$$\frac{u_{j+2} - u_{j-2}}{4\triangle x} = \frac{\partial u}{\partial x} + \underbrace{\frac{4}{6}\frac{\partial^3 u}{\partial x^3}(\triangle x)^2 + O(\triangle x)^4}_{\text{4th order accurate}} \quad (8.7)$$

A simple manipulation to eliminate the 2nd order error terms from equations 8.6 and 8.7 yields,

$$\frac{\partial u}{\partial x} = \frac{4}{3}\left(\frac{u_{j+1} - u_{j-1}}{2\triangle x}\right) - \frac{1}{3}\left(\frac{u_{j+2} - u_{j-2}}{4\triangle x}\right) + O(\triangle x)^4 \quad (8.8)$$

which is fourth-order accurate for the first derivative. Note, however, that two things happen for the higher-order accurate FDMs:

- Computer programming algorithms gets complicated
- treatments near boundaries get complicated and errors increase

Similarly, higher-order derivatives can be derived using the same Taylor series expansion in equations 8.4 and 8.5. For example, the second derivative can be approximated by subtracting 8.5 from 8.4 and realizing,

$$(u_{j+1} - u_j) - (u_j - u_{j-1}) = u_{j+1} - 2u_j + u_{j-1} = \frac{\partial^2 u}{\partial x^2}(\triangle x)^2 + O(\triangle x)^4 \quad (8.9)$$

Now let us think about the temporal derivatives. Using forward difference,

$$\frac{\partial u}{\partial t} = \frac{u_{n+1} - u_n}{\triangle t} \qquad \text{Forward difference} \qquad (8.10)$$

$$\frac{\partial u}{\partial t} = \frac{u_{n+1} - u_{n-1}}{2\triangle t} \qquad \text{Central difference} \qquad (8.11)$$

The centered difference algorithm given in equation 8.11 is called the "leap-frog" scheme as it requires communicating between the next and previous time-steps from the current one; it also requires storing spatial information at three time-steps and communicating with them for the temporal derivative computation.

Clearly, when the system of equations is discretized, there are two different ways to evaluate the spatial and temporal gradients at a particular time-step at a particular grid point. These are known as Explicit and Implicit Schemes. In the explicit scheme, all the values (derivatives) in the *new* time-step are calculated from the values at previous time-steps. In the implicit scheme, values at the *new* time-step depend on the spatial gradients of values at the new time-step. In this case, one writes the equation for each grid point and solves the whole set simultaneously. Clearly, this requires much larger storage space in the memory than the explicit scheme.

So, there are many ways to formulate the same physical set of equations in the numerical/computational domain. An obvious question is whether these various formulations can be evaluated using the same common performance criterion? One of the simple ones is to check for linear computational instability, which is given by choice of the grid size and time-step according to the speed of the physical process that you want to resolve. Mathematically,

$$\triangle t \leq \triangle x / c_{max} \qquad (8.12)$$

where c_{max} is the maximum wave propagation speed. The above equation 8.12 represents the so-called Courant–Friedrichs–Lewy (CFL) condition, which explicitly means that an advected quantity must not cross one grid interval in one-time step.

Clearly, if the time-step constraint is relaxed, it will result in the explosive growth of small errors, which will lead to larger errors and unrealistic simulations in the longer-term integration. Problems of numerical noise are a serious issue in numerical modeling exercises. Non-linear terms might lead to non-linear numerical instability or rapid increase of energy at very small scales. In general, implicit methods have performed better than explicit methods in terms of numerical stability and long-term simulations.

We discussed the Finite Difference Method in detail here, as many of today's models use this technique (in some modified form). However, there have been other numerical methods of discretization of analytical equations such as finite element method (FEM) and finite volume method (FVM). Both FEM and FVM are based on an unstructured grid system; a typical structure is a triangular mesh. So, they are suitable for complex and irregular geometries and boundaries. FDM is very easy to implement and a regular grid system allows for certain computational and storage advantages. However, FEMs allow for finer and well-resolved subdomains (with finer elements) within a large domain, which might be better for a complex region with varying topography or islands. For the FEM, the partial differential equations (PDEs) to be solved are written for the elements. The spatial gradients are needed to be redefined based on the element's shape and generally expressed as a sum of shape functions with different amplitudes. This formulation is based on the well-known Galerkins' method (Galerkin, 1915; Zienkiewicz & Taylor, 2005) which used the idea of the variation of parameters to a functional space to convert the continuous equations to a discrete set for numerical implementation. On the other hand, for the FVM, the equations are written in divergence form and integrated over the volume. One then applies Gauss's theorem and converts the volume integral over the divergence to a surface integral across the boundaries. This transformation leads to calculating fluxes of dependent variables across the boundary of the cells, which simplifies the differential equations substantially. For FEM and FVM applications in ocean models please see Table 10.1 in Chapter 10 and more Advanced books on numerical modeling (Mazumder, 2015; Neill & Hashemi, 2018; Rapp, 2016).

8.5   THE BASIC EQUATIONS

As derived in Section 2, the final seven equations called the primitive equations are the three momentum equations 2.28, 2.29, 2.30, and two tracer (temperature and salinity) equations 2.31 and 2.32, the continuity equation 2.33, and the equation of state 2.34. Let us rewrite them explicitly:

$$\frac{\partial u}{\partial t} = -u\frac{\partial u}{\partial x} - v\frac{\partial u}{\partial y} - w\frac{\partial u}{\partial z} - \frac{1}{\rho}\frac{\partial p}{\partial x} + fv + F_x \quad x\text{-momentum,} \quad (8.13)$$

$$\frac{\partial v}{\partial t} = -u\frac{\partial v}{\partial x} - v\frac{\partial v}{\partial y} - w\frac{\partial v}{\partial z} - \frac{1}{\rho}\frac{\partial p}{\partial y} - fu + F_y \quad y\text{-momentum,} \quad (8.14)$$

$$\frac{\partial w}{\partial t} = -u\frac{\partial w}{\partial x} - v\frac{\partial w}{\partial y} - w\frac{\partial w}{\partial z} - \frac{1}{\rho}\frac{\partial p}{\partial z} + g \quad z\text{-momentum,} \quad (8.15)$$

$$\frac{\partial T}{\partial t} = -u\frac{\partial T}{\partial x} - v\frac{\partial T}{\partial y} - w\frac{\partial T}{\partial z} + Q +$$

$$K_H\left(\frac{\partial^2 T}{\partial x^2} + \frac{\partial^2 T}{\partial y^2}\right) + K_z\frac{\partial^2 T}{\partial z^2} \qquad \text{Temp,} \qquad (8.16)$$

$$\frac{\partial S}{\partial t} = -u\frac{\partial S}{\partial x} - v\frac{\partial S}{\partial y} - w\frac{\partial S}{\partial z} - E + P + R +$$

$$K_{HS}\left(\frac{\partial^2 S}{\partial x^2} + \frac{\partial^2 S}{\partial y^2}\right) + K_{zS}\frac{\partial^2 S}{\partial z^2} \qquad \text{Salt,} \qquad (8.17)$$

$$\rho = \rho(S, T, p) \qquad\qquad\qquad\qquad\quad \text{State,} \qquad (8.18)$$

$$\frac{\partial u}{\partial x} + \frac{\partial v}{\partial y} + \frac{\partial w}{\partial z} = 0 \qquad \text{Continuity.} \qquad (8.19)$$

We will now call these the "primitive equations" for the seven "primitive" variables u, v, w, t, s, p, and ρ. The time-dependent equations (8.13, 8.14, 8.15, 8.16, and 8.17) are called the "prognostic" equations. The continuity equation 8.19 and the equation of state (8.18 are called the "diagnostic" equations. Note that equation 8.19 can be used to derive the vertical velocity w if the horizontal velocities u, v are known at a time. Similarly, the equation of state 8.18 can give the density ρ if the temperature, salinity, and pressure are known at a grid point. Also, if $\frac{\partial w}{\partial t} \approx 0$, then equation 8.15 reduces to the hydrostatic balance; otherwise, a non-hydrostatic model would need to be invoked. The models we described here are mostly hydrostatic models.

8.6 THE MODEL EQUATIONS

The primitive equations discussed so far have been developed on the (x, y, z) coordinate system. However, the Earth is spherical, and thus it makes sense to develop the general modeling equations on a spherical grid or $(\lambda, \theta, \text{and } z)$ coordinates, where λ is the longitude, θ is the latitude and z is the local vertical. We follow the classical development of the model equation based on the seminal work published in the GFDL Technical Report by Cox (1984). This was indeed one of the first comprehensive guides to ocean modeling equations and solution procedure after early developments of ocean models by Bryan (1969b) Cox (1970), and Semtner (1986). We encourage all students of ocean modeling to read through the Cox (1984) report to understand the details of the modeling framework and its very complex demands. Here we choose to describe salient features of the modeling equations and the solution procedure. We also follow the same notations as in the Technical Report by Cox (1984) for compliance.

In the spherical coordinate, we define the following quantities first.

$m = \sec\theta$ $1/\cos(\text{latitude})$

$n = \sin\theta$ $\sin(\text{latitude})$

$f = 2\Omega\sin\theta$ Ω is 1 revolution per day

$u = a\dot{\lambda}/m$ a is the radius of the earth

$v = a\dot{\theta}$ a · (dot) over a variable x denotes $\dfrac{\partial x}{\partial t}$ (8.20)

In the spherical coordinate, the advective operator for any scalar variable (μ) becomes

$$\Gamma(\mu) = ma^{-1}\left[\underbrace{(u\mu)_\lambda}_{\frac{\partial(u\mu)}{\partial x}} + \underbrace{(v\mu m^{-1})_\theta}_{\frac{\partial(v\mu)}{\partial y}}\right] + \underbrace{(w\mu)_z}_{\frac{\partial(w\mu)}{\partial z}} \qquad (8.21)$$

See the parallels of each term in equation 8.21 with corresponding components of the advective operator in the (x, y, z) system (equation 2.3). The quotient ma^{-1} arises because of the projection of u and v on the spherical coordinate axes, as given in equation 8.20.

So, the horizontal equations of motion can be written as

$$u_t + \Gamma(u) - fv = -ma^{-1}(p/\rho_0)_\lambda + F^u \qquad (8.22)$$

$$v_t + \Gamma(v) + fu = -a^{-1}(p/\rho_0)_\theta + F^v \qquad (8.23)$$

where the reference density ρ_0 is taken as unity. Note that u is the west-east zonal velocity; v is the south-north meridional velocity and w is the upward-downward vertical velocity.

The pressure p was rewritten decomposing into two parts (a surface pressure and a part that is dependent on the density variation in the vertical using the hydrostatic relationship.

$$p(z) = p^s + \int_z^0 g\rho\,dz \qquad (8.24)$$

The continuity equation is simply given by (when the scalar $\mu = 1$),

$$\Gamma(1) = 0 \qquad (8.25)$$

The temperature and salt conservation equations can be easily written as a "tracer" equation as follows.

$$T_t + \Gamma(T) = F^T \qquad (8.26)$$

It is interesting to note that using the above "tracer" formulation (equation 8.26, one can add multiple other variables such as Carbon-14 or Tritium or any other lagrangian particles which advect using the velocity fields and the equation can be closed by a suitable F^T term (would be zero for a conservation equation) on the RHS. Temperature and salinity are called active tracers (as they are constrained by the equation of state) and the others are called passive tracers in the sense that they just utilize the flow field for their movement.

The equation of state is simply

$$\rho = \rho(\Theta, S, z) \tag{8.27}$$

where Θ is the potential temperature (See Section 7.2.1).

These seven equations 8.22 through 8.27 are the primitive equations in the spherical coordinate system. Note that the tracer equation 8.26 represents both temperature and salinity equations. Now, since the pressure was subdivided in two parts in equation 8.24, the temporal derivatives of the horizontal velocities in the horizontal equations of motion can be broken up as a part dependent on surface pressure only and a part that is dependent on the density variation in the vertical. Mathematically, we can then write,

$$u_t = u_t' - ma^{-1} p^s{}_\lambda \tag{8.28}$$

$$v_t = v_t' - a^{-1} p^s{}_\theta \tag{8.29}$$

where

$$u_t' = -\Gamma(u) + fv - mga^{-1} \int_z^0 \rho_\lambda dz' + F^u \tag{8.30}$$

$$v_t' = -\Gamma(v) - fu - a^{-1} \int_z^0 \rho_\theta dz' + F^v \tag{8.31}$$

These two equations are unique in the sense that if we know the velocities at a given instance, the RHS of these equations are known and that automatically tells us about the temporal derivatives of the velocity components u' and v' to evaluate these components in the next time-step. This is where the uniqueness of Cox's approach comes. Let us look into this in a bit more detail.

As mentioned in Chapter 7, the ocean can be thought to be made up of an infinite layers in the vertical. The velocity profile $u(z)$ can then be composed of a mean barotropic (depth-independent, average) velocity ($\bar{u}$) and a depth-dependent baroclinic velocity ($\hat{u}(z)$). Dropping the "z"s for convenience, we can define

$$u = \bar{u} + \hat{u} \tag{8.32}$$

$$v = \bar{v} + \hat{v}$$

where

$$\bar{u} = \frac{1}{H} \int_{-H}^{0} u dz' \qquad (8.33)$$

$$\bar{v} = \frac{1}{H} \int_{-H}^{0} v dz' \qquad (8.34)$$

Now taking the mean of the equation 8.28 one gets

$$\overline{u_t} = \overline{u'_t} - \overline{ma^{-1}p^s_\lambda} \qquad (8.35)$$

$$\overline{u'_t} = \overline{u_t} + \overline{ma^{-1}p^s_\lambda} \qquad (8.36)$$

since the surface pressure is already depth-independent!

Using this relationship (8.36) with the time-derivative of equation 8.32 we get

$$
\begin{aligned}
\hat{u}_t &= u_t - \overline{u_t} \\
&= u'_t - ma^{-1}p^s_\lambda - \overline{u_t} && \text{using 8.28} \\
&= u'_t - (\overline{u_t} + ma^{-1}p^s_\lambda) \\
&= u'_t - \overline{u'_t} && \text{using 8.36} \qquad (8.37)
\end{aligned}
$$

Both of the terms on the RHS of equation 8.37 are determined by the density variation-dependent equations given in equations 8.30 and 8.31. These equations define the internal modes of the velocities. Now what is left is to solve the mean or the external mode $\bar{u}$. This can be solved in the following way using the surface pressure.

Under the "rigid-lid" condition, we can define the depth-independent volume transport streamfunction (much like the Sverdrup balance derivation in Chapter 4) as follows.

$$\bar{u} = -(aH)^{-1}\psi_\theta;$$

$$\bar{v} = m(aH)^{-1}\psi_\lambda \qquad (8.38)$$

Using vertical integration like Sverdrup and Stommel solutions and taking cross-derivatives to eliminate the surface pressure, we can obtain the following relationship,

$$curl_z(\bar{v}_t, \bar{u}_t) = ma^{-1}\left[\bar{v}_{t\lambda} - (\bar{u}/m)_{t\theta}\right] \qquad (8.39)$$

It is easier to appreciate these equations 8.38 and 8.39 in cartesian coordinates (x, y, z) as

$$\bar{u} = -\frac{1}{H}\Psi_y;$$

$$\bar{v} = \frac{1}{H}\Psi_x \tag{8.40}$$

and vertically integrating the above two equations 8.40, cross differentiating and utilizing the surface pressure equations similar to 8.36 in Cartesian coordinates, the external mode prognostic equation is

$$\frac{\partial}{\partial x}\left(\frac{1}{H}\frac{\partial\Psi_t}{\partial x}\right) - \frac{\partial}{\partial y}\left(\frac{1}{H}\frac{\partial\Psi_t}{\partial y}\right) = \frac{\partial\overline{v_t'}}{\partial x} - \frac{\partial\overline{u_t'}}{\partial y} \tag{8.41}$$

The Forcing terms become

$$F^u = A_{MV}\,u_{zz} + A_{MH}\nabla_H^2 u \tag{8.42}$$

$$F^v = A_{MV}\,v_{zz} + A_{MH}\nabla_H^2 v \tag{8.43}$$

$$F^T = A_{TV}\,T_{zz} + A_{TH}\nabla_H^2 T \tag{8.44}$$

To summarize we now have the final equations of numerical modeling as follows: two internal mode equations; two tracer equations; and one external mode equation – these are five prognostic equations. Then we have the continuity equation and the equations of state – these are the two diagnostic equations.

The external model equation is an equation of the transport stream function ψ (see Chapter 4 to recall its definition) the spatial derivatives of which will give you the two barotropic velocity components ($\bar{u}$ and $\bar{v}$).

These equations are generally solved using a numerical algorithm on a finite 3-D ocean grid, subject to the following boundary conditions.

- At lateral walls: $u, v, T_n = 0$
- At surface, $z = 0$:

$$\rho_0 A_{MV}(u_z, v_z) = (\tau_x, \tau_y)$$
$$A_{TV}(T_z) = Q_{surface}$$
$$w = 0$$

here τ_x and τ_y are the zonal and meridional components of the surface stress, and Q is a flux through the surface for the particular tracer.

- At bottom, $z = -H$:

$$\rho_0 A_{MV}(u_z, v_z) = (\tau_{xB}, \tau_{yB})$$
$$T_z = 0$$
$$w = -\left(u\frac{\partial H}{\partial x} + v\frac{\partial H}{\partial y}\right)$$

- At lateral walls:

$$\Psi_x = \Psi_y = 0$$

- and Ψ is one constant on land boundaries around the domain and is a set of different constants on island boundaries as appropriate. Subscript "B" is used to denote the bottom counterparts of stress variables.

While rewriting the continuous equations in finite difference, finite element or finite volume form may be done in any of several different ways with different and more advanced ways (e.g., with higher accuracy and to utilize massively parallel node architecture for computing power); but certain integral constraints must be maintained during the solution of the initial value problem. These constraints are cast as overall integrals for mass conservation (whatever gets in must equal whatever gets out) within a cell, over the basin, and over the volume; and for conservation of energy and to provide insulation for the tracers (temperature and salt) except at the surface or at the river mouths.

Let the basin be divided into cells with interfaces lying along common planes with constant latitude, longitude, and depth, as in Figure 8.3.

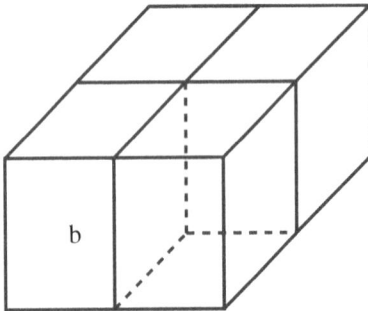

$$\sum_{b=1}^{6} V_b A_b = 0$$

Mass Conservation for each cell

Figure 8.3: A cell in a numerical model adjacent to the others. Mass needs to be conserved along with other conservative quantities in each of these cells.

Then these constraints are:

- Mass conservation within each cell (with six sides, $b = [1...6]$) with velocity V_b, and area A_b for each side,

$$\sum_{b=1}^{6} V_b A_b = 0$$

- Basin-wide integral, I' of any conserved quantity, q must remain unchanged by the advective process. So, if there are N cells in a closed basin

$$\frac{dI'}{dt} = -\sum_{n=1}^{N}\sum_{b=1}^{6} q_b V_b A_b = 0$$

- Volume integral I'' of the square of q remains unchanged over the domain. This means that the variance of the quantity or the energy contained by the thermohaline variables (T, S) remains constant in an integral sense, unless forced by external inputs. Assuming Q_n to be the average of q within a cell n,

$$\frac{dI''}{dt} = -2\sum_{n=1}^{N}\sum_{b=1}^{6} q_b Q_n V_b A_b = 0$$

- In the conservation of energy equation, kinetic energy gained or lost through the pressure term of the momentum equation should be balanced by the potential energy loss or gain through the advection terms.
- Since an insulating boundary condition exists for the tracers, everywhere except at the surface,

$$\int_{vol} F^T dV = \int_A \eta dA$$

8.7   THE COMPUTER ALGORITHM

The computer program back in the 1980s for the Cox model (Cox, 1984) was simple enough and instructive. It is described briefly here. Programming has become more complex and involved with new algorithms and new techniques of treating non-linearities and turbulence parameterization and boundary treatment, and many other components. However, the basic idea of the ocean modeling algorithm can be understood from the simple flow chart given in Figure 8.4.

The numerical implementation of the Cox model (Cox, 1984) given in Figure 8.4 can be described as follows. Each subroutine does a specific function. The model begins with **Ocean** calling **STEP** at every time-step. **STEP** gets the Initial fields at the first time-step and updates the forcing fields in every subsequent time-step (this is the RHS of the primitive equations). Once the whole domain is specified by every variable at all grid points, the internal mode or the baroclinic component of the velocity is solved by **CLINIC** in a row-by-row computation for the spatial domain.

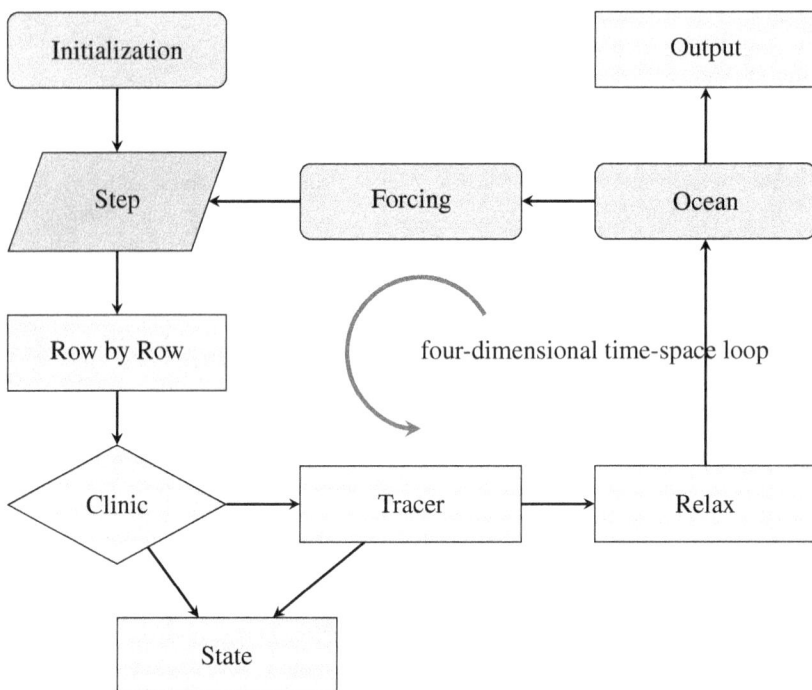

Figure 8.4: A first example of a program flow for ocean modeling.

CLINIC uses the density computation (Equation of **STATE**) at every time step and sends its velocity information to **TRACER** for active and/or passive tracers. This completes the internal mode velocity and temperature and salinity computations at all grid-points throughout the domain, which are then utilized to compute the external mode (barotropic stream function) using the Relaxation scheme (**RELAX**). This is the routine which utilizes the wind stress curl at the surface. The external and internal components of velocities are added together by the ocean at every time-step and sent to the output for storage and back to **STEP** for the next time-step computation on the whole domain again. Note that there are other routines for boundary condition, forcing, and surface pressure calculation at every time-step, which are not shown.

Finally, the reader is referred to the latest version of the GFDL model (MOM6) to gain familiarity with the new developments of the rather simple modeling system described in this chapter. Over the last four decades, extensive development of this simple architecture has led to a fully nested, coupled, interdisciplinary climate system modeling framework at GFDL and at NCAR and all over the world. Interested readers are requested to look at those websites for MOM6 and CMIP6.0. Some of the other models will be described in the next chapter to extend the idea of "knowing the models" rather than using the model as a black box. So, please read on.

8.8 CONCLUSION

In this chapter, we first discussed the purpose of modeling and the general approach to modeling with a description of an early ocean model. Then we discussed some details about the Finite Difference Method, a numerical method that is easy to comprehend. The modeling equations were cast from a spherical coordinate perspective. A basic setup of the model equations for numerical integration over a three-dimensional grid was presented and a standard algorithm using the barotropic/baroclinic mode splitting technique was discussed in detail. The basic modeling equations and the computer algorithm are discussed in some detail to give the student a fair idea of the process of numerical modeling operations. The model requires a set of initial and forcing fields, it had flat levels in the vertical and the effects of friction were parameterized by Ekman-like Eddy viscosity and diffusivity coefficients.

This chapter is kept simple enough to understand on purpose. Most of the recent modeling efforts build on the above methodology while providing tremendous sophistication of computer algorithm (e.g., for solving the spatial and temporal gradients at every time-step; maintaining integral constraints to meet conservation properties as accurately as possible). It is thus critical to understand the very basics of numerical modeling to develop and improve upon that basis for attaining higher levels of algorithmic efficiency as more and more computing power and speed become available in the future.

FURTHER READING

Bryan, K. (1997). A numerical method for the study of the circulation of the world ocean. *Journal of Computational Physics*, *135*(2), 154–169.

Cox, M. D. (1970). A mathematical model of the Indian Ocean. In *Deep Sea Research and Oceanographic Abstracts*, *17*(1), 47–75. Elsevier.

Cox, M. D., & Bryan, K. (1984). A numerical model of the ventilated thermocline. *Journal of Physical Oceanography*, *14*(4), 674–687.

Cushman-Roisin, B., & Beckers, J. (2011). *Introduction to geophysical fluid dynamics: Physical and numerical aspects*. Elsevier.

Haidvogel, D. B., & Beckmann, A. (1999). *Numerical ocean circulation modeling*. World Scientific.

Kantha, L. H., & Clayson, C. A. (2000). *Numerical models of oceans and oceanic processes*. Elsevier.

Mazumder, S. (2015). *Numerical methods for partial differential equations: Finite difference and finite volume methods*. Academic Press.

Neill, S. P., & Hashemi, M. R. (2018). *Fundamentals of ocean renewable energy: Generating electricity from the sea*. Academic Press.

Pond, S., & Pickard, G. L. (1983). *Introductory dynamical oceanography*. Gulf Professional Publishing.

Rapp, B. E. (2016). *Microfluidics: Modeling, mechanics and mathematics*. William Andrew.

Komodo Island, Indonesia

9 Turbulence and Eddies

> I do not know. [summarizing his life's work]
>
> ———————————————
>
> Joseph-Louis Lagrange
> (1736–1813)

OVERVIEW

In the last chapter, we introduced you to numerical ocean modeling with a set of basic equations and its numerical implementation with the finite difference method to solve for the gradients at every time-step. The model requires a set of initial and forcing fields, it was set up with flat levels in the vertical and the effects of friction were parameterized by Ekman-like eddy viscosity and diffusivity coefficients.

In this chapter, we focus on the idea of eddy parameterization, a difficulty that stems from the fact that the ocean is a turbulent fluid and this turbulence is pervasive over a large number of spatial and temporal scales that cover multiple oceanic phenomena (see Figure 1.5). So, this chapter discusses Turbulence and eddy parameterization together. Beginning with the challenges of understanding turbulence, we discuss the challenges for modeling this aspect of fluid flow and discuss the so-called *closure problem*. The well-known Kolmogorov theory of turbulence is briefly discussed to highlight different approaches to closure modeling. Relation to vertical and horizontal mixing of turbulent eddies are then discussed in the light of a one-dimensional mixed-layer model.

9.1 TURBULENCE AND EDDY VISCOSITY

To understand the difficulty with the turbulence and eddy viscosity (and eddy diffusivity) let us dive into the process of turbulence briefly. We understand that the eddy viscosity is a very different (but somewhat similar) phenomenon than the molecular viscosity and that they must be connected in the fluid system. How are they connected? And if we understand that, can we use that connecting pathway to develop a scheme that would then yield better results in numerical models than what a single value of eddy viscosity would yield? This is the subject of this section.

DOI: 10.1201/9780429347221-9

9.1.1 TURBULENCE

We have seen that when the scales of velocities and lengths are "large," then the non-linear terms in the left-side of the equations of motion (the advective terms) dominate the balance when compared to the molecular viscosity terms on the right side. The ratio of these two terms is called the Reynolds Number (R_e). Let us see how this happens. A typical advective term is $u\frac{\partial u}{\partial x}$ and the molecular viscosity term is $v\frac{\partial^2 u}{\partial x^2}$. So, choosing orders of velocity as U and length as L, the ratio of the two terms is simply $\frac{UU/L}{vU/L^2}$ or $R_e = UL/v$.

If we use the Gulf Stream as an example of a turbulent current, $U = 1\,\mathrm{ms}^{-1}$, and $L = 100\,\mathrm{km} = 10^5\,\mathrm{m}$; with a $v = 10^{-6}\,\mathrm{m^2 s^{-1}}$, we get a $R_e = 10^{11}$ – a hugely turbulent flow field. A typical Gulf Stream eddy would be on similar orders of magnitude of Reynolds number too! So, if the viscosity to describe these currents and eddies are appropriate for such large R_e values and called the "eddy viscosity," as was discussed in Chapter 4. How does the energy finally dissipate from such large eddy-scales down to the molecular level? There must be a connection between the two.

How do we think about such a turbulent fluid and understand its evolution? One of the ways to think about such a flow system is to characterize the flow into its mean and fluctuating components and treat the fluctuating parts as the ones that evolve rapidly, extract energy from the mean flow and develop eddies and other interesting features different from the mean flow. The average or mean flow is then determined as a time-averaged flow and the total velocity is given by $u = \bar{u} + u'$, where the average is denoted by the overbar, the prime denotes fluctuating part, and by definition, $\overline{u'} = 0$.

Now, you can substitute $u = \bar{u} + u'$ and similar forms for v and w and other variables in the equations of motion (8.13, 8.14, 8.15). You can average the resulting equations to get equations for the mean flow and the fluctuations. It is easy to see that since the average of the fluctuations is zero, the pressure gradient, Coriolis, and friction terms (with molecular viscosity) reduces to the terms containing mean velocities and mean of pressure and density, without any fluctuating components.

However, the only terms that are left with fluctuating components are the non-linear advection terms on the left side. Let us look at these terms and follow them on the left side of the x-component equation,

$$(\bar{u} + u')\frac{\partial(\bar{u} + u')}{\partial x} + (\bar{v} + v')\frac{\partial(\bar{u} + u')}{\partial y} + (\bar{w} + w')\frac{\partial(\bar{u} + u')}{\partial z}$$

It is left to the reader to see that terms like $\bar{u}\frac{\partial \bar{u}}{\partial x}$ can be absorbed with the local derivative $\frac{\partial \bar{u}}{\partial t}$ to complete the full equation of motion for the average flow without any fluctuating component in the total derivative of the mean flow.

Now, think about the continuity equation and use $u = \bar{u} + u'$ to yield two separate and similar continuity equations for mean and fluctuating components. Use the continuity equation for the fluctuating part to add (which does not change anything as it is zero) to the non-linear part of the mean-flow equation of motion, to yield the following,

$$\frac{d\bar{u}}{dt} = -\alpha\frac{\partial \bar{p}}{\partial x} + 2\Omega\sin\phi\bar{v} - 2\Omega\cos\phi\bar{w} + v\left(\frac{\partial^2\bar{u}}{\partial x^2} + \frac{\partial^2\bar{u}}{\partial y^2} + \frac{\partial^2\bar{u}}{\partial z^2}\right)$$
$$-\frac{\partial\overline{u'u'}}{\partial x} - \frac{\partial\overline{u'v'}}{\partial y} - \frac{\partial\overline{u'w'}}{\partial z} \tag{9.1}$$

While you do that, please remember that time averaging is required so that non-linear mixed terms like the following vanish.

$$\overline{u'\frac{\partial\bar{u}}{\partial x}} = \overline{\bar{u}\frac{\partial u'}{\partial x}} = 0 \tag{9.2}$$

Which then make the non-linear terms with fluctuations become like below

$$\overline{u'\frac{\partial u'}{\partial x}} + \overline{v'\frac{\partial u'}{\partial y}} + \overline{w'\frac{\partial u'}{\partial z}} = \frac{\partial}{\partial x}\overline{u'u'} + \frac{\partial}{\partial y}\overline{u'v'} + \frac{\partial}{\partial z}\overline{u'w'} \tag{9.3}$$

since $\nabla \cdot \vec{u'} = 0$ due to continuity.

The equation 9.1 is the Reynolds equation in x-direction named after Osborne Reynolds, who first derived this equation. The last three terms represent the impact of turbulence on the mean flow. These terms are the divergence of what is called the Reynolds stresses, which are given by $\rho\overline{u'u'}$, $\rho\overline{u'v'}$, and $\rho\overline{u'w'}$. You can see the similarity of these stresses with those that were described in Chapter 2 for Newtonian stress tensors. These are nine fluctuating stresses on a fluid element due to perturbations of the mean flow. The next question is then how do we solve these equations now?

9.1.2   THE CLOSURE PROBLEM

So, now we see that the turbulent fluxes can be represented by the three new terms in the forms like $\overline{u'u'}$, $\overline{u'v'}$, $\overline{u'w'}$, etc. But these are new variables (u', v', w') in addition to the seven primitive variables and we need additional equations to solve for these fluctuating terms! We need a complete set of equations (as many equations as variables) to arrive at a unique solution. But it seems that we cannot find it ever. This is **THE** problem of theoretical turbulence as we know it. Well, Why exactly is there a problem? Why can't we get new equations?

Let us try and go one step further to see this problem in a little more detail. Following Vallis (2000), we consider a model non-linear system as below,

$$\frac{du}{dt} + uu + ru = 0 \tag{9.4}$$

where r is a constant, and $u = \bar{u} + u'$, i.e. with mean and fluctuating components. Now take the average of the above equation 9.4 to get,

$$\frac{d\bar{u}}{dt} + \overline{uu} + r\bar{u} = 0 \tag{9.5}$$

The problem is that the term $\overline{uu}$ cannot be determined from just the mean $\bar{u}$ because it has the fluctuating part u' within it. Note that $\overline{uu} = \bar{u}\bar{u} + \overline{u'u'} \neq \bar{u}\bar{u}$. Using this relation in equation 9.5 yields the following:

$$\frac{d\bar{u}}{dt} + \bar{u}\bar{u} + r\bar{u} = -\overline{u'u'} \tag{9.6}$$

So, it is actually the correlation terms on the right-side such as $\overline{u'u'}$ that cannot be determined. Why not?

Well, to solve for $\overline{u'u'}$, we need to go one order up. Let's try that. Let us multiply equation 9.4 by u and take the average. This will lead to an equation for u^2 as follows,

$$\frac{1}{2}\frac{d\overline{u^2}}{dt} + \overline{uuu} + r\overline{u^2} = 0 \tag{9.7}$$

Substituting $u = \bar{u} + u'$, we get the following:

$$\frac{1}{2}\frac{d\overline{u'u'}}{dt} + \frac{1}{2}\frac{d\bar{u}\bar{u}}{dt} + \overline{u'u'u'} + \bar{u}\bar{u}\bar{u} + 3\bar{u}\overline{u'u'} + r\bar{u}\bar{u} + r\overline{u'u'} = 0 \tag{9.8}$$

What just happened? Now this equation 9.8 has a "triple correlation" term, which needs to be solved (or unknown). To solve for $\overline{u'u'}$, we need to know $\overline{u'u'u'}$.

Yes, we can try to go one order up. We will run into a "quartic" correlation. So on and on. This will go on ad infinitum!

This is the problem of closure! We can never get to close the system of equations with fluctuations or turbulence. Not only has this problem not been solved, but it also is not clear whether there exists a closed-form solution at all.

It is probably possible to observe the fluctuations and come up with those fluctuating terms quantitatively locally, but that would not work in all situations. We can probably express the Reynold's stresses in terms of mean flow variables and their gradients based on our physical understanding of how turbulence might work on the "average" or with the "mean" flow. This approach is called "*parameterization.*" By *parameterization*, we mean that we can express the turbulence quantities in terms of parameters which we can observe more readily and/or determine from our equations, for example, the mean velocity variables and their gradients. This is the problem of "closure." We do not have a formal way to close the system with new equations for the fluctuating components. So, we need to "parameterize" their effects on the mean flow and represent them in terms of the variables of the mean flow. One of the popular ways is of closing such hierarchical systems is to make assumptions about the relationship between two consecutive orders of correlation terms. For example, one could assume that

$$\overline{uuuu} = \alpha \overline{uuu}\,\overline{u} + \beta \overline{uu}\,\overline{uu} \tag{9.9}$$

where α and β are set parameters based on physical and/or observational understanding.

Realize that the eddy viscosity thus is a derived parameter of this closure problem and it would depend on the mean flow $(\overline{u}, \overline{v}, \overline{w})$ and their gradients. It does not depend on the property of the fluid (water), unlike the molecular viscosity.

9.1.3 A SIMPLE EDDY VISCOSITY

Let us go back to the Reynolds Equation 9.1. The molecular viscosity terms $(v \frac{\partial^2 \overline{u}}{\partial x^2})$ can be written as $\partial(v \partial \overline{u}/\partial x)/\partial x$. Similar to the molecular friction formulation (recall Newton's law of friction and the construct in Chapter 2, equation 2.15), a first approximation form of the Reynolds stress would be to express it with a constant times the gradient of the mean flow. This constant would be called the "eddy viscosity." Mathematically, this is simply,

$$-\overline{u'u'} = K_{vx} \frac{\partial \overline{u}}{\partial x} \tag{9.10}$$

where K_{vx} is the eddy viscosity, and its value is much larger than the value of the molecular viscosity v. Similarly,

$$-\overline{u'v'} = K_{vy} \frac{\partial \overline{u}}{\partial y}; \quad -\overline{u'w'} = K_{vz} \frac{\partial \overline{u}}{\partial z} \tag{9.11}$$

This is also doable for the fluctuations due to turbulence for the heat equation and a similar "eddy diffusivity" (K_d) can be defined with the following relationship for the buoyancy (b) variable,

$$-\overline{u'b'} = K_d \frac{\partial \overline{b}}{\partial x} \tag{9.12}$$

So, the first of the three non-linear fluctuation terms in the Reynold's equation becomes $\partial(\overline{u'u'})/\partial x = \partial(K_{vx} \frac{\partial \overline{u}}{\partial x}/\partial x) = K_{vx} \frac{\partial^2 \overline{u}}{\partial x^2}$, by taking K_{vx} outside of the derivative. Similarly, the other two terms become $K_{vy} \frac{\partial^2 \overline{u}}{\partial y^2}$ and $K_{vz} \frac{\partial^2 \overline{u}}{\partial^2 z}$.

Now, it is easy to absorb the molecular viscosity within the eddy viscosity by using equation 9.11 in the Reynold's equation 9.1 and close the system as follows for the mean flow,

$$\frac{d\bar{u}}{dt} = -\alpha\frac{\partial\bar{p}}{\partial x} + 2\Omega\sin\phi\bar{v} - 2\Omega\cos\phi\bar{w} + A_x\frac{\partial^2\bar{u}}{\partial x^2} + A_y\frac{\partial^2\bar{u}}{\partial y^2} + A_z\frac{\partial^2\bar{u}}{\partial z^2} \quad (9.13)$$

Note that we are only considering the x-momentum equation here for simplicity.

Here, $A_x = K_{vx} + v$, $A_y = K_{vy} + v$ and $A_z = K_{vz} + v$. Unlike molecular viscosity, the eddy viscosity can be different in different directions since it is a flow-dependent characteristic. The units of K, A, and v are L^2T^{-1}, i.e. m^2s^{-1}.

In a similar manner, one could derive the Reynolds averaged buoyancy equation. See Cushman-Roisin and Beckers, 2011 for details.

9.1.4 THE KOLMOGOROV THEORY OF TURBULENCE

The typical value of molecular viscosity (v) is $10^{-6}\,m^2s^{-1}$. The typical value for the eddy viscosity in the above simple approximation would be about $10^{+6}\,m^2s^{-1}$ to make the non-linear (now frictional) terms balance those of inertial or Coriolis or pressure-gradient terms. Note that the value of eddy viscosity depends on the scales that you are trying to resolve. Eddy diffusivities are scale-dependent.

So, thinking about the process of turbulence, a first-order question that arises is that of the connection. What is the mechanism by which large-scale energy from the eddies is transferred to the molecular viscous scales? This is turbulence. At the fundamental level, a turbulent fluid is a collection of eddies of different sizes and strengths, interacting at multiple scales within and outside each other, transferring energy and fluxes between them. In fact, interactions occur between nearby scales, wherein vortex stretching by a slightly larger eddy transfers energy to one slightly smaller (see Kundu et al., 2012 for a discussion). Two variables characterize this process: the eddy-length scale (d) and the orbital velocity of the eddy (u^*). The intensity of the turbulent eddy field can be measured by the **turbulent kinetic energy** (TKE) or E.

In general, the TKE (E) is defined as the mean kinetic energy per unit mass associated with the eddies in the turbulent flow. So, mathematically,

$$E = \frac{u'^2 + v'^2 + w'^2}{2} \quad (9.14)$$

Here u', v', w' are the fluctuating components (eddy-part) of the total velocity fields as described in Section 9.1.1. The unit of TKE is Joules/kg $= m^2s^{-2}$.

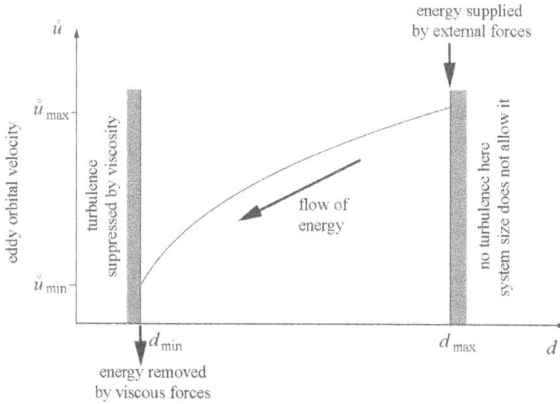

Figure 9.1: The turbulent Energy Cascade from large-scale eddies to the molecular level. According to this theory, the energy fed by external forces excites the largest possible eddies and is gradually passed to ever smaller eddies, all the way to a minimum scale where this energy is ultimately dissipated by viscosity. Republished with permission of Elsevier, from "Introduction to Geophysical Fluid Dynamics: Physical and Numerical Aspects," Cushman-Roisin, Benoit, 2009; permission conveyed through Copyright Clearance Center, Inc.

We first assume that the turbulent field is stationary (statistically unchanged in time), homogeneous (uniform in space), and isotropic (no preferred direction). We also assume that the fluid is stirred at large scales and dissipated at small scales. The turbulent energy cascade from large scale to small scale was theorized by Kolmogorov (1941) (Frisch & Kolmogorov, 1995; Kolmogorov, 1991) and we will follow his ideas here. See Figure 9.1. The range of eddy diameters (along the x-axis) increases as their characteristic velocities increases. The longest existing eddy diameter provides the largest eddy orbital velocity. The shortest eddy would have the lowest orbital velocity. In the "inertial sub-range" in between, neither forcing nor dissipation is important, and non-linear terms govern the dynamics.

The idea is that energy moves downscale. Energy gets passed on from larger scales to smaller scales, to even smaller scales, and then the energy dissipates to the molecular level. The input of energy happens at the largest of scales and the viscous energy dissipation is at the smallest scales of molecular level. So, what is the rate of dissipation? This is the most powerful assumption of the Kolmogorov theory. Since the turbulence is statistically steady, this dissipation rate ε is assumed to be set by the rate at which energy is input at large scales. The rate at which energy is transferred downscale is independent of eddy size; i.e., ε is defined to be the rate of energy loss at small scales. Otherwise, some eddies of a given size would grow or shrink in numbers over time. In fact, it is the rate at which energy is supplied at the largest scales – the same rate at which energy cascades from larger to smaller scales till it reaches the molecular level and it is the exact same rate at which energy dissipates at the smallest scales due to viscosity. This dissipation rate ε is the turbulent kinetic energy dissipation rate and has the unit of energy per unit mass per unit time, or $m^2 s^{-2} s^{-1}$, or $m^2 s^{-3}$. See the definition of TKE in equation 9.14, which has the unit of energy per unit mass or $m^2 s^{-2}$.

So, now, we have defined a few very important variables for the turbulence process: the TKE (E) and its dissipation rate ε. Remember that there is also the eddy size (d) and eddy characteristic velocity $(u*)$ involved in the process.

Since the energy cascade is across the eddy sizes, what might the spectrum look like? Assuming forcing and dissipation scales are not important, the Energy (E) must be a function of the eddy size $(d = 1/\kappa$; wavenumber is $\kappa)$ and dissipation rate (ε). This spectrum is assumed to be universal, i.e. the same for any turbulent flow. However, Landau (1944) maintained that "the variation of ε over times of the order of the period of the larger eddies should be different for different flows, and the results of such averaging therefore cannot be universal." At the same time, the universality seems to hold for most observational data, and so the generality has been generally accepted.

Let us see what the character of this spectrum by dimensional analysis is. First, by definition, the dissipation rate depends on the energy and the timescale of the eddy. So,

$$\varepsilon \approx \frac{u^{*2}\kappa}{\tau_\kappa} \tag{9.14a}$$

The energy is nothing but square of the characteristic velocity, or

$$u^{*2}(\kappa) = \phi_E(\kappa)\kappa \tag{9.14b}$$

Now, the eddy timescale can be thought as,

$$\tau_\kappa = \frac{1/\kappa}{u^*(\kappa)} = \left[\kappa^3 \phi_E(\kappa)\right]^{-1/2} \tag{9.14c}$$

Thus utilizing the last expression (c) in the first expression (a) and then using (b) in that resulting expression, we get the following,

$$\varepsilon \approx \frac{u^{*2}(\kappa)\tau_\kappa}{}$$

$$= \frac{\left[\phi_E(\kappa)\kappa\right]\left[\kappa^3 \phi_E(\kappa)\right]^{-1/2}}{}$$

$$= \phi_E(\kappa)^{3/2}\kappa^{5/2} \tag{9.15}$$

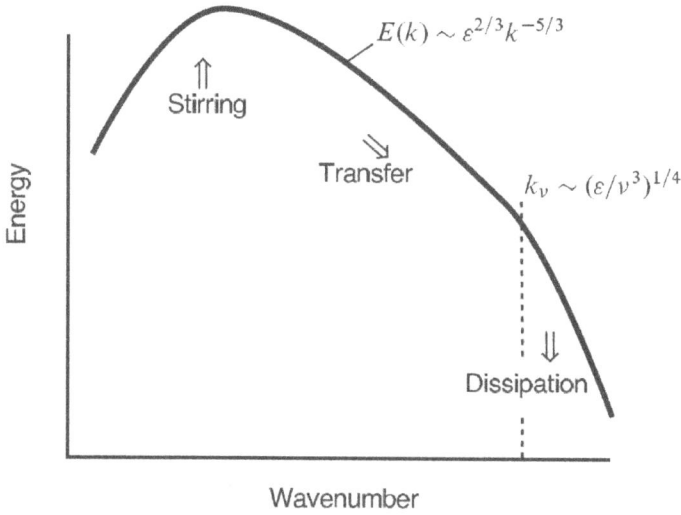

Figure 9.2: A schematic Turbulence Spectrum of the TKE with the fall of −5/3 in the inertial subrange. Schema of the energy spectrum in three-dimensional turbulence, in the theory of Kolmogorov. Energy is supplied at some rate ε; it is cascaded to small scales, where it is ultimately dissipated by viscosity. There is no systematic energy transfer to scales larger than the forcing scale, so here the energy falls off. Republished with permission of Cambridge University Press, from "Atmospheric and Oceanic Fluid Dynamics," Geoffrey K. Vallis, 2006; permission conveyed through PLS clear.

which leads to the Kolmogorov spectra equation,

$$\phi_E(\kappa) = C\varepsilon^{2/3}\kappa^{-5/3} \tag{9.16}$$

where C is a constant. A schematic Kolmogorov spectrum is shown in Figure 9.2, where the fall of the spectral energy (TKE) follows the well-known slope of −5/3rd.

If we think of dimensions of these variables as follows,

$$[E] = L^2 T^{-2}$$
$$[\phi_E(\kappa)] = L^3 T^{-2}$$
$$[\varepsilon] = L^2 T^{-3}$$
$$[\kappa] = L^{-1}$$
$$[d] = L \tag{9.17}$$

then the Kolmogorov spectra equation 9.16 is satisfied with both sides having a dimension of $L^3 T^{-2}$.

Utilizing the dimensional analysis with variables in equation 9.17, we also can get some qualitative behaviors as follow. First, note that the characteristic velocity is related to the size of the eddy as $u^* = A(\varepsilon d)^{1/3}$. This implies that larger dissipation would lead to larger eddy speeds and larger eddies are also associated with faster speeds.

The credence for the Kolmogorov turbulence theory has been observationally validated by many over the years. See Pope (2001), Tennekes and Lumley (2018), and Cushman-Roisin and Beckers (2011) for examples. The cascading nature of turbulence makes the closure problem real and in some sense infinitely impossible to solve! On the other hand, the limited number of variables $(E, \varepsilon, d(\kappa), u*)$ provides a basis for inventing reasonable closure schemes with a targeted process-oriented and physically meaningful manner. There are more than 200 closure schemes in practice for different applications of solving turbulent flow situations in various configurations such as high speed flows, atmospheric flows, hydraulics, etc.

For more details on eddy viscosity and diffusivity and their closure schemes that depend on fluid flow (rather than properties) the reader is referred to advanced texts (mentioned above) on this subject. A typical way to represent the horizontal eddy viscosity to incorporate the effects of sub-grid scale processes in a numerical model (with grids of $\triangle x \times \triangle y$) is proposed by Smagorinsky, 1963.

$$A = \triangle x \triangle y \sqrt{\left(\frac{\partial u}{\partial x}\right)^2 + \left(\frac{\partial v}{\partial y}\right)^2 + \frac{1}{2}\left(\frac{\partial u}{\partial y} + \frac{\partial v}{\partial x}\right)^2} \tag{9.18}$$

9.1.5   APPROACHES TO CLOSURE MODELING

Now that we understand the basics of turbulence and the problem of closure (due to non-linear terms being related to correlations of Reynold's stresses), we can begin to appreciate some of the different approaches to arrive at a closure model based on additional physical assumptions of the flow.

We begin our search for parameterization assuming that the turbulence is a local problem, i.e. the energy cascade is purely local, which is also known as the *one-point closure model*. We assume that the mean flow has a length scale l_m, and since the velocity fluctuations are not resolved explicitly, the turbulent energy is extracted by dissipation from the mean flow itself with eddy viscosity (K) at the rate of ε. In this situation, the Reynolds number with the mean flow velocity (u_m) and the length scale (l_m) should be on the order of unity, if the viscosity is considered to be eddy viscosity itself. This means that the following relation holds for the one-point closure,

$$\frac{u_m l_m}{v_E} \approx 1 \tag{9.19}$$

Since the dissipation rate is energy over time, or

$$\varepsilon = \frac{u_m^2}{\text{eddy-timescale}}$$
$$= \frac{u_m^2}{l_m/u_m}$$
$$= \frac{u_m^3}{l_m} \tag{9.20}$$

If we rewrite this in terms of the TKE ($E = u_m^2/2$), we get another relationship for the dissipation rate,

$$\varepsilon = C\frac{\kappa^{3/2}}{l_m} \tag{9.21}$$

where C is a calibration constant.

So, knowing κ and ε would allow us to determine the eddy viscosity (remember that this is the goal!). In the process, we evaluate the length-scale l_m, over which dissipation (mixing) happens through equation 9.21. This is why l_m is also called the "mixing-length."

Note that if we could design a way to define the length and velocity scales (l_m, u_m), we would be able to get to determine the eddy viscosity (v) as well. This was first done by Ludwig Prandtl, who considered a case of a horizontal mean flow (u_m) with vertical shear $\frac{\partial u}{\partial z}$ only. He assumed that the only correlation for Reynolds stresses that is non-zero is that between the horizontal and vertical fluctuations, or $\overline{u'w'} \neq 0$. (See Figure 9.3.) The reasoning behind $\overline{u'w'} \neq 0$, is the following. When you move the fluid parcel upward (or downward), as in Figure 9.3, it initially keeps its original horizontal momentum (ρu), so relative to the fluid surrounding it at its new location, it has relatively less (or more) momentum. Thus, for upward displacement, $w' > 0$, and $u' < 0$, and for

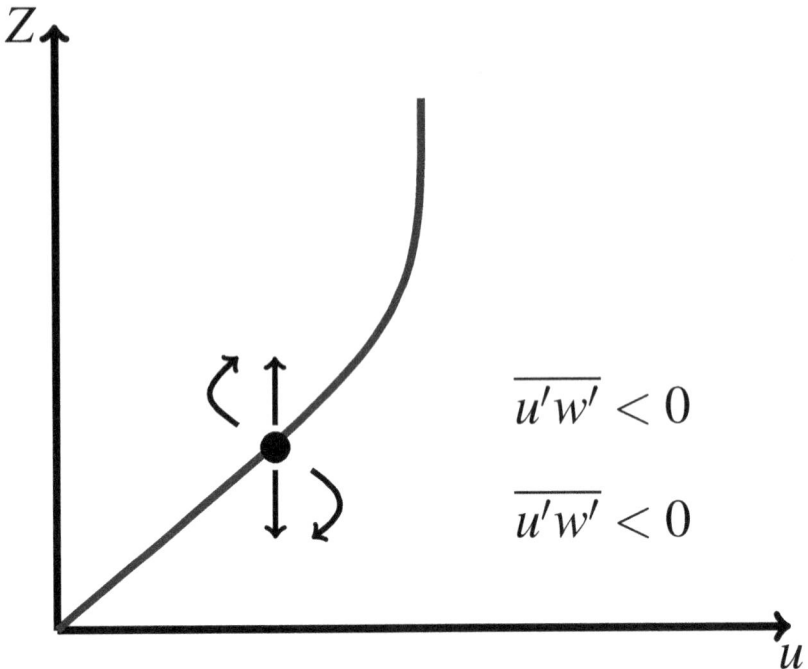

Figure 9.3: Prandtl's mixing length formulation. The fluid is assumed to have a horizontal mean flow with a vertical shear only. The non-zero correlations for Reynold's stresses are between u' and w' only.

downward displacement $w' < 0$, and $u' > 0$. After some time, the particle's excess (or deficit) momentum diffuses into the surrounding fluid, either speeding it up and slowing it down, and thus the turbulent eddy has increased the rate at which momentum has diffused through the fluid.

Furthermore, he assumed that the fluctuations are random, and thus the correlation relationship is given by:

$$\overline{u'w'} = r\sqrt{\overline{u'^2}}\sqrt{\overline{w'^2}}$$

Furthermore, assuming that the velocity fluctuations are proportional to the velocity scale (u_m) of the flow, the Reynold's stress terms can be simply written as,

$$\overline{u'w'} = Cu_m^2$$

.

Now, as the eddy viscosity is defined by

$$\overline{u'w'} = -v_E \frac{\partial \bar{u}}{\partial z},$$

and, $v_E = u_m l_m$, since R_e at this scale is 1, the characteristic velocity and the eddy viscosity can be expressed as follows, which depends on the mixing length and the velocity shear of the flow,

$$u_m = l_m \left| \frac{\partial \bar{u}}{\partial z} \right| \tag{9.22}$$

$$v_E = l_m^2 \left| \frac{\partial \bar{u}}{\partial z} \right| \tag{9.23}$$

The absolute values are included to ensure positive eddy viscosity, and therefore a downgradient advective flux. Note that the higher the shear, the higher the viscosity, or turbulent diffusion. This confirms the idea that increased shear will cause increased turbulence. For more exposition on this model, see more advanced texts in Further Reading.

Now, let us look at a more complicated model which takes into account the fluid's "memory," in contrast with Prandtl's model of instantaneous flow. To do this, one seeks governing equations with time-derivatives or looks for prognostic equations instead of diagnostic equations. Two of the most important variables to relate to the eddy viscosity are the TKE (E) and the dissipation rate (ε).

First, let us remember that the conservation equations are valid for the mean flow as well as for the fluctuations separately. So, we can use the equation

$$\frac{\partial u'}{\partial x} + \frac{\partial v'}{\partial y} + \frac{\partial w'}{\partial z} = 0$$

to aide in developing the prognostic equation as a constraint on the fluctuations. We also note that the Turbulent kinetic energy is simply,

$$E = \frac{u'^2 + v'^2 + w'^2}{2} \tag{9.24}$$

Using the difference between the momentum equations for the mean flow and the Reynolds equations (with the fluctuations) we can get three equations for the evolution of the fluctuations u', v', w'. Multiplying the u' evolution equation with u', v' equation with v' and w' equation by w' and adding them up and then using the fluctuations conservation equation and equation 9.24, we can arrive at the following prognostic equation for the TKE (E),

$$\frac{dE}{dt} = P_s + P_b - \varepsilon - \underbrace{\left(\frac{\partial q_x}{\partial x} + \frac{\partial q_y}{\partial y} + \frac{\partial q_z}{\partial z} \right)}_{\nabla \cdot \vec{q}} \tag{9.25}$$

where P_s and P_b are a collection of terms with multiple combinations of Reynolds stresses and gradients of the mean flow. P_s is generally the collection of shear-dependent terms and is called the "shear production" term. P_b is called the "buoyancy production" term and is given by $-\overline{\rho' w'} g/\rho_0$. The third term is the dissipation rate and it involves all of terms of the molecular viscous dissipation by turbulent motions. The terms like $q_x, q_y,$ and q_z contain the pressure and velocity fluctuations and the spatial gradients of the TKE (E) (which signifies redistribution of TKE).

The next step in this process is to model all of these terms in terms of mean flow variables. Some of these "parameterizations" are shown above. This process is not very complicated but could be tedious and special care is needed to close all the gaps with physically meaningful assumptions. This is where the closure models are different, challenging, and innovative. For example, of detail methodology, the reader is referred to more advanced textbooks such as Cushman-Roisin and Beckers (2011) and others in the Further Reading section.

To end the discussion on the approaches on Closure modeling, it suffices to say that other closure models with two governing equations (*two-point closure models*) have been sought to avoid the prescriptive mixing-length type one-equation models. Two governing equations for k and l_m, or for k and ε have been sought. Once you have two variables determined, the third could be easily obtained using the diagnostic relationship between them, as discussed in this section. There is also a very popular model with k and kl_m as the variable set (Mellor & Yamada, 1982), which we will discuss later.

Let us now talk about vertical mixing and then return to some modeling frameworks.

9.1.6   RELATION TO VERTICAL MIXING

Mixing is an irreversible process by which waters of different densities diffuse into a common water mass at the molecular level.

Turbulence is essentially motion at small length scales, and it is this motion that deforms ambient gradients (that are usually small), creating within them large gradients – this is stirring. Molecular diffusion proceeds at a rate that is dependent on the gradients in the fluid (e.g. $\kappa \nabla^2 T$). Since the turbulent stirring has increased the gradients in the fluid, molecular diffusion acts faster, leading to more diffusion. It is this diffusion that is the irreversible part of the mixing. So, in the end, turbulence leads to reduced gradients, but only by first increasing the gradients so that molecular diffusion can act more quickly. See Figure 9.4 (upper panel).

This has important consequences for the chemical and biological gradients, especially in the surface layers (euphotic zones where there is light available for photosynthesis; see Chapter 13 later). If you think of vertical mixing of warm lighter water from above going below and colder, denser water rising above, then this mixing process would raise the gravitational potential energy of the system. See Figure 9.4 (bottom panel) for a schematic representation of the time-evolution of this process from a stratified system to a mixed layer.

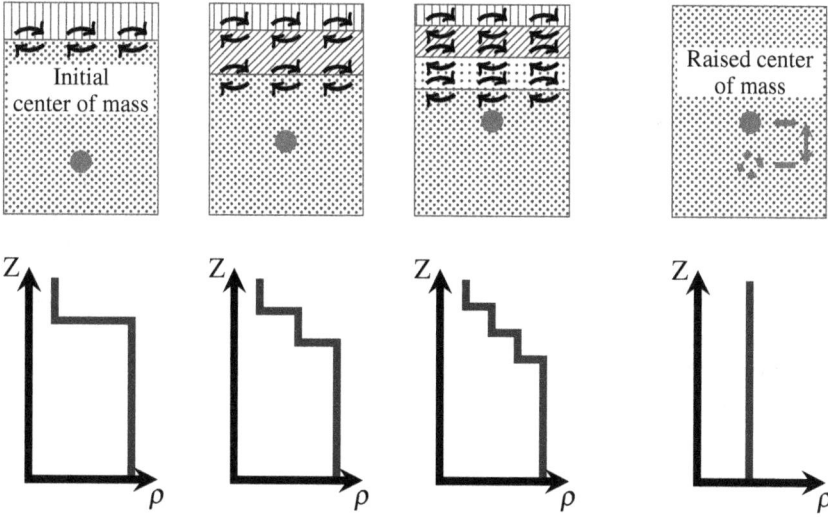

Figure 9.4: Turbulent induced mixing and reduction of density gradient in stages. Figure credit: Adrienne Silver.

Mixing is parameterized the same way as we did for the viscosity above. Consider the thermodynamic equations rather than the momentum equations. For example, take the simpler advection-diffusion equation below.

$$\frac{\partial \rho}{\partial t} + \vec{u} \cdot \nabla \rho = \kappa \nabla^2 \rho \tag{9.26}$$

Now, we can perform Reynold's decomposition with mean and fluctuations for ($u = \bar{u} + u'$; and $\rho = \bar{\rho} + \rho'$) and follow through the steps as in Section 9.1.1. Clearly, you are going to end in Reynold's stresses like $\overline{w'\rho'}$, which will need to be related to the vertical gradient of the density through the eddy diffusivity. Mathematically,

$$\overline{w'\rho'} = -\kappa_e \frac{\partial \bar{\rho}}{\partial z} \tag{9.27}$$

This means that the Reynolds fluxes are downgradient turbulent eddy flux. Substituting this in the advection-diffusion equation, one gets,

$$\frac{d\rho}{dt} = \kappa_e \frac{\partial^2 \rho}{\partial z^2} \tag{9.28}$$

This equation shows how the change of density with time is linked to the curvature of density in the vertical (the second derivative on the right-side) through the eddy diffusivity term, which is related to the turbulent correlations! There is the closure problem and hence the need for closure schemes again.

In short, it is the eddy viscosity and eddy diffusivity terms in the equations of motion that are related to the turbulence correlation terms (of Reynolds stresses) and these latter terms are needed to be solved through some closure schemes for better simulations and prediction through numerical modeling.

In oceanic numerical models, in addition to the first-order Pacanowski-Philander scheme described later in Chapter 10, three other vertical mixing schemes with different closure schemes are very popular. These are: (i) the Mellor-Yamada Model; (ii) the K-Profile Parameterization; and (iii) the Generic Large Scale (GLS) model. We will discuss the basic parameterizations of these three as part of the ROMS modeling effort in Chapter 10. Note again that the purpose of these schemes is to determine appropriate eddy viscosity and diffusivity values (or profiles) for the flow condition at that time and use it for the time-integration.

9.2   TURBULENCE AND MIXED LAYER

The surface layer of the ocean is a region where the forcing from the wind and heat from the sun play important roles in governing the circulation and modify the water masses by changing their density structure. Naturally, this upper layer is called the "Mixed layer" as turbulence mixes the waters through the non-linear terms derived above. The viscosity and diffusivity in this turbulent region are critical elements for the success of numerical models. Both coefficients (viscosity and diffusivity) are related to the turbulence in this surface mixed layer and a considerable research effort has focused on parameterizing turbulence in the mixed layer to connect it to the layers below.

Let us think about the process of "mixing." Mixing happens when different water masses of different densities are mixed together. Lighter water goes down; heavier water comes up; you pour milk in the coffee or tea and stir – i.e. create turbulence which mixes the two different density fluid, etc. So, increased turbulence results in enhanced mixing. This happens as the initial density gradient that existed in the system reduces. At the same time, the heavier fluid coming up and lighter fluid going down – that effect raises the potential energy of the system. Where does that energy come from? Well, it is the turbulent kinetic energy (Remember the Reynold's stresses and non-linear terms) that cascades energy down the scales (from eddy scale to molecular viscous scales). So, let's think about this. We need mean velocity gradients (remember the Reynolds stresses are parameterized in terms of mean velocity gradients and the coefficient of viscosity) to act on the density gradients to reduce them in favor of turbulence or mixing. So, there is a competition between the density gradient and velocity gradient in the vertical, which can be captured by their ratio to quantify vertical mixing. Now, the vertical gradient of density is already characterized by the buoyancy frequency (N) and called the static stability criteria (remember Section 7.2.1). This is given by,

$$N^2 = -\frac{g}{\rho}\frac{\partial \rho}{\partial z}\qquad(9.29)$$

If the density gradient $\frac{\partial \rho}{\partial z} > 0$ within the mixed layer (z +ve upward), then the mixed layer is unstable. The unstable water column would allow for enhanced velocity fluctuations in the vertical (similar to convection). If turbulence persists, then mixing will continue and move the density toward uniformity within the mixed layer. Positive stability will inhibit the fluctuations and keep the surface layer stratified.

How do we characterize the contribution of velocity gradients on the turbulence in relation to the density gradient? Lewis Fry Richardson, a famous English Meteorologist, who first came up with the idea of numerical weather prediction (see later in Chapter 14), came up with a non-dimensional quantification of this relative measure of density gradient versus velocity gradient (flow shear). This is now known as the Richardson Number (R_i) and is given as follows,

$$R_i = N^2 / \left(\frac{\partial u}{\partial z}\right)^2 = -\frac{\frac{g}{\rho}\frac{\partial \rho}{\partial z}}{(\frac{\partial u}{\partial z})^2}\qquad(9.30)$$

Note that if $\frac{\partial v}{\partial z} \neq 0$, then the denominator of the right-side of the Richardson Number can be replaced by $(\frac{\partial V_H}{\partial z})^2$ or $(\frac{\partial u}{\partial z})^2 + (\frac{\partial v}{\partial z})^2$.

It is evident from equation 9.30 that when the value of R_i is greater than unity, stability dominates velocity gradients, and turbulence is limited. However, as the value of R_i decreases below 1, turbulence increases considerably, and mixing will increasingly reduce density gradients. Laboratory experiments by Miles (1961) have shown that a stratified shear flow (which has both $N \neq 0$ and $\frac{\partial u}{\partial z} \neq 0$) is actually stable if $R_i > 0.25$ everywhere in the flow. Please see LeBlond and Mysak (1981) for a proof of this theorem and supporting laboratory experiments.

One way to think about the R_i is that it is the ratio of "static stability" (defined by N) and "dynamic stability (or instability)" (defined by the square of the velocity gradient, coming from fluctuations or the Reynold's stresses or the non-linear terms in the equation of motion. So, in the mixed layer, when $R_i < 0.25$, dynamic instability will take over the static stability and flow will be completely turbulent and mixed.

9.2.1 A MIXED-LAYER MODEL
- THE PRICE-WELLER-PINKEL (PWP) MODEL

Based on the above ideas, Price et al., 1986 developed a so-called "bulk mixed-layer model," now known as the PWP model. This model is a one-dimensional bulk model in the sense that it integrates the model prognostic equations over the mixed layer and balances quantities over the entire mixed layer. There are three essential stability processes in this model that are applied to the one-dimensional mixed layer (defined by its depth only) and the layer below to allow turbulence and mixing to play out through the upper ocean and below. Once the surface boundary forcing of wind, heat flux, and other factors (precipitation, evaporation, river, etc.) are applied to the model fields, at a time-step, we have a vertical profile of density. The first process test for the PWP model is to check for static stability. This is done by checking whether

$$\frac{\partial \rho}{\partial z} < 0 \qquad\qquad (9.31)$$

If this condition is not satisfied between the top two layers, the waters are mixed (all properties are mixed) and then checked for static stability with the next level below. This is continued downward until the condition in equation 9.31 is satisfied. This is the first guess of the mixed-layer depth.

The second criterion in PWP is that the mixed layer is defined as the minimum depth required to maintain a bulk Richardson Number (R_{ib}) of the well-mixed layer to be above a critical value of 0.65. This bulk Richardson number condition is given by,

$$R_{ib} = \frac{g \triangle \rho h}{\rho_0 (\triangle V)^2} \geq 0.65 \qquad\qquad (9.32)$$

This condition determines whether the layer below the mixed layer is to be entrained in the mixed layer or not. If $R_{ib} < 0.65$, then the layer below is entrained in the mixed layer, the mixed-layer depth is increased, the mixed-layer is homogenized, and the value of R_{ib} is recalculated. This process goes on till we reach a particular depth at which the condition in equation 9.32 is satisfied ($R_{ib} > 0.65$). This value of 0.65 was fixed by studying the deepening of the mixed-layer during a storm by using different turbulence energy conversion models such as the Dynamics Instability Model and the Turbulent Erosion Model by Price et al., 1978 and comparing with real data. They found that the lower bound of the mixed-layer depth was best when the bulk Richardson Number reaches 0.65, with realistic variations being in the range of 0.4 to 0.8.

In the final step, PWP calculates the gradient Richardson Number (R_{ig}, same as R_i as given in equation 9.30) for mixing below the mixed-layer. This is the standard condition for shear flow stability given by,

$$R_{ig} = \frac{\frac{\delta \rho}{\delta z}}{\rho_0 (\frac{\partial v}{\partial z})^2} \geq 0.25 \qquad (9.33)$$

The value of R_{ig} is calculated for all layers below the mixed layer down to the bottom of the profile. If any level with $R_{ig} < 0.25$ is found, the model partially mixes the two neighboring levels (above and below the interface level) to a value of $R_{ig} = 0.30$. This process again continues till $R_{ig} \geq 0.25$ for all levels below the mixed layer.

After this mixing process is completed in step 3, one can go back to step 2 and recalculate the depth of the mixed layer based on R_{ib} criteria. And then the process continues till we arrive at a continuous variation of stability measure (in terms of R_{ib} and R_{ig} through the base of the mixed layer. Effectively, this process reduces the sharp gradients at the base of the mixed layer.

One easy way to think about the PWP model is the three-step process oriented toward redistribution and maintenance of stability between the layers: (i) static stability throughout the water column; (ii) mixed-layer stability by making sure $R_{ib} \geq 0.65$ within the mixed-layer; and (iii) shear-flow stability by making sure $R_{ig} \geq 0.25$ below the mixed-layer.

CONCLUSION

ocean is a turbulent fluid. This turbulence is pervasive over multiple eddy scales and cascades energy from larger eddies to molecular levels. Turbulence is quantified by Reynolds stresses which are related to the fluctuating velocity components about the mean flow. The non-linearity of the Reynolds stresses leads to a closure problem that is not yet solvable. The well-known Kolmogorov theory of turbulence was briefly discussed to highlight different approaches to closure modeling. Multiple ways of dealing with this closure problem exist, which have been briefly discussed here. This process is called eddy turbulence parameterization for viscosity and diffusivity. Relation to vertical and horizontal mixing of turbulent eddies are then discussed in the light of a one-dimensional mixed-layer model known as the PWP model. Different closure schemes and eddy parameterization techniques constitute the uniqueness of various ocean models, some of which will be discussed in the next chapter.

FURTHER READING

Batchelor, C. K., & Batchelor, G. (2000). *An introduction to fluid dynamics*. Cambridge University Press.

Cushman-Roisin, B., & Beckers, J. (2011). *Introduction to geophysical fluid dynamics: Physical and numerical aspects*. Elsevier.

Frisch, U., & Kolmogorov, A. N. (1995). *Turbulence: The legacy of A.N. Kolmogorov*, 296 pp. Cambridge University Press.

Haidvogel, D. B., & Beckmann, A. (1999). *Numerical ocean circulation modeling*. World Scientific.

Kolmogorov, A. N. (1991). The local structure of turbulence in incompressible viscous fluid for very large Reynolds numbers. *Proceedings of the Royal Society of London. Series A: Mathematical and Physical Sciences*, *434*(1890), 9–13.

Kundu, P., Cohen, I., & Dowling, D. (2012). *Fluid mechanics*, 920 pp. Academic Press, New York, NY.

Pedlosky, J. (2013b). *Ocean circulation theory*. Springer Science & Business Media.

Tennekes, H., & Lumley, J. L. (2018). *A first course in turbulence*. MIT Press.

Thorpe, S. A. (2007). *An introduction to ocean turbulence* (Vol. 10). Cambridge University Press, Cambridge.

Vallis, G. K. (2017). *Atmospheric and oceanic fluid dynamics*. Cambridge University Press.

Poipu Beach, Kaua'i

10 Multiscale Ocean Models

> No human investigation can be
> called real science if it cannot be
> demonstrated mathematically.
>
> ─────────────────────────
> Leonardo da Vinci (1452–1519) in
> Treatise of Painting (1651)

OVERVIEW

The last two chapters discussed the simple modeling architecture, high-lighted the finite difference method for numerical implementation and outlined the challenges of modeling the viscosity for a turbulent-fluid like ocean. This chapter describes a number of multiscale ocean models with their individual characteristics in terms of the adapted numerical solution schemes, turbulent closure models and vertical coordinate choices. Together, these triage of attributes define the kernel of a numerical model. We provide a list of models for easy reference first. Then we discuss the idea of coordinate choices from a generalized perspective. Finally, we describe a selected set of models with their equation set ups and discuss the attributes briefly.

10.1 A BRIEF BACKGROUND

Scales in both time and space matter. Processes occur in different spatial and temporal scales. These processes interact with each other over multiple scales in both time and space. A single process might also span over multiple time and spatial scales. So, how do you resolve such processes? What setup do you need to model these scales?

We have come to address these processes in various different setups and thus there have been a number of different ways to model the ocean. Numerical modeling of the ocean followed the field of numerical weather prediction (NWP) as envisioned by Richardson in 1922 in his futuristic Book (Richardson, 1922), later formalized by Charney et al., 1950, and then by Norman Phillips in the 1950s (Phillips, 1954, 1956). See Chapter 13 for the story of evolution of NWP. Numerical Ocean Modeling was first formalized by Bryan and Cox (Chapter 8) and then flourished from simple simulation models to advanced data-assimilative four-dimensional modeling systems in the 80s and 90s. With the explosive advances in computing technology in the last two decades, the computer models are essentially available as black-boxes (with examples to run,

DOI: 10.1201/9780429347221-10

verify, and apply to a user problem) and many of these models have become part of a larger community resource framework. Some of these newer ocean and atmospheric modeling systems are now being made available in Python, Amazon Web Services, and being ported through Google for Societal Good and Microsoft's AI platform for various applications (Bolton & Zanna, 2019; Krasnopolsky, 2007).

10.2   MULTISCALE MODELS

Due to the increased availability of high-performance computers (HPC) in the late 80s and early 90s, numerical ocean models became freely available to multiple groups worldwide. Development occurred at a very fast rate. HPC, cheaper data storage, cloud distribution, faster internet, all of these allow both observations and model simulations in real-time nowadays. Many of these ocean models are available in open-source architectures worldwide. It is thus important to realize and understand the basics of these models and their capabilities before we start using one of these as an opaque system. Three attributes of numerical models are: (i) numerical methods or the solution algorithms; (ii) turbulence and closure schemes and (iii) generalized vertical coordinates. (See Figure 10.1.) In Chapter 8, we discussed one of the many numerical methods (the finite difference method) in detail. We leave the reader to explore the other methods (FEM, FVM, Spectral methods) through more appropriate texts. In Section 9.1.1, we discussed the turbulence and closure schemes. In this chapter, we will first discuss the generalized vertical coordinates. We will then describe these characterizations for a number of modeling systems so that you can appreciate the difference between these models when you select them for modeling your own scientific problem.

Table 10.1 lists a number of general-purpose much-used three-dimensional ocean circulation models. These are listed in their chronological development order.

Table 10.1: An Example Set of Ocean Models

Model	Horizontal Coordinate	Vertical Coordinate	Reference	Other
MOM	(λ, θ) (lon,lat)	Z (Flat Levels)	Bryan (1969); Bryan and Cox (1972); Cox (1984)	Rigid-lid
HOPS	(x, y)	Single/Double Sigma (with One/Two Critical Levels)	Spall and Robinson (1989)	Rigid-lid
				Continued on next page

Table 10.1 – continued from previous page

Model	Horizontal Coordinate	Vertical Coordinate	Reference	Other
POM	(x,y)	Terrain Following (η included)	Blumberg and Mellor (1987)	Free Surface
POP	Similar to MOM		Dukowicz and Smith (1994)	Parallel Architecture
ROMS	(η, ζ)	Stretch Coordinate (s)	Song and Haidvogel (1994)	Free Surface
MITGCM	(x,y)	z-coordinate	Marshall et al. (1997a, 1997b)	Non-hydrostatic
HYCOM	$(\sigma, z,$ and ρ in Vertical)		Chassignet et al. (2007)	Global Operational Set Up
MSEAS	Similar to HOPS		Haley and Lermusiaux (2010)	Uncertainty prediction
Navy layered models	Layered Formulation		Wallcraft and Moore (1997)	Operationally Efficient (Layer ρ, Thick, Vol Trans)
FVCOM	Finite Volume		Chen et al. (2006)	Unstructured Grid
QUODDY	Finite Element Grid		Lynch-Werner (1987, 1991)	Regional Applications
MOM6	Finite Volume Core		Adcroft et al. (2019)	2020 Evolution from MOM
ADCIRC	2/3D unstructured FEM		Luettich et al. (2018)	Storm Surge, Tides, Larval Transport, etc.
SWAN	3rd-generation Wave Model		Booji et al. (1999)	Coastal and inland waters
DELFT3D	Boundary-fitted	σ	Baracchini et al. (2020)	Waves, Water Quality, Sediment Transport, etc.
ECOM3D	Similar to POM		Blumberg and Mellor (1987)	Estuarine Application with Sediment Transport

Other than the models listed above, there are Lagrangian models, oil-spill models, and interdisciplinary and climate models. The latter two types of models will be discussed in Chapter 12. We refer the reader to some of the specialized references for Lagrangian and oil spill models.

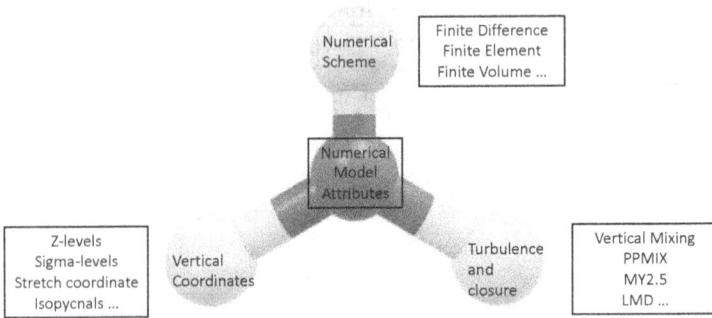

Figure 10.1: The chemistry of a numerical ocean model. Modeling attributes are shown as the three different elements of a complex oceanic system that are connected together so that the model (as a system) works seamlessly on a computer and solves the non-linear dynamics in a prognostic manner (with all oceanic variables).

10.3   GENERALIZED VERTICAL COORDINATES

As apparent from the list of ocean models in Table 10.1, one of the key difference between different models is the way they treat the vertical coordinate. Some use the height z, some use a non-dimensional sigma coordinate which is a function of z, some use the density ρ. In all cases, the new variable in the vertical is a monotonic function of the height z. So, it is worthwhile to learn about the generalization of the vertical coordinate and what happens to the equations of motion when we use a vertical coordinate other than the "z"-coordinate.

Consider a general vertical coordinate "r", which is a monotonic function of "z". To get the equations of motion in the (x,y,r,t) system, we need to make the variable z a function of (x,y,r,t). Effectively, $z(x,y,r,t)$ becomes a dependent variable in a system where the independent variable set is now (x,y,r,t) and not (x,y,z,t).

Any dependent variable $A(x,y,z,t)$ in the original coordinate system can now be described as a function of the form $A(x,y,z(x,y,r,t),t)$ in the new coordinate system. The new vertical partial derivative in the new coordinate system is given by the chain rule of calculus,

$$\frac{\partial A}{\partial r} = \frac{\partial A}{\partial z}\frac{\partial z}{\partial r} \qquad (10.1)$$

and similarly,

$$\frac{\partial A}{\partial z} = \frac{\partial A}{\partial r}\frac{\partial r}{\partial z} \qquad (10.2)$$

Now, a partial derivative with respect to any other independent variable s, where s could be any one of (x,y,t) can be written as

$$\left.\frac{\partial A}{\partial s}\right|_r = \left.\frac{\partial A}{\partial s}\right|_z + \frac{\partial A}{\partial z}\left.\frac{\partial z}{\partial s}\right|_r \tag{10.3}$$

$$\left.\frac{\partial A}{\partial s}\right|_r = \left.\frac{\partial A}{\partial s}\right|_z + \frac{\partial A}{\partial r}\frac{\partial r}{\partial z}\left.\frac{\partial z}{\partial s}\right|_r \quad \text{using equation 10.2} \tag{10.4}$$

We now use ∇ for the partial operator $\frac{\partial}{\partial s}$, i.e. $\nabla = (\frac{\partial}{\partial x}, \frac{\partial}{\partial y})$ in the new coordinate system. We also use vector notation to represent A by any velocity component $\vec{v}$ in the new coordinate system, i.e. $\vec{v} = (u, v, \Omega)$, where Ω is the velocity component in the (x, y, r) space. It is easy to see that equation 10.4 can be rewritten as

$$\nabla_z A = \nabla_r A - \frac{\partial A}{\partial r}\frac{\partial r}{\partial z}\nabla_r z \tag{10.5}$$

$$\nabla_z \cdot \vec{v} = \nabla_r \cdot \vec{v} - \frac{\partial \vec{v}}{\partial r}\frac{\partial r}{\partial z}\nabla_r z \tag{10.6}$$

Note that ∇_z in the above equation denotes horizontal gradients with respect to the "fixed z" coordinate. And similarly for ∇_r.

Now, let us see how the vertical velocity (w) gets transformed in the new vertical coordinate system.

$$w = \frac{dz}{dt} \tag{10.7}$$

$$= \frac{dz(x, y, r(x,y,z,t),t)}{dt} \tag{10.8}$$

$$= \left.\frac{\partial z}{\partial t}\right|_r + \left.\frac{\partial z}{\partial x}\right|_r \left.\frac{\partial x}{\partial t}\right|_r + \left.\frac{\partial z}{\partial y}\right|_r \left.\frac{\partial y}{\partial t}\right|_r + \left.\frac{\partial z}{\partial r}\right|_r \left.\frac{\partial r}{\partial t}\right|_r \tag{10.9}$$

$$= \left.\frac{\partial z}{\partial t}\right|_r + \vec{v}\cdot\nabla_r z + \dot{r}\frac{\partial z}{\partial r} \tag{10.10}$$

where the dot notation means $\frac{d}{dt}$.

The total derivative (Chapter 2) in z-coordinate can also be expressed in terms of derivatives in the "r" coordinate system as follows.

$$\frac{dA}{dt} = \frac{\partial A}{\partial t}\bigg|_z + \vec{v} \cdot \nabla_z A + w \frac{\partial A}{\partial z} \tag{10.11}$$

$$= \frac{\partial A}{\partial t}\bigg|_r + \vec{v} \cdot \nabla_r A + \left(w - \vec{v} \cdot \nabla_r z - \dot{r}\frac{\partial z}{\partial r} \right) \frac{\partial A}{\partial r}\frac{\partial r}{\partial z} \tag{10.12}$$

$$= \frac{\partial A}{\partial t}\bigg|_r + \vec{v} \cdot \nabla_r A + \dot{r}\frac{\partial A}{\partial r} \tag{10.13}$$

Here $\vec{v}$ is the horizontal velocity components and ∇_z is the horizontal gradient as before.

In the above, to arrive at equation 10.12 from 10.11, we used the relationship in equation 10.2 and added and subtracted similar terms. Then, to arrive at equation 10.13 we used the vertical velocity expression given in equation 10.10 with equation 10.12.

Note that the total derivative in the new coordinate system (x, y, r, t) has an extra term $\left(\dot{r}\frac{\partial A}{\partial r} \right)$ compared to the total derivative in (x, y, z, t) system given in equation 2.2 due to the dependence of r on z.

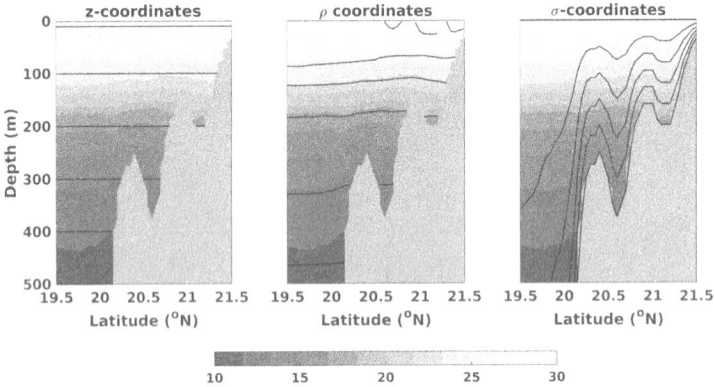

Figure 10.2: Levels for different vertical coordinates. Left – Flat staircase levels for z-coordinate. Middle – Density coordinate system levels following isopycnals. Right – sigma coordinate with terrain-following levels. The background is shaded with temperature field. This particular section was prepared from real observations for section ($19.5°$N–$21.5°$N) along the $87.5°$E longitude in the Bay of Bengal. Figure Credit: Sourav Sil.

A number of vertical coordinate representations are shown in Figure 10.2. Different models use different vertical coordinate systems (see later in this chapter).

Let us apply the above transformation to the hydrostatic equation and see what happens. The hydrostatic equation 2.30 is simply written as,

$$\frac{\partial p}{\partial z} = -g\rho \qquad (10.14)$$

transforming to the "r" coordinate

$$\frac{\partial p}{\partial r}\frac{\partial r}{\partial z} = -g\rho \qquad (10.15)$$

using the chain rule

$$\frac{\partial p}{\partial r} = -g\rho\frac{\partial z}{\partial r}$$

$$\frac{\partial p}{\partial r} = -\rho\frac{\partial gz}{\partial r} \qquad (10.16)$$

The horizontal pressure gradient terms in the momentum equations become

$$\nabla_r p = \nabla_z p + \frac{\partial p}{\partial z}\nabla_r z$$

$$\nabla_z p = \nabla_r p - \frac{\partial p}{\partial z}\nabla_r z$$

$$\nabla_z p = \nabla_r p + \rho\nabla_r gz \quad \text{using equation 10.16} \qquad (10.17)$$

One can use such derivations to arrive at newly transformed equations of motion for their own choice of coordinate system in the vertical (as well in the horizontal). The equations become more complicated than the simpler (x,y,z,t) system equations; but offer advantages for representing and resolving complex oceanic and atmospheric processes. So, such motivations have driven many groups to design new systematic transformations to uncover new dynamics of such complex processes. The result is a number of unique and elegant set mathematical representations of the equations of motion and their constraints and boundary conditions. These are our present-day set of different "Numerical Ocean Models." We discuss only a few of them in this chapter and refer the reader to advanced texts such as Haidvogel and Beckmann (1999), Kantha and Clayson (2000), and others (Hecht & Hasumi, 2013; Jacobson & Jacobson, 2005) listed in the Further Reading section for more.

10.4 HARVARD OCEAN PREDICTION SYSTEM (HOPS)

In the mid-to-late seventies two very important milestones in physical oceanography were reached. One was the appreciation that the vast ocean is full of eddies, a number of well-organized western boundary currents and a lot of finer-scale features in between. The second was the advent of satellite oceanography and recognition of the existence of such mesoscales in the open ocean which persist for days and weeks with slowly evolving dynamics. In other words, the multiscale nature of the ocean was visible, without going to the ocean and surveying and mapping the in-situ data – through the infra-red eyes of the satellites. (See Figure 1.12). In the early 80s, several groups realized the need for applying ocean models in a regional setting to understand mesoscale physical processes and for region-specific applications. The idea of predicting the ocean (much like weather) over a period of days to weeks seemed to have become a realistic goal to achieve. The Harvard Oceanography Group was one of the pioneers in developing a regional ocean model in Cartesian coordinates Spall and Robinson (1990). This modeling system is now known as the Harvard Ocean Prediction System (HOPS).

One of the primary requirements for regional modeling was that it ought to resolve the topography well so that the topographic interactions on both horizontal and vertical are well resolved. However, just increasing the horizontal and vertical resolution in a (x, y, z) modeling system would demand too much of computing power and even if that is possible, it might then create spurious gradients near the bottom, which would make simulations unrealistic because the actual bottom is more continuous than even a highly resolved grid. In other words, the idea was a topography following vertical coordinate transformation might actually be better than a highly resolved staircase topography. However, this is arguable as both approaches have their own pros and cons, which will not be discussed here.

First, let us briefly consider the evolution of the horizontal grid system. Figure 10.3 shows different configurations for gridding using the FDM. Early models had used the "A" grid, in which the velocity components, u and v were located at the same location as temperature. This configuration renders a relatively poor solution in terms of accuracy. The "B" grid has the velocity components at the vertices of the grid boxes with the tracer variables (t and s) in the center of the grid cell. This improved the accuracy of the solution, but became computationally extensive. The "C" grid has the u and v components at different locations for easier gradient computations across the grid boxes. The "D" and "E" grids are more advanced and the reader is encouraged to decipher them in more advanced texts or on their own.

Let us come back to the vertical coordinates. It was one of the major developments for a class of models that redefined the vertical coordinate in terms of topography. One of the first of these was the hybrid coordinate system as designed by Spall and Robinson (1990) as a simple transformation of variables in the vertical for all model levels below a prescribed depth called the "critical depth" z_c,

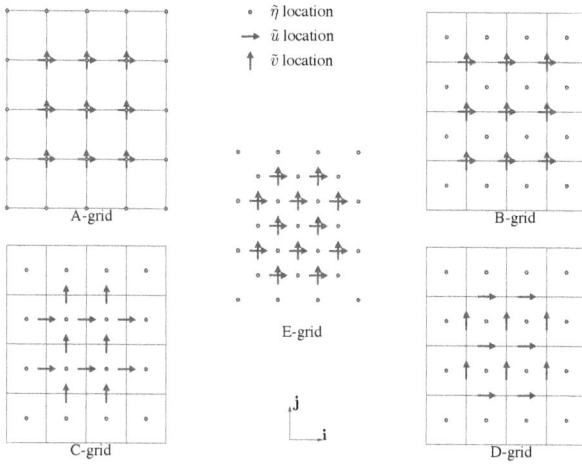

Figure 10.3: Arrangement of variables on different Arakawa grids. The five Arakawa grids. On the A–grid, the variables $\tilde{\eta}$, $\tilde{u}$, and $\tilde{v}$ are collocated, and staggered on other grids, called B–, C–, D– and E–grids. Note that the E–grid (center) has a higher grid-point density than the other grids for the same distance between adjacent nodes. Republished with permission of Elsevier, from Cushman-Roisin, B. (2009). *Introduction to geophysical fluid dynamics: Physical and numerical aspects*, Cushman-Roisin, Benoit, 2009; permission conveyed through Copyright Clearance Center, Inc.

$$\sigma(x,y) = \frac{z - z_c}{H'} \tag{10.18}$$

$$H' = H(x,y) - z_c, \; z_c \leq z \leq H(x,y) \tag{10.19}$$

In the above formulation, z increases downwards. See Figure 10.2 for an example of different representations of vertical coordinates (z, ρ, σ) for a particular section in the Bay of Bengal.

This redefinition of the vertical coordinate (σ) allows the model's surfaces to follow the bathymetry $(H(x,y))$ below the critical depth (z_c) and the surfaces above the critical depth to be horizontal.

Now, because of this redefinition, the primitive equations would change. Let us first see how a simple transformation from z-level to σ-level happen when we assume,

$$\sigma = \frac{z}{H(x,y)} \tag{10.20}$$

It is the simplest transformation in z. Note the difference of this formulation from that given by HOPS (equation 10.19). It just says that the new vertical coordinate σ is simply a non-dimensional representation of the depth (z). At the surface $z = 0$, $\sigma = 0$; and at the local bottom $z = H(x,y)$, $\sigma(x,y) = 1$. The bottom topography is a surface of unity in the σ coordinate system. All other model surfaces would follow the bathymetry $H(x,y)$. This is why it (σ) is also called the "terrain-following" coordinate system.

Using the development in Section 10.3, let us see what happens to the hydrostatic equation and the momentum equations.

In this case, the new coordinate, $r = \sigma = z/H$. So,

$$\frac{\partial r}{\partial z} = \frac{\partial \sigma}{\partial z} = \frac{1}{H}$$

$$\frac{\partial z}{\partial r} = H \qquad (10.21)$$

So, the hydrostatic equation using the σ transformation becomes,

$$\frac{\partial p}{\partial \sigma} = -\rho g \frac{\partial z}{\partial r} = -\rho g H$$

$$\text{or,} \quad \frac{\partial p}{\partial \sigma} + \rho g H = 0 \qquad (10.22)$$

Similarly, the momentum equations simply becomes (using equation developed earlier),

$$\frac{d\vec{v}}{dt} + f v + \frac{1}{\rho}\nabla_r p - \underbrace{\frac{1}{\rho}\nabla_r(gz)}_{-\frac{1}{\rho H}\frac{\partial p}{\partial \sigma}\nabla_\sigma z} = 0 \qquad (10.23)$$

So, with the Spall and Robinson (1990) σ coordinate redefinition (equation 10.19), and using what we learned in Section 10.3 about how momentum equations get modified as shown above (equation 10.23), the primitive equations become as follows,

$$u_t + \frac{u}{H'}(H'u)_x + \frac{v}{H'}(H'u)_y + w(u)_\sigma - fv = -\frac{1}{\rho_0}\left(P_x - g\sigma H'_x\rho\right) + F_m$$

(10.24)

$$v_t + \frac{u}{H'}(H'v)_x + \frac{v}{H'}(H'v)_y + w(v)_\sigma + fu = -\frac{1}{\rho_0}\left(P_y - g\sigma H'_y\rho\right) + F_m$$

(10.25)

$$P = H' \int_0^\sigma \rho\, d\sigma$$

(10.26)

$$\frac{1}{H'}(H'u)_x + \frac{1}{H'}(H'v)_y + w_\sigma = 0$$

(10.27)

$$T_t + \frac{u}{H'}(H'T)_x + \frac{v}{H'}(H'T)_y + w(T)_\sigma = 0$$

(10.28)

$$S_t + \frac{u}{H'}(H'S)_x + \frac{v}{H'}(H'S)_y + w(S)_\sigma = 0$$

(10.29)

$$\rho = \rho(T, S, \sigma)$$

(10.30)

Note that, if you set $H' = z_c$, then we get back the standard Cartesian primitive equations for depths above the critical depth.

The solution procedure was very similar to the one described in Chapter 8 for the GFDL MOM (Cox, 1984) model. Please see Spall and Robinson (1990) for details on the solution procedure. Over the last few decades, HOPS has been adopted and transformed by the group at MIT and is now part of a larger framework called the MIT Multidisciplinary Simulation, Estimation, and Assimilation System (MSEAS; Haley and Lermusiaux, 2010), which is discussed later in Section 10.9.

10.5   THE PRINCETON OCEAN MODEL (POM)

The Princeton Ocean Model (POM) was developed during the early to late eighties as well. It has a terrain-following coordinate system similar to HOPS, but with a free surface. Explicitly, the vertical σ coordinate was defined by

$$\sigma = \frac{z - \eta}{H + \eta}$$

(10.31)

where $D = H + \eta$ is the total local water depth, with $H(x,y)$ being the bottom topography and $\eta(x,y,t)$ is the surface elevation. So, σ varies from $\sigma = 0$ at the free surface where $z = \eta$ to $\sigma = -1$ at the bottom, where $z = -H$.

A full exposition of the governing equation can be found in the works of Phillips (1957) and Blumberg and Mellor (1980, 1987).

The model also uses curvilinear orthogonal coordinates in the horizontal and an "Arakawa C-grid" scheme like MOM6 (original MOM used the "B-grid") and HOPS in a finite difference formulation like discussed in Chapter 8.

The POM has been used by many groups around the world for its simplicity to run (it still runs as a Fortran code and allows the user to add features). It is presently archived by Dr Tal Ezer at Old Dominion University and has a wave sub-model (evolves the free surface) and wet-dry capability. These last two attributes allow the ocean model to be used for inundation and storm surge studies and to be linked/coupled with other coastal/estuarine hydrodynamics/sediment transport models.

10.5.1 VERTICAL MIXING IN POM

One of the most important contributions of the POM modeling system was its implementation of a second moment turbulent closure sub-model within the numerical scheme to compute the vertical mixing coefficients. This turbulent closure scheme was developed by Mellor (1973) and modified later by Mellor and Yamada (1974, 1982). This turbulence model has been important in simulating mixed layer dynamics in the upper ocean. This scheme has been later adopted in multiple modeling systems including ROMS and HyCOM.

10.5.1.1 Mellor-Yamada 2.5 Closure scheme

The Mellor-Yamada closure scheme is a hierarchy of closures that allows for increasing orders of complexity. The one that is more popular and implemented in POM, ROMS, and HyCOM is the one called "Level 2.5." This closure scheme is a two-equation closure model as defined in Section 9.1.1 before – the variables are the turbulent kinetic energy ($E = q^2/2$) and a product of TKE with a length scale ($q^2 l$). The one-dimensional TKE equation is given by:

$$\frac{D}{Dt}\left(\frac{q^2}{2}\right) - \frac{\partial}{\partial z}\left[K_q \frac{\partial}{\partial z}\left(\frac{q^2}{2}\right)\right] = P_s + P_b - \xi_d \qquad (10.32)$$

where P_s is the shear production, P_b is the buoyant production and ξ_d is the dissipation of TKE, given by

$$P_s = K_m \left[\left(\frac{\partial u}{\partial z}\right)^2 + \left(\frac{\partial v}{\partial z}\right)^2\right] \qquad (10.33)$$

$$P_b = -K_s N^2 \qquad (10.34)$$

$$\xi_d = \frac{q^3}{B_1 l} \qquad (10.35)$$

Here B_1 is a constant.

The equation for the product variable (TKE times l) is given by,

$$\frac{D}{Dt}(lq^2) - \frac{\partial}{\partial z}\left[K_l \frac{\partial lq^2}{\partial z}\right] = lE_1(P_s + P_b) - \frac{q^3}{B_1}W \qquad (10.36)$$

W is the wall proximity function:

$$W = 1 + E_2 \left(\frac{l}{\kappa L} \right)^2$$

$$L^{-1} = \frac{1}{\zeta - z} + \frac{1}{H - z} \tag{10.37}$$

Once you have determined q and l from the above, the vertical viscosity and diffusivities can be obtained as below (with a background ambiance):

$$K_m = qlS_m + K_{m_{\text{background}}}$$

$$K_s = qlS_h + K_{s_{\text{background}}} \tag{10.38}$$

$$K_q = qlS_q + K_{q_{\text{background}}}$$

The stability coefficients S_m, S_h, and S_q are obtained from a set of algebraic equations which are known functions of (l, N^2, q) and a number of predetermined constants. See Mellor and Yamada (1982), Cushman-Roisin and Beckers (2011), and ROMS website for these expressions.

10.6   REGIONAL OCEAN MODELING SYSTEM (ROMS)

One of the well-known ocean modeling systems today is the regional ocean modeling system (ROMS). It has a generalized vertical coordinate system that stretches the vertical coordinate (s) with depth so that the bottom essentially flattens and smooths out the sharp topographic gradients at $z = h(x, y)$. It also allows for better-resolved surface mixed layers even in the deeper oceanic regions.

The s-coordinate system was developed by Song and Haidvogel (1994), where the vertical coordinate z is a function of s as follows.

$$z = \zeta(1 + s) + h_c s + (h - h_c)C(s) \quad -1 \le s \le 0 \tag{10.39}$$

where h_c is a critical depth (either a minimum depth or a shallow depth) above which we more resolution is desired. Song and Haidvogel (1994) designed a form of C(s) as

$$C(s) = (1 - b)\frac{\sinh(\theta s)}{\sinh \theta} + b\frac{\tanh[\theta(s + \frac{1}{2})] - \tanh(\frac{1}{2}\theta)}{2\tanh(\frac{1}{2}\theta)} \tag{10.40}$$

Here "θ" and "b" are surface and bottom resolution control parameters. The usual ranges are $0 - 20$ for θ, and $0 - 1$ for b. It is easy to see from Equation

10.40 that for $s = 0$, $C(s) = 0$, and then from 10.40, $z = \zeta$. Similarly, for $s = -1$, $z = h$.

Some of the properties of this generalization are given below.

- If we let $\theta - > 0$, and use L'Hopital's rule ($limit_{x->0} \frac{f(x)}{g(x)} = limit_{x->0} \frac{f(x)}{g(x)}$), we get back the sigma coordinate equation from 10.39

$$z = (\zeta + h)(1 + s) - h \qquad (10.41)$$

or, $s = \frac{z - \zeta}{h + \zeta}$. This is the exact equation that is used for POM vertical coordinates as indicated earlier in equation 10.31.
- As θ increases, the resolution increases above h_c.
- For $b = 0$, the resolution increases to the surface only as $\theta - > 0$.
- For $b = 1$, the resolution is distributed between surface and bottom with increasing θ.

There are some other interesting and unique properties of this generalization and the reader is referred to the papers by Song and Haidvogel (1994), the book by Haidvogel and Beckmann (1999), and the ROMS user manual from their website given in Table 10.1. A good exercise is to program this function with real bathymetry for a section of your favorite ocean and see how the model levels changes with different choices of θ and b. The stretched coordinate formulation then modifies the primitive equations considerably. In a general sense, s is now a function of (x,y,z), and z is a function of (s,x,y).

Let us see what happens to the equations of motion. After substituting for appropriate derivatives, one can arrive at the following dynamical equations for ROMS

$$\frac{\partial u}{\partial t} + \vec{u} \cdot \nabla u - fv = -\frac{\partial \phi}{\partial x} - \left(\frac{g\rho}{\rho_0}\right)\frac{\partial z}{\partial x} - g\frac{\partial \zeta}{\partial x} + F_u + D_u \qquad (10.42)$$

$$\frac{\partial v}{\partial t} + \vec{u} \cdot \nabla v + fu = -\frac{\partial \phi}{\partial y} - \left(\frac{g\rho}{\rho_0}\right)\frac{\partial z}{\partial y} - g\frac{\partial \zeta}{\partial y} + F_v + D_v \qquad (10.43)$$

$$\frac{\partial T}{\partial t} + \vec{u} \cdot \nabla T = F_T + D_T \qquad (10.44)$$

$$\frac{\partial S}{\partial t} + \vec{u} \cdot \nabla S = F_S + D_S \qquad (10.45)$$

$$\rho = \rho(T, S, P) \qquad (10.46)$$

$$\frac{\partial \phi}{\partial s} = \left(\frac{-gH_z\rho}{s}\right) \qquad (10.47)$$

$$\frac{\partial H_z}{\partial t} + \frac{\partial H_z u}{\partial x} + \frac{\partial H_z v}{\partial y} + \frac{\partial H_z \Omega}{\partial z} = 0 \qquad (10.48)$$

where, the redefined velocity vector $\vec{u} = (u, v, \Omega)$ and the advection operator is given by

$$\vec{u} \cdot \nabla = u \frac{\partial}{\partial x} + v \frac{\partial}{\partial y} + \Omega \frac{\partial}{\partial \Omega} \tag{10.49}$$

In the above equations, the F and D terms are for forcing and diffusion for mass (T, S) and momentum (u, v).

Now, the vertical velocity in s-coordinate system is given by

$$\Omega(x, y, s, t) = \frac{1}{H_z} \left[w - (1+s) \frac{\partial \zeta}{\partial t} - u \frac{\partial z}{\partial x} - v \frac{\partial z}{\partial y} \right] \tag{10.50}$$

which also yields,

$$w = \frac{\partial z}{\partial t} + u \frac{\partial z}{\partial x} + v \frac{\partial z}{\partial y} + \Omega H_z \tag{10.51}$$

The last equation is generally used to convert from s-coordinate vertical velocity (Ω) to z-coordinate vertical velocity (w).

In the s-coordinate system, the boundary condition for the vertical velocity simplifies to $\Omega = 0$.

In the horizontal direction, ROMS also uses a boundary-following coordination system to better resolve the irregular coastal boundaries. This is done by introducing a curvilinear orthogonal coordinate transformation in the horizontal. The new horizontal coordinates $\varepsilon(x, y)$ and $\eta(x, y)$ (this η is not the vertical height as was for POM in the above section) are defined in such a way that the relationship between the horizontal arc length to the differential distance is given by:

$$(ds)_\varepsilon = \left(\frac{1}{m} \right) d\varepsilon (ds)_\eta = \left(\frac{1}{n} \right) d\eta \tag{10.52}$$

Note that $m(\varepsilon, \eta)$ and $n(\varepsilon, \eta)$ are the scale factors which relate the differential distances $(\triangle\varepsilon, \triangle\eta)$ to the physical arc lengths. See Batchelor (2000) for details of how to compute the operators curl, grad and div for such a transformation.

Once the above transformations are applied to the primitive equations, the final equations are obtained for the ROMS dynamical model. See the ROMS manual for details. They are similar to the ones above except in (ε, η) space with appropriate scale factors (m, n, mn) applied to appropriate terms. It is a worthwhile exercise for you to carry this out algebraically.

The numerical solution for ROMS is carried out on the Arakawa-C grid (Figure 10.3) and standard vertical gridding (in s-coordinate, but stretched in z-coordinate as in Figure 10.4).

The ROMS system has some additional unique improvements compared to its predecessors. Some of them are listed below.

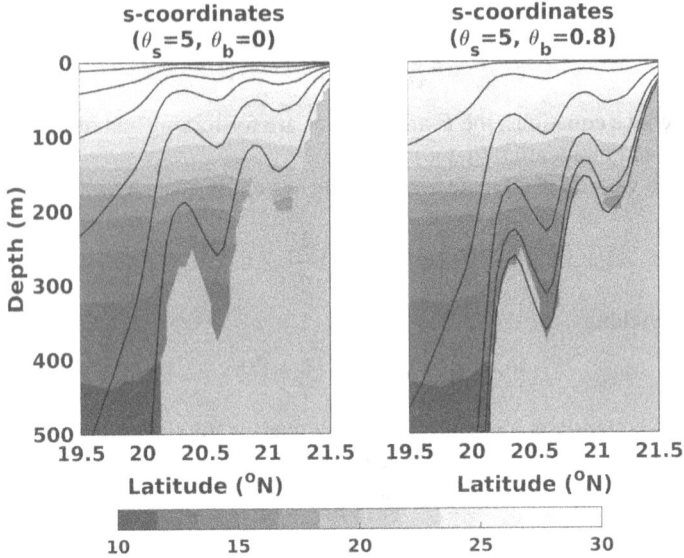

Figure 10.4: ROMS vertical coordinate with different s-coordinate parameters for the same section in Figure 10.2. Figure Credit: Sourav Sil.

- Several options for vertical mixing
  - Brunt-Vaisala frequency-based mixing $(1/N)$.
  - Large-McWilliams-Doney interior scheme
  - Pacanowsky-Philander closure scheme
  - Mellor-Yamada 2-level closure
  - Mellor-Yamada 2.5 level closure
  - Convective, double-diffusive, local K-profile mixing schemes
  - Bottom boundary layer (Styles and Glenn (2000) formulation)
- Masking for coastlines and islands
- River input
- Sea-ice module (Budgell, 2005)

Due to the complicated transformations and a number of approximations it is important to maintain the stability of solutions during time integration. At the same time, for computational efficiency, the primitive equations for momentum are solved using a split-explicit time-stepping scheme between barotropic and baroclinic computations. A number of fast barotropic time-stepping is done within a slowly evolving baroclinic time-step. Such computational advantage is accomplished by employing innovative numerical algorithm developments to preserve exact volume conservation and consistency preservation for the tracer equations by Shchepetkin and McWilliams (2003, 2005).

Additionally, ROMS has incorporated a number of data-assimilation methodologies within its broader distribution network. Some of these are mentioned below and are definitely worth further reading. We will discuss some basics of data-assimilation later in this book, but leave the more involved and advanced state of developments to the experts and their works below.

- Four-dimensional variational methods – Both strong and weak constraints are available (Arango et al., 2006; Di Lorenzo et al., 2006)
- Adjoint model sensitivities (Moore et al., 2004, 2006) – which provide four separate models and capabilities to couple to different interdisciplinary (biogeochemistry) and atmospheric models in a Earth System Modeling Framework (ESMF). Please see the ROMS website for details of these options.

10.6.1   VERTICAL MIXING IN ROMS

One of the most common closure schemes, the MY2.5 was discussed earlier as it evolved with the development of POM. Three other common vertical mixing schemes are briefly discussed below. These are all different closure schemes, the rationale of which was described in Section 9.1.1 earlier. Note that all of the equations below were needed to have been transformed for the ROMS stretched coordinate system in their implementation.

10.6.1.1   Philander and Pacanowski – PPMIX

This scheme, as originally used for MOM and has the value of being simple and instructive. Here we look into the vertical mixing parameters, namely, the coefficients of vertical eddy viscosity (ν) (in the momentum equations) and the vertical eddy diffusivity (κ) (in the temperature and salt equations). Experiments have shown that they vary anywhere between 1 and 100 cm^{2}s^{-1} in different oceans and also are very small within the depths of the ocean. Early measurements (Crawford and Osborn, 1979; Osborn and Bilodean, 1980) suggested that the mixing is strongly influenced by shears of the mean currents. This shear dependence was empirically fit to the eddy coefficients by Robinson (1966) and Jones (1973) as follows.

$$\nu = \frac{\nu_0}{(1 + \alpha R_i)^n} + \nu_b \tag{10.53}$$

$$\kappa = \frac{\nu}{(1 + \alpha R_i)} + \kappa_b \tag{10.54}$$

here, the Richardson number is

$$R_i = \frac{\beta g T_z}{\left(\frac{\partial u}{\partial z}\right)^2 + \left(\frac{\partial v}{\partial z}\right)^2} \tag{10.55}$$

and the coefficient of thermal expansion (β) for water is

$$\beta \approx 8.75(T+9) \times 10^{-6}.$$

In the above expressions, T is the potential temperature ($^\circ C$), u and v are the horizontal velocity components, g is gravity, v_b and κ_b are the background viscosity and diffusion, and v_0, α, and n are adjustable parameters.

It was in the eighties that Pacanowski and Philander (1980) had used the above formulation in an equatorial model of thermocline and showed how a choice of $\kappa = 0.1$ cm^{2}s^{-1} could sustain a thermocline after 2 years of integration instead of a value of $\kappa = 1.0$ cm^{2}s^{-1} which would spread the thermocline at mid-depth reducing the strength of the equatorial undercurrent. With a set of values of $v_0 = 50$ cm^{2}s^{-1}, $n = 2$, $\alpha = 5$, they were able to get the best results for the equatorial Pacific using a MOM model with a (40 km $\times$ 70 km) resolution.

Later developments using a similar mixing schemes as the PPMIX in many other oceans have yielded successful simulations for particular processes; however a global value does not yield an overall satisfactory performance, neither do we expect it for such a non-linear turbulent fluid system. Efforts are underway for investigating a variable range of v and κ formulation in different oceans and for different regions.

In the meantime, a number of other ways of parameterizing the vertical mixing have been developed, which include turbulence closure scheme (Mellor-Yamada; discussed earlier), subgridscale parameterization, Large Eddy Simulations (LES) etc. These are discussed in more detail by Kantha and Clayson, 2000 and briefly outlined below. It is instructive for the student to go back to the earlier section on POM and reread the vertical mixing parameterization outlined for POM (the MY2.5 scheme) now to understand the continuity of development of vertical mixing schemes.

10.6.1.2   K-Profile Parameterization – LMD94

The ROMS setup implemented the KPP scheme developed by Large et al. (1994) (LMD94), which uses separate parameterization schemes for turbulence closure in the surface boundary layer (SBL) and within the interior below. The boundary layer depth (h_{SBL}) is first calculated using a boundary layer similarity theory. This parameterization is finally matched at the interior with a separate closure scheme that accounts for effects of local shear, internal waves, and double-diffusion.

For the layers (σ) within the SBL, the viscosity and diffusivity coefficients are assumed to be a product of the length scale, h_{SBL} and the turbulent velocity scale $u_x(\sigma)$ and a non-dimensional shape function $G_x(\sigma)$. The subscript "x" denotes the variable of interest (momentum or temperature or salinity). Mathematically, this formulation is:

$$A_x(\sigma) = h_{SBL}u_x(\sigma)G_x(\sigma) \qquad (10.56)$$

For details on how to calculate the above terms, please see Large et al. (1994) and the ROMS website for Vertical mixing.

The interior scheme is based upon adding the different effects of different turbulence-generating mechanisms. These are shear mixing (based on gradient Richardson number), double diffusion and internal wave (treated as a constant background). For the specific formulation and implementation details please see the ROMS website.

10.6.1.3   Generic Length Scale

There is a more general two-equation turbulence closure model developed by Umlauf and Burchard (2003). This generic model was implemented by Warner et al. (2005) for ROMS and can be tuned to behave like the MY2.5 scheme or other two-equation closure models including $(E - El; E - \varepsilon; E - \omega)$. Note that the TKE is denoted by E here, in many other books, TKE is denoted by the symbol κ, which is used for wavenumber here. Please see the ROMS website and the references above for details.

10.7   MITGCM

One of the more futuristic developments in ocean model in the twentieth century was the development of a general circulation model (GCM) at MIT. This model was developed by Marshall et al. (1997a, 1997b). It uses the full incompressible, Bousinesq, non-hydrostatic equations on a sphere. A major advancement was the implementation of the non-hydrostatic vertical momentum equation compared to its predecessors. It required the solution algorithms to be computationally demanding which was met by elegant and sophisticated numerical algorithms (matrix solvers) in a massively parallel processing architecture.

Here we outline the full three-dimensional primitive equations below. Compare the equations developed and discussed throughout the text from Chapter 2 through now (and even later for HYCOM) with the ones below to see and think about what and why approximations are made to these governing equations to simplify and develop particular ocean models applicable to a particular oceanic region, process and purpose. All of the models are solving the nonlinear Navier-Stokes Equations, with different approximations and algorithms and geometries.

These equations were originally given by Charney and Flierl (1981) in spherical coordinates and are repeated below.

$$\frac{1}{(R+z)\cos\theta}\frac{\partial u}{\partial\phi} + \frac{1}{(R+z)\cos\theta}\frac{\partial}{\partial\theta}(v\cos\theta) +$$
$$\frac{1}{(R+z)^2}\frac{\partial}{\partial z}\left[(R+z)^2 w\right] = 0 \quad (10.57)$$

$$\frac{du}{dt} - \frac{uv\tan\theta}{(R+z)} - 2\Omega v\sin\theta + 2\Omega w\cos\theta + \frac{uw}{(R+z)} =$$
$$-\frac{1}{(R+z)\cos\theta}\frac{\partial p}{\partial\phi} + \frac{\partial}{\partial z}\left(K_M\frac{\partial u}{\partial z}\right) +$$
$$A_M\left[\nabla^2 u + \frac{(1-\tan^2\theta)u}{(R+z)^2} - \frac{2\sin\theta}{(R+z)^2\cos^2\theta}\frac{\partial v}{\partial\phi}\right] \quad (10.58)$$

$$\frac{dv}{dt} - \frac{u^2\tan\theta}{(R+z)} - 2\Omega u\sin\theta + \frac{vw}{(R+z)} =$$
$$-\frac{1}{(R+z)}\frac{\partial p}{\partial\theta} + \frac{\partial}{\partial z}\left(K_M\frac{\partial v}{\partial z}\right) +$$
$$A_M\left[\nabla^2 v + \frac{(1-\tan^2\theta)v}{(R+z)^2} + \frac{2\sin\theta}{(R+z)^2\cos^2\theta}\frac{\partial u}{\partial\phi}\right] \quad (10.59)$$

$$\frac{dw}{dt} - \frac{u^2}{(R+z)} + 2\Omega u\cos\theta - \frac{v^2}{(R+z)} =$$
$$-\frac{\partial p}{\partial z} - \frac{\rho}{\rho_0}g + \frac{\partial}{\partial z}\left(K_M\frac{\partial w}{\partial z}\right) +$$
$$A_M\left[\nabla^2 w + \frac{(1-\tan^2\theta)v}{(R+z)^2} - \frac{2\sin\theta}{(R+z)^2\cos^2\theta}\frac{\partial u}{\partial\phi}\right] \quad (10.60)$$

where the total derivative of any variable X is expressed by

$$\frac{dX}{dt} = \frac{\partial X}{\partial t} + \frac{1}{(R+z)\cos\theta}\frac{\partial}{\partial\phi}(uX) + \frac{1}{(R+z)\cos\theta}\frac{\partial}{\partial\theta}(v\cos\theta X) +$$
$$\frac{\partial}{\partial z}(wX) = 0 \quad (10.61)$$

The Laplacian $\nabla^2 X$ is given by

$$\nabla^2 X = \frac{1}{(R+z)^2 \cos^2 \theta} \frac{\partial^2 X}{\partial \theta^2} + \frac{1}{(R+z)^2 \cos \theta} \frac{\partial}{\partial \theta} \left(\cos \theta \frac{\partial X}{\partial \theta} \right) \quad (10.62)$$

The equation of state is simply in terms of potential temperature,

$$\rho = \rho(\Theta, S, p) \quad (10.63)$$

and the tracer equations in (Θ, S) can be given by

$$\frac{d\Theta}{dt} = \frac{1}{\rho_0 c_p} \frac{\partial I_s}{\partial z} + \frac{\partial}{\partial z} \left(K_H \frac{\partial \Theta}{\partial z} \right) + A_H \nabla^2 (\Theta) \quad (10.64)$$

$$\frac{dS}{dt} = \frac{\partial}{\partial z} \left(K_H \frac{\partial S}{\partial z} \right) + A_H \nabla^2 (S) \quad (10.65)$$

where, I_s is the solar insolation, and the rest of the symbols have their usual meaning used in this book previously. Note that the equations use a "modified spherical coordinate," in that the radial coordinate is replaced by $(R+z)$ instead of R, ϕ is the longitude and θ is the latitude.

In their non-hydrostatic model implementation, Marshall, et al. (1997) started with the above set of equations and then neglected the tendency, advective and diffusion terms for the vertical velocity. The meridional component of the Coriolis acceleration was retained in the "z-momentum" equation. This makes the pressure "p" a prognostic variable to be solved at every time step, which is given by its Laplacian,

$$\nabla^2 p = \xi_p; \text{ and } \nabla p \cdot n = 0 \quad (10.66)$$

Note the procedural difference of this solution setup with those in MOM, HOPS, POM and ROMS discussed earlier.

The RHS of this equation 10.66, or the term ξ_p is complicated and left for the reader to look up Marshall et al. (1997a). In fact, solving this equation is computationally demanding and a clever decomposition (see Marshall et al., 1997a, 1997b) into three pressure components was done. These are: (i) the rigid-lid pressure p_s; (ii) the hydrostatic pressure, p_h; and (iii) the non-hydrostatic pressure, p_{nh}. The hydrostatic pressure can be easily solved from the standard diagnostic hydrostatic equation. The surface pressure or the pressure with rigid-lid can be solved by the usual barotropic set up, solving the two-dimensional Poisson equation (the external mode in some of the previous models discussed). The non-hydrostatic part, p_{nh} can be solved in a three-dimensional Poisson equation, which is more complicated. Please see Marshall et al. (1997a, 1997b) and Kantha and Clayson (1995), and Haidvogel and Beckmann (1999) for a more detailed discussion on their solution procedure.

It is sufficient to end the discussion on the MITGCM here by stating that it is one of the more popular models when a non-hydrostatic model is required for process representation. It also offers unique computational advantages due to its more advanced algorithms for matrix solving and can be run in both hydrostatic and non-hydrostatic modes, separately or in a combined time-delayed and nested regional modes. MITGCM also has components that do data assimilation and creates reanalysis fields or Ocean State Estimations such as ECCO (estimating the circulation and climate of the ocean) and SOSE (Southern Ocean State Estimate) (see MITGCM website for details).

10.8   THE HYBRID COORDINATE MODEL (HYCOM)

For an adiabatic fluid, the density (actually, potential density) is conserved. Furthermore, since density increases monotonically with depth, it makes sense to treat the density ρ as the vertical coordinate. This system (x, y, ρ) is called the "isopycnal" coordinate system. It has some specific advantages, which will become clearer as we delve into the equations. The Hybrid Coordinate system of models (HYCOM) uses all three coordinates, z, σ, and ρ in different parts of the ocean to resolve the coastal regions, mixed layer, and isopycnal layers to the bottom. See hycom.org for more details.

Following Section 10.3, we can now set $r = \rho$. Let us look at the continuity equation first. It becomes (the reader is encouraged to do this themselves):

$$\partial_t \partial_\rho z + \nabla_r \cdot \partial_\rho z \vec{v} + \frac{\partial}{\partial \rho}(\partial_\rho(z\dot{\rho})) = 0 \qquad (10.67)$$

Here, $\partial_\rho = \frac{\partial}{\partial \rho}$.

Note that $\dot{\rho}$ is non-zero only when diabatic terms (which ones? – reader to think) force the flow to cross isopycnals. This means that the continuity equation can serve as a "layer thickness equation" of constant density!

The hydrostatic equation becomes the following.

$$\frac{\partial p}{\partial \rho} + \rho \frac{\partial}{\partial \rho} gz = \frac{\partial p}{\partial \rho} + \frac{\partial}{\partial \rho} \rho gz - gz$$

$$= \frac{\partial}{\partial \rho} \rho M - gz = 0 \qquad (10.68)$$

where, $M = p/\rho + gz$, called the "Montgomery potential." Look at its similarity with the idea of specific volume anomaly discussed in Chapter 2.

The momentum equations become (in vector form)

$$D_t \vec{v} + fV + \frac{1}{\rho}\nabla_\rho\, \rho M = 0 \qquad\qquad (10.69)$$

Isopycnal models are ideal for lateral transfer processes in the absence of adiabatic processes. The topography is represented generally by a piecewise-linear form with vanishing layer thicknesses, which provides a very smooth model topography.

Much like ROMS, HYCOM has also implemented similar vertical mixing schemes (MY2.5 and LMD being the common ones). Note that all of those closure equations above were needed to have been transformed for the HYCOM hybrid coordinate system in their implementation. Additionally, HYCOM has the option of using PWP for its surface mixed layer, if the user chooses so. Please see the HYCOM website for details on these special numerical implementations.

10.9   MIT – MSEAS

The MIT Multidisciplinary Simulation, Estimation, and Assimilation System (MSEAS; Haley and Lermusiaux, 2010, Haley Jr et al., 2015) is the present evolution of the HOPS multiscale modeling apparatus for simulating multi-resolution 2-way nested re-analyses fields. MSEAS modeling capabilities include implicit two-way nesting for realistic tidal-to-mesoscale dynamics with hydrostatic primitive equations (PEs) and a nonlinear free-surface (Leslie et al., 2008) as well as a finite-volume code for non-hydrostatic dynamics and incubation of new methods (Ueckermann & Lermusiaux, 2012).

This system has been used for fundamental research and for realistic tidal-to-basin-scale primitive-equation simulations in varied ocean regions (e.g. Leslie et al., 2008; Onken et al., 2008). Ocean applications include mesoscale and regional dynamics (Gangopadhyay et al., 2011; Lermusiaux et al., 2011), internal tide interactions (Kelly & Lermusiaux, 2016; Kelly et al., 2016), ecosystem dynamics (Lermusiaux et al., 2011), monitoring (Lermusiaux, 2007), and acoustics (Duda et al., 2019; Lam et al., 2009; Xu et al., 2008).

Recent developments focused on MSEAS implementation for adaptive modeling, adaptive data assimilation, and adaptive sampling following the ideas presented by Lermusiaux (2007) toward building an intelligent Autonomous Ocean Observing System (Lermusiaux et al., 2017). Adaptive modeling can be viewed as the intelligent allocation of computational resources over the space of stochastic dynamical models. Adaptive sampling considerations are complex and might have several choices based on availability of past observations and evolution of the dynamical system. One of the hallmarks of the MSEAS development was to design a set of "Dynamically Orthogonal" (DO) primitive equations, which are used for optimum path planning for autonomous vehicles (Subramani et al., 2017) given a background flow field as well as being adaptive to the evolving flow field. The DO evolution equations for uncertainty prediction in an evolving stochastic model subspace allows for estimating joint Bayesian inference of the state, equations, geometry, boundary conditions and initial conditions. Such developments are at the cutting-edge of applying Bayesian learning of stochastic dynamics models to incorporate into both adaptive modeling and adaptive sampling for future ocean modeling and prediction exercises in real-time. To learn more about this evolving system please see the website at http://mseas.mit.edu/.

10.10 MOM6

MOM6 is the latest evolution of the original MOM from GFDL over the last five decades. The model equations are the layer-integrated vector-invariant form of the hydrostatic primitive equations (either Boussinesq or non-Boussinesq) (Adcroft et al., 2019; Griffies, 2018). Effectively, MOM6 implements the flux-form equations of motion in height coordinates as one vector-invariant momentum equation, thickness (continuity) equation, two tracer equations (potential temperature and salinity) and the equation of state. For details of the mathematical formulation of MOM6, please see Adcroft et al. (2008), Adcroft et al. (2019), Adcroft and Hallberg (2006), and White et al. (2009).

MOM6 uses the structured Arakawa C-grid for the spatial staggering of variables (see Section 10.4). MOM6 also allows for the horizontal grids to be spherical, tripole, regional, or cubed sphere. The vertical coordinate is Lagrangian in that the interfaces between the layers are free to move up and down with time. The interfaces have target densities or depths depending on your choice of the vertical coordinate (ρ, or z). The finite volume core uses an arbitrary Lagrangian-Eulerian (ALE) framework with general vertical coordinates (Bleck, 2002; White et al., 2009). MOM6 also uses the Piecewise Parabolic Method (PPM) for the advection scheme (Carpenter Jr et al., 1990; Colella & Woodward, 1984). PPM allows for a variable to be uniquely prescribed within the cell by its two edge values and be conservative. For more details please see the references and the website mentioned in this section.

In MOM6, it is common to have at least four different timesteps (from more frequent to less frequent): (i) the barotropic timestep (which solves the 2-d linear momentum and integrated continuity equations); (ii) the baroclinic (momentum dynamics) timestep (which solves the Lagrangian dynamics with 3-d stacked shallow water equations); (iii) the tracer timestep (which solves the column physics, or the tracer advection, thermodynamics and mixing); and (iv) the remapping interval (which remaps all variables in the ALE framework to bring the vertical coordinate back to target specification). There can also be a forcing timestep on which model coupling occurs. For this model's code availability, installation procedure, and implementation details, please see the MOM6 documentation website at https://mom6.readthedocs.io/en/dev-gfdl/about.html.

CONCLUSION

In this chapter, we described a number of multiscale ocean models including MOM, HOPS, POM, ROMS, and HyCOM with their individual characteristics in terms of the adapted numerical solution schemes, turbulent closure models, and vertical coordinate choices. Four different vertical coordinate systems (terrain following – σ, hybrid, stretched – s and density – ρ) are discussed. Three different turbulence closure schemes (MY 2.5, PPMix, and K-Profile parameterization) are discussed. This completes the discussion on modeling at this level. The curious mind is requested to look into more advanced texts and modeling group websites for older research papers and keep looking for state-of-the-art work in model attributes that are constantly evolving for the better.

FURTHER READING

Cushman-Roisin, B., & Beckers, J. (2011). *Introduction to geophysical fluid dynamics: Physical and numerical aspects.* Elsevier.

Di Lorenzo, E., Moore, A. M., Arango, H. G., Cornuelle, B. D., Miller, A. J., Powell, B., Chua, B. S., & Bennett, A. F. (2007). Weak and strong constraint data assimilation in the inverse regional ocean modeling system (ROMs): Development and application for a baroclinic coastal upwelling system. *Ocean Modelling, 16*(3–4), 160–187.

Haidvogel, D. B., Arango, H., Budgell, W. P., Cornuelle, B. D., Curchitser, E., Di Lorenzo, E., Fennel, K., Geyer, W. R., Hermann, A. J., Lanerolle, L., et al. (2008). Ocean forecasting in terrain-following coordinates: Formulation and skill assessment of the regional ocean modeling system. *Journal of Computational Physics, 227*(7), 3595–3624.

Haidvogel, D. B., & Beckmann, A. (1999). *Numerical ocean circulation modeling.* World Scientific.

Hecht, M. W., & Hasumi, H. (2013). *Ocean modeling in an eddying regime* (Vol. 177). John Wiley & Sons.

Jacobson, M. Z., & Jacobson, M. Z. (2005). *Fundamentals of atmospheric modeling.* Cambridge University Press.

Kantha, L. H., & Clayson, C. A. (2000). *Numerical models of oceans and oceanic processes.* Elsevier.

Large, W. G., McWilliams, J. C., & Doney, S. C. (1994). Oceanic vertical mixing: A review and a model with a nonlocal boundary layer parameterization. *Reviews of Geophysics, 32*(4), 363–403.

Marshall, D. P., & Tansley, C. E. (2001). An implicit formula for boundary current separation. *Journal of Physical Oceanography, 31*(6), 1633–1638.

Mellor, G. L., & Yamada, T. (1982). Development of a turbulence closure model for geophysical fluid problems. *Reviews of Geophysics, 20*(4), 851–875.

Moore, A. M., Arango, H. G., Di Lorenzo, E., Cornuelle, B. D., Miller, A. J., & Neilson, D. J. (2004). A comprehensive ocean prediction and analysis system based on the tangent linear and adjoint of a regional ocean model. *Ocean Modelling*, 7(1–2), 227–258.

Spall, M. A., & Robinson, A. R. (1989). A new open ocean, hybrid coordinate primitive equation model. *Mathematics and Computers in Simulation, 31*(3), 241–269.

Spall, M., & Robinson, A. (1990). Regional primitive equation studies of the GS meander and ring formation region. *Journal of Physical Oceanography, 20*, 985–1016.

Hanakapi'ai Beach, Kaua'i

11 Simulation and Prediction

> When a butterfly flutters its wings
> in one part of the world, it can
> eventually cause a hurricane in
> another.

Edward Norton Lorenz
(1917–2008)*

OVERVIEW

In the last three chapters, we discussed a number of important attributes of multiple numerical models including MOM, HOPS, POM, ROMS, HYCOM, and MITGCM. We realize that three main elements of numerical models are (i) the numerical algorithm; (ii) the turbulence closure scheme; and (iii) the vertical coordinate system.

In this chapter, we will discuss the process of simulation and prediction using such numerical models. Two important aspects are: (i) grids and domain set up and (ii) initialization schemes. These are discussed in detail here. Techniques of optimal interpolation (OI) and data assimilation are briefly outlined which are required for both simulation and prediction. Three example simulations for North Atlantic, Bay of Bengal, and Brazil Current System are presented.

We finally discuss the art of prediction and the challenge of predicting a non-linear system such as an ocean basin or a coastal region in some detail. Examples of predictions using numerical models by various agencies that are being used for monitoring and other applications are mentioned in closing this chapter.

11.1 CONTEXT OF SIMULATION AND PREDICTION

Once you identify a model and a process to address, you need to accurately compute the system's response (for that process) using the dynamical numerical model to a set of inputs. There are typically two ways that are followed by researchers in oceanography: (i) simulation and (ii) prediction. Both use the modeling framework. However, let us define their characteristics to help set the stage for the next few chapters.

---

*Credit/ Source: AZQuotes.com, Wind and Fly LTD, 2021. https://www.azquotes.com/quote/1077951, accessed September 22, 2021.

DOI: 10.1201/9780429347221-11

A simulation computes the model response (output) using the input data and initial conditions.

A prediction computes the model response at some specific time in the future using the current as well as the initial condition.

In this definition, a simulation is characterized by the fields it generates over a given length of time utilizing all the information of its forcing and boundary conditions during the simulation time. A simulation provides the best field estimate from the dynamical adjustment of the known observations and known forcing during the simulation period. In contrast, the prediction does not have the information of its forcing and boundary conditions from the future. So, it is a combination of four elements that provides the future field estimates: (i) the dynamical model's capability to adjust the observations with forcing at initialization; (ii) the appropriateness of the dynamical model to represent the process (and the variables representing the responses); (iii) the predictability of the natural system under consideration; and (iv) the fidelity of the forcing fields which are being used to force the ocean model.

Both simulation and prediction need the initial condition (primitive variables at the initial time) to start the prognostic dynamical model system. Thus initialization is a critical need for both a successful simulation and prediction. It is important to realize that "Ocean prediction is an initial value problem." However, simulation is a "data and model adjusted field estimation problem." A number of times in literature, the boundary between simulation and prediction becomes blurred and it is good to keep in mind the different elements that effect the "model run." A model run, be it simulation or prediction, follows a number of steps. An organizational flowchart for a typical model run is shown in Figure 11.1.

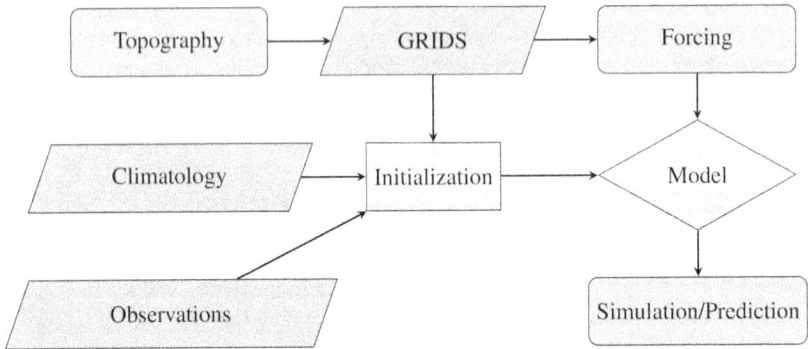

Figure 11.1: A typical organizational chart for an ocean model run.

11.2   GRID AND MODEL SETUP

Once you have selected a model, you need to select a region or a basin appropriate for your problem. A model setup requires a thoughtful choice of horizontal and vertical grid spacing and time-steps of integration to resolve the process or phenomenon of interest. The space-time coverage of the processes and the model's ability to represent and resolve such processes should match. A good literature survey on the regional models in the region of interest and their process studies or previous simulations would help the reader to arrive at some possible alternative modeling systems to choose from.

Once some ideas on these basic parameters ($\triangle x$, $\triangle y$, and $\triangle t$) are formed, it is time to set up the "model domain" with bathymetry and coastline. There are excellent choices for a basic bathymetric dataset from the ETOPO database prepared and made available by Smith and Sandwell, 1997. These are now available at different resolutions, for 5 minutes (ETOPO5), 2 minutes (ETOPO2), and 1 minute (ETOPO1). Each model allows its own GRIDS (or its equivalent) routine to have the user perform the following operations:

- set up the horizontal extent and resolution,
- define the vertical discretization,
- extract topography from a gridded, evenly spaced data set,
- extract multiple sets of nested topography from the parent for nested domain configurations,
- condition the topography for improved simulations.

This initial step is what defines the model configuration – spherical or *x-y-z* coordinate, sigma or density or hybrid coordinate, number of levels, and the model's ability to resolve mesoscale (or other scales) and the computational demand of the planned simulation. It is critical to spend some time and to iterate on the choice of configuration before running the final simulations. More often than not, a configuration and the initial condition (described next) and the boundary condition set are run for a few time-steps to see how the model is adjusting to the mass and velocity fields and the system is in a proper numerically stable state. Some of these numerical issues and adjustment problems are discussed in detail by more advanced books on numerical modeling (Haidvogel & Beckmann, 1999; Kantha & Clayson, 2000). The user manuals of each modeling group website also discuss these choices and sometimes offer default choices to move ahead with a process study.

11.3   MODEL INITIALIZATION

After the model is set up, all of the primitive variables (T, S, ρ, p, u, v, w) have to be assigned at every model grid point at time $t = 0$. This is the initial condition of mass and momentum that a model needs to carry forward the integration

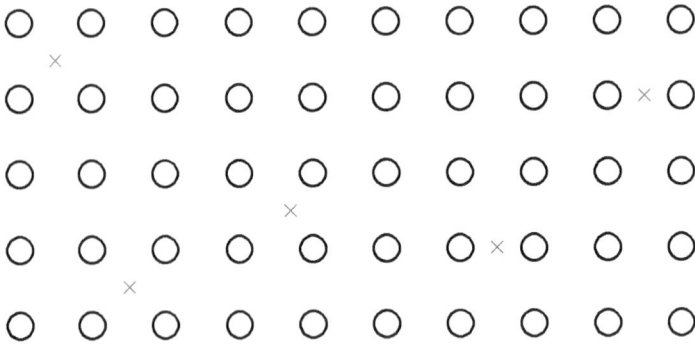

Figure 11.2: A typical initialization mapping issue is to go from an irregular set of observations to a regular gridded field.

using the model equations for each simulation time-step. It is often the situation that we do not have enough observations to map the three-dimensional mass or momentum fields at the initial time See Figure 11.2 for a typical situation of the need for mapping irregularly spaced observation on a regular model grid. Remember the first model of the Indian Ocean by Cox, 1970? He used a uniform temperature and salinity field with vertical variation based on a few observations. He also used a seasonally varying wind forcing. His grid size was $4°$, with seven vertical levels. He started from such a mass field with zero motion ($u = v = w = 0$) and he had to integrate for over 200 years of simulation time for the fields to adjust and produce some currents. More details are left for the reader to explore from his paper. It is amazing that back in 1970, with so little computing power and so few observations being available, his model simulations produced a seasonally reversing Somali Current as a vindication of the idea of numerical modeling to study the oceanic features of circulation. This success was primarily due to the formulation of equations of motion and novel design of the numerical methodology employed to solve the prognostic system, as shown in Figure 8.4 and discussed in Chapter 8 for the first ocean model (MOM).

We have come a long way in the last five decades (1970–2020).

11.3.1 CLIMATOLOGY

Beginning in the early 80s, we have access to a gridded product called climatology. Let us explore the evolution of this product.

The World Ocean Atlas (WOA) is a data product of the Ocean Climate Laboratory of the National Oceanographic Data Center (U.S.). The WOA consists of a climatology of fields of in situ ocean properties for the World

Ocean. It was first produced in 1994 (Levitus & Boyer, 1994) (based on the earlier Climatological Atlas of the World Ocean, Levitus, 1982), with later editions at roughly 4-year intervals in 1998 (Antonov, 1998), 2001 (Conkright et al., 2002), 2005 (Boyer et al., 2005), 2009 (Levitus et al., 2010), 2013 (Levitus et al., 2013) and 2018 (Locarnini et al., 2018).

The fields that make up the WOA dataset consist of objectively analyzed global grids at $1°$ spatial resolution. The fields are three-dimensional, and data are typically interpolated onto 33 standardized vertical intervals from the surface (0 m) to the abyssal seafloor (5500 m). In terms of temporal resolution, averaged fields are produced for annual, seasonal, and monthly time scales. The WOA fields include ocean temperature, salinity, dissolved oxygen, apparent oxygen utilization (AOU), percent oxygen saturation, phosphate, silicic acid, and nitrate. Early editions of the WOA additionally included fields such as mixed layer depth and sea surface height.

In addition to the averaged fields of ocean properties, the WOA also contains fields of statistical information concerning the constituent data that the averages were produced from. These include fields such as the number of data points the average is derived from, their standard deviation and standard error. A lower horizontal resolution ($5°$) version of the WOA is also available. The WOA dataset is primarily available as compressed ASCII, but since WOA2005, a netCDF version has also been produced. Standardized intervals are at 0, 10, 20, 30, 50, 75, 100, 125, 150, 200, 250, 300, 400, 500, 600, 700, 800, 900, 1000, 1100, 1200, 1300, 1400, 1500, 1750, 2000, 2500, 3000, 3500, 4000, 4500, 5000, 5500 m.

The World Ocean Database (WOD) is a collection of scientifically quality-controlled ocean profile and plankton data that includes measurements of temperature, salinity, oxygen, phosphate, nitrate, silicate, chlorophyll, alkalinity, pH, pCO_2, TCO_2, Tritium, ^{13}C, ^{14}C, ^{18}O, Freon, Helium, ^{3}He, Neon, and plankton (Locarnini et al., 2018). The WOD was first conceived as a way to provide reproducibility for the WOA series of gridded fields. The WOA series is a continuation of the Climatological Atlas of the World Ocean (Levitus, 1982), a set of global one-degree gridded climatological mean fields of oceanographic variables at standard depth levels in the ocean to be used, among other things, as initial and boundary conditions for coupled climate models.

A relatively new effort on creating regional climatology for different targeted regions of the world ocean has been initiated at the national Centers for Environmental Information (NCEI/NOAA). These contain six decadal climatologies from 1955 to 2012 with annual, seasonal, and monthly time resolutions. The fields contain both temperature and salinity at the standard 33 depth levels and are based on the WOD archive of observations spanning more than 100 years of data and incorporating new data which were not accessible previously. Some of these regions are: (i) Southwest North Atlantic; (ii) GIN Seas; (iii) Northeast pacific; (iv) Northern North Pacific;

(v) Northwest Atlantic; (vi) Nordic Seas and Northern North Atlantic; (vii) Arctic; (viii) East Asian Seas; and (ix) Gulf Of Mexico. Please see the website https://www.nodc.noaa.gov/OC5/regional_climate/. Many of these fields are available at $1°, 1/4°$, and $1/10°$ resolution, which are made with a 3-pass objective analysis (see next section).

11.3.2 OBSERVATIONS FOR MODELING SYSTEMS

A modeling effort requires observations for multiple purposes. These include initialization, assimilation, validation of simulations, and verification of forecasts. In this section, we highlight some of these types of data, their observational platforms and data products and their sources that are available to the community. It is to be appreciated how far we have come from the scarce data regime in the 1970s to the present day (2020s) with modern information technology. However, the ocean is such a vast place, that there is always the need for more data, so the model's ability to bridge the data gaps is tested more often in a data-model, simulation-prediction framework.

For example, many times, a modeling effort is warranted for process studies and simulations that are carried out with initial condition, which utilizes available data as well as climatology. Since available observations cannot provide all the information needed at all the high-resolution grid points of the model domain, a process called "Objective Analysis" is invoked, which meld the new observations with background climatology (or similarly gridded field from another model run). We will explain this methodology, which is also used for "data assimilation" later. Here let us explore some of these observational opportunities for initialization.

11.3.2.1 In situ Observations

If you are on board collecting data and/or modeling for someone who is collecting data, you have a basic set of observations from the standard suite of instrumentation such as CTD, XBT, fluorometer, current meters, pH meter, oxygen sensor. Let us look at some of these briefly here.

- CTD – This is well-established instrumentation for measuring the conductivity, temperature, and depth at a location by lowering a probe that samples and senses the water throughout the water column as it goes down and comes up in a vertical line. The downward and upward casts are processed in real-time to yield the temperature and salinity profiles $T(z)$ and $S(z)$ at a particular location on board a vessel at a particular time (x, y, t).
  - Depth – The CTDs actually measure pressure which is a proxy for Depth. Although the relationship between pressure

and depth is complicated (compressibility of water makes pressure depth dependent at greater depths); to a large extent, each one meter (3.28 ft) of water is approximately equivalent to one decibar of pressure. The water pressure at the surface is zero decibar. So, at 100 m of depth, the pressure should be 100 db.

- Temperature – The temperature is measured in degree Celsius (°C) with high accuracy (better than 0.0005°C. In the ocean, the temperature ranges from -2°C in the Polar waters to about 35°C in the Equatorial surface waters.

- Salinity – The salinity is the amount of dissolved salt in the water, and is measured in practical salinity units (PSU), which is a variant of the older PPT (kg salt per kg of water in parts per thousand). The PSU is based on a consistent standard and equation that works on a single ion mix to fit the real salinity of diluted North Atlantic seawater, instead of the previous equation based on a mix of all oceans. See Siedler and Peters, 1986 for more details. For a CTD, the process works as follows. In electric circuits, metals are known for their high conductivity of electric currents as opposed to glass. As the sea water contains salt and minerals, it has high conductivity compared to freshwater, which has none. Conductivity also depends on temperature – higher the temperature, higher the conduction. So, by measuring the conductivity, temperature, and pressure of water, one can find its salt content, or the salinity in PSU. The salinity in the open oceans ranges from 28 (near Polar regions) to about 35 in the Ocean basins; however, the salinity of about 37 PSU is observed in the Mediterranean, off of Brazil and Arabian Sea regions.

Two common types of CTDs are: (i) profiler (from ships) and (ii) moored. There is also a newer, less accurate version for hand-held lighter CTDs and smaller attachable sensors to be used with fishing gears. Technology is opening new opportunities with the so-called micro-CTDs.

- Current Meters – These are used to measure currents at depths and there are four types of current meters.

  - Mechanical – CMs are the rotary type and called the Rotary Current meter (RCM). They use a propeller to detect the water velocity and a vane to determine the direction of the flow. This older but proven technology is still in use by the Alfred Wegener Institute to monitor the Arctic inflow through Fram Straits.

- Acoustic – ADCP The Acoustic Doppler Current Profiler (ADCP) measures the current velocities over a depth range using the Doppler effect of sound waves scattered back from particles within the water column. They use a ceramic transducer to emit or receive the sound waves. They use the travel time, which requires at least two acoustic signals, one upstream and one downstream. The average speed of water can be determined from the time to travel from the emitter to the receiver in both directions. The 3-D water velocities can also be determined using multiple paths.
- Electromagnetic Induction – This approach has been used in the Florida Straits from 1980 till recently utilizing the electromagnetic induction in the submerged telephone cable. The idea is to exploit the physical property of the ions in the seawater to follow the ocean currents. Using the Law of Induction (Faraday's Law), the variability of the averaged horizontal flow can be determined by measuring the induced electric currents. See Chave et al. (2012) for details.
- Tilt CM – There are two types of Tilt current meters – Floating or Sinking. The idea here is to use a sub-surface buoyant housing (a floating tilt meter) that is anchored either to the sea floor with a flexible line or hanged from an attachment point on a mooring line (a sinking tilt meter). The housing then tilts as a function of its shape, buoyancy, and the current velocity. Since the characteristics of the housing are known, the velocity can be determined by measuring the angle of the housing and the direction of the tilt. A sinking TCM is used with moorings or docks, while the floating type is used at the bottom of an anchor or with lobster traps, among other possibilities.

11.3.2.2 Buoys

There are permanent buoys maintained by different countries for monitoring oceanographic as well as meteorological parameters. For example, there are NDBC Buoys all around different countries in the Global Ocean, and the data from these are available in near-real time, making them very useful for operational modeling systems. Some of the useful websites to look into are given below.

- USA – National Data Buoy Center (NDBC) is part of the US National Weather Service and probably keeps the most comprehensive

collection of buoy observations worldwide in collaboration with many countries around the world. https://www.ndbc.noaa.gov/
· United Kingdom Live Buoy Observations
(https://www.eldoradoweather.com/buoy/United%20Kingdom/buoy-xhtml.php)

After the 2004 Tsunami, a global effort was launched to monitor the wave heights across the Globe in a systematic way and a fleet of new DART (Deep-Ocean Assessment and Reporting of Tsunamis) buoys were launched. This is a global network of tsunami detecting buoys funded, monitored, and maintained by many agencies including the Global Drifter Program under the auspices of the Global Ocean Observation System (https://www.aoml.noaa.gov/global-drifter-program/).

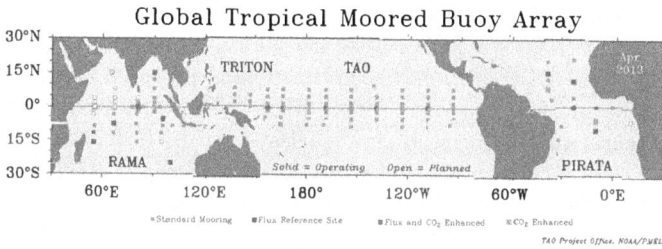

Figure 11.3: Global Tropical Moored Buoy Array. TAO in the Pacific, PIRATA in the Atlantic, and RAMA in the Indian Ocean (Source: http://www.pmel.noaa.gov /tao/oceansites/images/map_lg.gif).

11.3.2.3   Global Arrays

There are three Global efforts along the equatorial regions on the Pacific, Atlantic, and Indian Oceans. This is a multi-national effort to provide data in real-time for ocean modeling and forecasting. The Global Tropical moored Buoy Array program has three components (see Figure 11.3), which include the TAO/TRITON array in the Pacific, the PIRATA in the Atlantic, and the RAMA in the Indian Ocean. The focus of the three systems is ENSO in the Pacific, inter-hemispheric dipole mode and hurricane activity in the Atlantic, and the monsoon and intraseasonal variability in the Indian Ocean. The program is a contribution to the Global Ocean Observing System (GOOS), the Global Climate Observing System (GCOS), and the Global Earth Observing System of Systems (GEOSS). Please see https://www.pmel.noaa.gov/gtmba/ for more information. It is worth exploring the excellent data visualization tools available on their websites to have a feel for real-time as well as past events, which are archived and sometimes animated and explained in peer-reviewed literature.

11.3.2.4 Regional Arrays

The Ocean Observatories Initiative (OOI) has housed seven regional observatories for different purposes and some of these data streams will be very valuable for process studies using models in the near future. It is important to be mindful of these activities from a data usage perspective. The observatories are briefly outlined below. Details with an excellent data portal are available at https://ooinet.oceanobservatories.org/.

- Coastal Endurance Array – The Endurance Array (44.5°N, 122.2°W) consists of a set of two cross-shelf moored array lines – the Oregon Line (also called the Newport Line) and the Washington Line (also called the Grays Harbor Line). Each line has three fixed sites going from the inner-shelf (25–30 m), to the shelf (80–90 m), out to the slope (500–600 m). All six sites have moorings with fixed sensors at various depths for physical, biological, and chemical measurements.
- Coastal Pioneer Array – Situated near the Northeast Shelfbreak (41°N, 69.4°W), this observatory has been functioning for more than five years now and is operational in real-time. The Pioneer Array is a multiscale multi-platform observational system. Its backbone is a frontal scale moored array with three surface moorings (anchored to the sea bottom) and seven profile moorings. The moored array is supplemented by six Gliders and two AUVs to resolve cross-shelf and along with front eddy fluxes due to frontal instabilities, wind forcing, and mesoscale variability. The Pioneer Array is scheduled to be relocated to the mid-Atlantic region in 2024–2025 to test its portable functionality.
- Global Argentine Basin – This particular initiative has been suspended for now for lack of funds. A network of moorings at a high latitude location in the South Atlantic within the Argentine Basin (42°S, 36.8°W). Observations include air-sea fluxes of heat, moisture and momentum, and physical-biological-chemical properties throughout the water column.
- Global Irminger Sea – A network of moorings at a high latitude location in the North Atlantic within Irminger Sea (59°N, 36.3°W). Observations planned similar to Argentine Basin.
- Global Southern Ocean – A network of moorings at a high latitude location within the Southern Ocean, southwest of Chile (56°S, 83.5°W). Observations planned are similar to those in the Argentine Basin.
- Global Station Papa – A network of moorings at a high-latitude location, Station Papa (50°N, 145°W) in North Pacific. Observations planned similar to Argentine Basin.
- Regional Cable Array – The first-of-a-kind cable array across a tectonic plate (the Juan De Fuca Plate) with the real-time observatory on the west coast of the USA off of Newport, Oregon.

To emphasize, in addition to physical sensors, the OOI includes sensors that measure key biogeochemical properties (pH, pCO_2, bio-optics, nitrate, dissolved oxygen) on both moored and mobile autonomous platforms across all its arrays in the Atlantic, Pacific, and Southern Oceans. It is time to take advantage of these data for use in models to carry out more interdisciplinary process-oriented studies.

11.3.2.5 ARGO Floats

Another global effort in collecting data for oceanographic and climate studies is the ARGO (the Array for Real-time Geostrophic Oceanography) program. This global program was an evolution after the success of the World Ocean Circulation Experiment (WOCE) in the 1990's and the Satellite Altimetry programs in the mid-80s to late 90s. It is a global array of autonomous profiling floats which collect real-time temperature and salinity profiles in the upper 2000 m of the ocean. As of September 17, 2020, there are 3924 floats in the Ocean (see Figure 11.4), collecting continuous observations and being archived by the program office at the Scripps Institute of Oceanography (https://argo.ucsd.edu). The data are available for researchers and community to use for understanding the world ocean and climate (thanks to years of hard work and persistence of so many people over the globe to make this happen).

During the WOCE program, Russ Davis from SIO and Doug Webb of Webb Research Corporation developed the Autonomous Lagrangian Circulation Explorer (ALACE) floats (R. E. Davis, 1991, 2005; R. Davis et al., 1992). ALACE floats extended the original technology of Swallow floats (J. C. Swallow, 1955, J. Swallow and Worthington, 1961; see Chapter 5), which used the principle of neutral buoyancy to follow currents at a particular pressure level. For a detailed history and development of neutrally buoyant floats, please visit the UK Argo site (https://www.ukargo.net/about/float_history/) and the University of Rhode Island site of Tom Rossby's group (http://www.po.gso.uri.edu/rafos/general/history/index.html). With the advances in GPS technology, the ALACE floats were tracked by satellites in real-time during WOCE and it was soon realized that they could carry temperature and salinity sensors during their ascent and transmit the profiles in real-time along with location information. By the end of the WOCE program, most of the ALACE floats were equipped with such sensors and they became Profiling ALACE (PALACE) floats (R. Davis et al., 2001).

The idea of converting the profile temperature and salinity data using the thermal wind relationship (equation 3.27) to geostrophic velocity was put forward simultaneously by Dean Roemmich of Scrips and by Ray Schmitt of Woods Hole in 1998. These proposals were supported by various US agencies to start the ARGO program. This program is now supported and maintained globally by 30 countries and has become a well-known international partnership

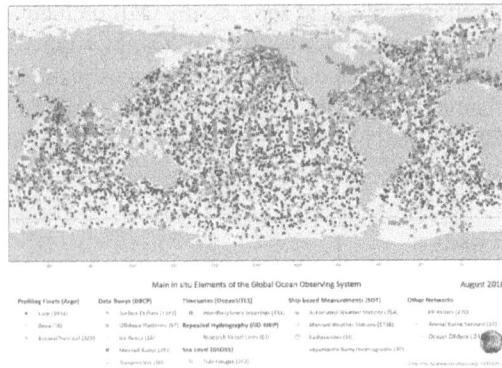

Figure 11.4: Global Coverage of multiple sets of observation systems including Argo Floats. Credit: NOAA/JCOMMS.

effort with an International Argo Steering team and monitored by the Technical Coordinator at JCOMMOPS located in IFREMER in Brest, France. See Figure 11.4 for coverage of Argo floats as of December 2020. There are three distinct Argo missions: Core Argo, BioGeoChemical Argo, and Deep Argo. Core Argo provides the temperature/salinity/pressure operationally; BGC Argo is in its pilot phase and is some of the variables that are being tested are: Oxygen, Nitrate, pH, Chlorophyll_a, suspended particles, and downwelling irradiance (see https://biogeochemical-argo.org/ for details). Deep Argo is also in the pilot phase and looking to break the challenges offered by going deeper than 2000 m with housing and sensors and possibly reach down to 4000 m, 6000 m, and to the full depth of our ocean, systematically! Visit the website for Deep Argo (https://argo.ucsd.edu/expansion/deep-argo-mission/) for some exciting news in the coming years!!

11.3.2.6   Drifters

Surface drifters are used for initializing, constraining (assimilation), and verification of real-time modeling for many regional models. The Global Drifter Program is an effort of NOAA's Global Ocean Observing System (GOOS) program. Its objectives are to maintain a global ($5° \times 5°$) gridded array of about 1300 surface drifting buoys. These satellite-tracked buoys provide observations of mixed-layer currents, sea surface temperature and salinity, atmospheric pressure, winds, and waves. The program office at AOML (https://www.aoml.noaa.gov/global-drifter-program/) has more information on data distribution and usage protocol. It is a shared effort by AOML, SIO, and Industry.

11.3.2.7 Survey Data

In the late eighties and early nineties, concerted efforts to seasonally survey the coastal ocean in a systematic way were given priority from a fisheries perspective. These twice-a-year surveys have been archived by the National Marine Fisheries Service and have become a major source of information of concurrent physical, chemical, and biological data for the coastal regions of the US and many other countries. Examples are

(i) MARMAP (Marine Resources Monitoring, Assessment and Prediction program) during 1977–1987, which is now part of the COPEPOD (Coastal and Oceanic Plankton Ecology, Production and Observation Database) https://www.st.nmfs.noaa.gov/copepod/data/us-05102/.

(ii) CALCOFI California Cooperative Oceanic Fisheries Investigations has about seventy years of data from Spring 1949 till date along the coast of California. https://www.calcofi.org/ccdata.html.

(iii) BCO-DMO The Biological and Chemical Oceanography Data management Office at the Woods Hole Oceanographic Institution hosts a vast amount of interdisciplinary data for community usage with the motto "It's not the size of the data ... it's the impact of the science."

The BCO-DMO website captures the utility of sharing the data very well with the following sentiment, "Ocean Researchers collect data using a variety of platforms, instruments and sensors. Each individual data set may be small, but when integrated and combined with laboratory experiments and model results, they enable BIG science."

Nothing can be closer to the truth than data. Data is the *Veritas* of the oceans. Models are tools that help understand the science that is present in the data.

11.3.2.8 Satellite Observations

Beginning with the success of SeaSAT in 1978, and followed by multiple experimental satellites to observe the oceans and their different surface parameters, we now have streams of satellite-derived data products. These include the sea surface temperature (SST – sensed by the infra-red bands in the spectrum), the sea surface height (SSH – measured by altimeter which resolves the height of local sea surface compared to the Geoid), the sea surface color (SSC – the reflectance in the visible range of the spectrum, which is a signature of the chlorophyll a at the surface of the ocean), the surface winds (speed and direction from scatterometer) and sea surface salinity (SSS – measured by radiometers which are able to detect subtle changes of microwave emission from the ocean surface due to changes in salinity). For description and details of the instruments and to learn about satellite oceanography (which is a vast field of merging of engineering and science and technology), please see some advanced books mentioned in Further Reading.

In this subsection, we highlight the accessibility of satellite data from various NASA sites and other joint university-laboratory consortium efforts. Processing, archiving, distributing of an enormous amount of satellite data requires huge resources and an extensive knowledge base from dedicated satellite oceanographers and technologists and meteorologists. Thanks to all of them and more, that we have a continuous stream of the dataset for use in numerical models and to understand the oceans and climate. Some of these efforts are mentioned below.

- SST – A series of NOAA satellites from NOAA-11 to NOAA-20 has the Advanced Very High-Resolution Radar (AVHRR) instrument, which resolves the SST at 1.1 km resolution over the Globe. Infrared observations cannot see through the cloud, so 3-day and 7-day composites are often used for model assimilation and simulations. There are also lower resolution microwave SST sensors that can see through clouds.
- SSH – Started in 1985 with GEOSAT and continuing through Topex/Poseidon and many present-day the US, European, Indian and Chinese missions. Continuous global SSH fields are available through multiple sites at multiple space-time resolutions (7 km along-track to 25-km gridded, daily to 10-day repeat cycle) from multiple satellites.
- SSC – NOAA-20, 9 km; multi-sensor daily data. MODIS. A similar problem like SST with clouds. Additionally, SSC can only be measured during the day.
- SSS – Relatively new technology. The first NASA mission, Aquarius (2011–2015) was successful in getting SSS at 150 km resolution. The recent NASA mission, Soil Moisture Active Passive (SMAP) is operating since May 2015. Its data is collected at 40 km resolution; however, due to the high-noise content of this data at 40 km, it is spatially averaged to be useful for science purposes at 60–70 km.
- Winds – Multiple wind products are available from multiple satellites. Some are: (i) wind speed from SAR; (ii) Vector winds (speed and direction) from Scatterometers on board SCATSAT-1, ASCAT, and SSM/I.
- Sea Ice – Both thermal (VIIRS) and microwave (AMSR-2) sensors are used to estimate sea ice temperature, concentrations and thickness.

- Surface Roughness – Synthetic Aperture Radar (SAR) maps the surface microwave radar reflectivity at very high resolutions (less than a meter to 100 m), and these can see through clouds. The data is used for obtaining surface wind speed and many other geophysical variables.

Many of the satellite products and fields are available through the following websites.

- NOAA Coastwatch – A great source of information for all satellite products available for US coasts. coastwatch.noaa.gov
- NASA/JPL – Jet Propulsion Laboratory https://podaac.jpl.nasa.gov
- NASA Earthdata – https://earthdata.nasa.gov/ See the different DAAC (Distributed Active Archive Center) for different regions.
- Asia-Pacific Data Research Center (APDRC) – The University of Hawaii hosts the APDRC of the International Pacific Research Center (IPRC), which is a great resource of all data (satellite and in situ) needs in the Asia-Pacific sector. http://apdrc.soest.hawaii.edu/
- COPERNICUS – It is the European Union's Earth Observation program. It offers information services based on satellite and in situ data. It collects, processes, archives, and serves a huge amount of satellite data from all sources in a synthesized gridded field format (netCDF), which is easy to use for the ocean and atmospheric modelers. See https://www.copernicus.eu/en and explore!

Note that the satellite-derived SST and SSH and SSC and other such gridded products are observations at the surface. Subsurface remains a mystery. Scientists use the surface observations in an assimilation algorithm and then the dynamical model ingests and adjusts the subsurface to the surface satellite data. Sometimes the surface properties also represent an integrated measure of the upper layer of the ocean, which can then be utilized with such understanding during analysis and for data assimilation in numerical models.

11.3.2.9 New Observational Platforms

In recent years, with the advances in information technology and satellite communication systems coupled with new battery technology, new instruments have been made possible to survey the ocean in real-time in addition to ship surveys. These include gliders, autonomous underwater vehicles (AUVs), HF radars, new oxygen and pH sensors, and turbulence (mixing) measurement devices (vertical microstructure profilers or VMPs) with higher accuracy than a decade back. See the blog at sirates.sites.umassd.edu for a recent cruise on salinity intrusion from warm-core rings to the shelf ocean across the shelf break in the western North Atlantic, which utilized CTDs, AUVs, VMP, ADCP, and other instruments including chlorophyll and nitrate sensors. For a review of new and recent observational instruments, please see Venkatesan et al. (2018). Assimilation of both Glider and HF Radar in numerical ocean models are topics of current research and the students are requested to look at the modeling websites for recent studies on this evolving subject.

Furthermore, there are innovative ideas with micro-sensors being used as tags on various species. A great example is to use turtles (which live for more than 100 years) as temperature sensors by tagging them with a small temperature sensor for climate change monitoring. For some recent fisheries and oceanographic instrumentation including futuristic efforts, see Wang et al. (2019).

11.3.3   SURFACE AND BOUNDARY FORCING FIELDS

The model runs (simulations or predictions) require specification of forcing fields including from winds (wind stresses), heat fluxes, precipitations, river runoff. These 2-D fields (except for Rivers, which are treated as coastal node buoyancy and flow input – or point observations) are available from "reanalysis fields." These are model-simulated fields using past data and atmospheric models to create a dynamically balanced and consistent state of the atmosphere over a long time period which can be used for the community for various needs at a later time.

Some of these reanalysis fields are available from the following websites:

- Research Data Archive from NCAR, USA https://rda.ucar.edu/
- NOAA NCEI has three sets of reanalyses (i) Climate Forecast System Reanalysis (CFSR) – a global reanalysis (a best estimate of the observed state of the atmosphere) of past weather from January 1979 through March 2011 at a horizontal resolution of 0.5°. (ii) North American Regional Reanalysis (NARR) – a regional reanalysis of North America containing temperatures, winds, moisture, soil data, and dozens of other parameters at 32 km horizontal resolution. (iii) Two global reanalyses of atmospheric data spanning 1948/1979 to the present at a 2.5° horizontal resolution (Reanalysis-1 / Reanalysis-2) https://www.ncdc.noaa.gov/data-access/model-data/model-datasets/reanalysis
- ECMWF climate reanalysis https://www.ecmwf.int/en/research/climate-reanalysis

11.4   OPTIMAL INTERPOLATION

There is a huge need for interpolation from irregularly spaced observations to a regular modeling or an analysis grid. There are several ways to achieve this.

11.4.1   THE CORRECTION METHOD

One of the early ideas was to create an estimate from a first guess based on multiple observations within a radius of influence. This follows the developments of Cressman, 1959 and Barnes, 1964. Mathematically speaking,

$$E_{i,j} = F_{i,j} + C_{i,j} \qquad (11.1)$$

i.e. the Estimate (E) equals the sum of the first guess (F) and the correction (C).

This correction is a distance-weighted mean of all grid-point differential values that lie within the area bounded by the influence radius.

mathematically speaking,

$$C_{i,j} = \frac{\sum_{j=1}^{N} W_s Q_s}{\sum_{j=1}^{N} W_s} \qquad (11.2)$$

where Q_s = the difference between the observed mean and the first-guess at the sth point in the influence area.

$W_s = e^{Er^2/R^2}$ for $(r \le R;)$ $W_s = 0$ for $r > R$; r = distance of observation from the gridpoint of the estimate; R = influence radius and E=4.

This technique was used for determining climatology by Levitus, 1982 at 1-degree resolution with a 300 km influence radius. Later in 1994 (Levitus & Boyer, 1994), with more data being available, a reduced R was applied with partially monthly climatology with the depth going down to only 250 m for the monthly climatologies. A 3-pass multipass analysis with progressively smaller influence radius (892 (321) km for pass 1, and 669 (267) km for pass 2 and 446 (214) km for pass 3) for $1°$ $(1/4°)$ was applied to create climatology at different time-scales (Boyer et al., 2005).

11.4.2 KRIGING

Named after a South African Mining Engineer, Danie G. Krige, who developed this technique to estimate distance-weighted average Gold grades in a reef. He wanted to obtain a most likely estimate of gold grade from the observations of a few bore holes. Mathematically, it is a method of interpolation based on Gauss-Markov theory assuming a Gaussian process whose prior covariances are known (Krige, 1951). Kriging is mostly used in spatial analysis (GIS) and computer modeling. It is a useful tool for data smoothing and mapping irregularly spaced data onto a regular grid. The mathematical formulation can be expressed as follows:

$$\hat{Z}(s_0) = \sum_{i=1}^{N} \lambda_i Z(s_i) \qquad (11.3)$$

where $Z(s_i)$ is the measured value at the ith location, λ_i is an unknown weight for the measured value at the ith location, s_0 is the interpolation location, and N is the number of measured values. There are different ways of determining the weights (λ_i) including an assumption of random variables and then applying the Gauss-Markov and/or Bayesian techniques for setting up a system of equations to determine their values. This method is gaining in popularity and it is worth looking into details of the techniques in the works by Cressie (1990) and Krige (1951).

11.4.3 OBJECTIVE ANALYSIS

A very well-established idea of projecting irregularly spaced data on a regular grid is to use statistics of the data for projection. How do we do this? Interpolation or extrapolation? Definitely, interpolation. We do not wish to extend our data beyond the boundaries of the data domain. Well, there are some exceptions in the case of known data properties, but those are with limitations, and we should be mindful of any extrapolation beyond the scope of the data domain. What else do we need? We have some observations or data. So we have some knowledge of data density. We can find out about the spatial scales of the data. We also have some knowledge about the temporal variation of the data (maybe a time series).

To use the statistics of the data, we need the mean, the standard deviation and the data distribution! Well, if we don't know the data distribution a priori, we can assume that the autocorrelation of the data is a Gaussian of the form (e^{-x^2/R^2}) where x_0 is the zero-crossing and R is the decay scale where the value of the autocorrelation becomes $1/e$ times the maximum value or the function e-folds! So, this is an important assumption – the correlation function is Gaussian - has a zero-crossing and a decay scale. Now this zero-crossing and decay scale can be in all three spatial directions (x, y, z) and in time (t).

So, now we are ready to formulate the problem of seeking the interpolated estimates on a regular grid from an irregular set of observations. This process is called objective analysis (OA).

An OA is a formulation on minimization of variance between data and estimate given a data set and its correlation functions.

The present algorithm discussed here is based on seminal work by Carter and Robinson (1987a), which was based on the original formulation of OA by Gandin (1965) and then developed for oceanography by Bretherton et al. (1976, July) and used by Clancy (1983) and A. Robinson and Leslie (1985) for ocean analysis.

Let us look into the statistical model (following Carter and Robinson, 1987a) briefly.

A linear estimator $\tilde{\theta}_x$ is sought formally as a combination of available observations (ρ_s) with coefficients (α_{xs}) which are yet to be determined.

$$\tilde{\theta}_x = \sum_{s=1}^{N} \alpha_{xs} \rho_s \qquad (11.4)$$

Consider a large number of different realizations of θ_x and observations ϕ_s, then the error variance for $\tilde{\theta}$ is

$$(\theta_x - \tilde{\theta}_x)^2 = \left(\theta_x - \sum_{s=1}^{N} \alpha_{xs}\rho_s\right)^2$$

$$= C_{xx} - 2\sum_{s=1}^{N} \alpha_{xs}C_{xs} + \sum_{r,s}^{N} \alpha_{xr}A_{rs}, \text{ where } A_{xs} = \overline{\phi_r\phi_s}$$

(11.5)

Here C_{xx} is the auto-correlation of the estimate, θ_x; and C_{xs} is the cross-correlation (covariance) of the estimate, θ_x with observation, θ_s. and After a little bit of algebraic manipulation (see Carter and Robinson (1987) and BDF76 for a fuller exposition of matrix conversions), one can arrive at the conclusion that the values of the coefficients that minimize this variance (or the difference between observations and estimates) are given by:

$$\alpha_{xr} = \sum_{r'} C - xrA_{r'r}^{-1}$$

(11.6)

which leads to

$$\tilde{\theta}_x = \sum_{s=1}^{N} C_{xr} \left(\sum_{s=1}^{N} A_{rx}^{-1}\phi_s\right)$$

(11.7)

C_{xr} is the covariance between the estimate and the observations, where

$$A_{rs} = \overline{\phi_r\phi_s} = C([X_r] - [X_s]) + \varepsilon^2\delta_{rs}$$

is the covariance Matrix between all pairs of observations (think about correlations)
and

$$C_{xr} = C(x - x_r)$$

C is the correlation of the observations, ϕ. This is where we can now assume homogeneous, stationary, and non-isotropic correlation function of the form

$$C(r) = C(\triangle x, \triangle y, \triangle t)$$

Now, one can define C_θ to be the autocorrelation for the series (vector) of observations and $C_{x\theta}$ as the cross-correlation between the estimates and the observations, i.e.

$$C_\theta = A_{\theta\theta}C_{x\theta} = A_{x\theta}$$

So now the general form can be written more specifically as

$$\hat{\theta} = C_{x\theta}C_\theta^{-1}\phi \qquad\qquad (11.8)$$

$$C_e = C_x - C_{x\theta}C_\theta^{-1}C_{x\theta}^T \qquad\qquad (11.9)$$

The last equation is for the error matrix. It is a very important component of the OA formulation. The error field is critical to assess the confidence in the initial field's sub-regions. The errors propagate during time integration (simulation or prediction) and so the regions with high initial error might be subjected to faster loss of predictability than those with less initial errors. In a real-time adaptive sampling scenario, the regions with higher errors might demand more data and hence better sampling to bring the solutions to a more stable overall state with data-assimilative forecasts.

Some of the correlation models usually implemented for initialization and assimilation purposes in ocean models are:

- A time-dependent model is $C = (1 - r^2)e^{-r^2/2}$, where $r^2 = \left(\frac{\triangle x}{L_x}\right)^2 + \left(\frac{\triangle y}{L_y}\right)^2 + \left(\frac{\triangle t}{T}\right)^2$

- A more general model with phase-speed in both directions is $C = (1 - r^2)e(-\alpha r^2)$ where $r^2 = \frac{1}{L^2}\left[(x - c_1t)^2 + (y - c_2t)^2\right]$

You can also construct your own correlation function or adapt to one of the above. The values of L_x, L_y, and T are the decay scales in space and time. The zero-crossings are generally assigned to a value beyond which you do not wish to extend the correlation function.

Note that the OA procedure described here is done for a scalar field (T, S, u, v, ψ), etc. For a vector field such as wind or current meter data, one applies the OA to the individual components and then recreates an objectively analyzed vector field. This leads to certain errors and a separate field of Vector OA has been developed with complex algebra, where the minimization is sought for the vector covariance between data and estimates.

An OA package is available through the ROMS server. Several modifications have been made to the original Mariano and Brown, 1992 OA package. A new Matlab version with bathymetry-dependent correlations using the level-set method (Agarwal & Lermusiaux, 2011) is available from the MSEAS website. This latter bathymetric OA is very useful for initialization of models with unstructured grid and with complicated island and bottom topographic features like the South China Sea and the Philippines. For some interesting modeling applications with the capabilities of the bathymetric OA see Lermusiaux et al. (2017) and the MSEAS website publications.

11.5   DATA ASSIMILATION

Now that we have a model grid and set up (Section 11.1), we have some data (Section 11.3) to initialize (Section 11.2) the model using an appropriate interpolation technique (11.3), it is time to run the model and generate some simulation or just run the model to predict the next few days or weeks or months!

Sure, we can do that and we will discuss some of such simulations in the next section. Before we do that, let us take a step back and think through the process of simulation and prediction. It is clear that we have some climatological data to initialize and then some initialization technique to map those data into the model grid and start the model. But, really, we might not have enough data to have much confidence in this initial field (to claim it as synoptic at a particular time) based on a climatology mean over so many years, which smooths out many high-resolution features due to the weighting distance or influence radius or decorrelation scales in space and time during interpolation. Then, once you get the prediction or the simulations, you have some more observations, which you can use to validate and verify and then go back and redo the simulations.

But there could be a possibility to use the new observations directly (or indirectly) in the model and have the model fields adjust to the data dynamically during integration. This idea of using data to be assimilated in the model fields in a mathematical framework led to the development of a vast new field called "Data Assimilation." It was originally conceived a problem of applying optimal estimation extending Gauss's original least-square problem to a time-dependent recursive solution. One of the first authoritative books on this subject was by Gelb, 1974, who defined the optimal estimation as follows. *"An optimal estimator is a computational algorithm that processes measurements to deduce a minimum error estimate of the state of a system by utilizing: knowledge of system and measurement dynamics, assumed statistics of system noises and measurement errors, and initial condition information."* Draw a parallel of the system to our ocean model, measurements to our observations, knowledge of the system, and observation as our correlation functions and make the noise and errors independent and uncorrelated – and we get back to something close to OA discussed before. The purpose of the OA was to initialize the model with data, the purpose of the DA is to use the data within the model's dynamical framework to have the model fields adjust to the few observations as a whole system.

Let us follow Gelb's illustration of the three different ways the data can be used by a system while performing a simulation or a prediction exercise. Figure 11.5 shows three types of estimation problems:

(i) Filtering – when all of the observations available up to the time of estimation is used by the assimilative modeling system (AMS).

Data Assimilation

(a) Filtering

Data Assimilation

(b) Smoothing

DA

No Assimilation

(c) Prediction

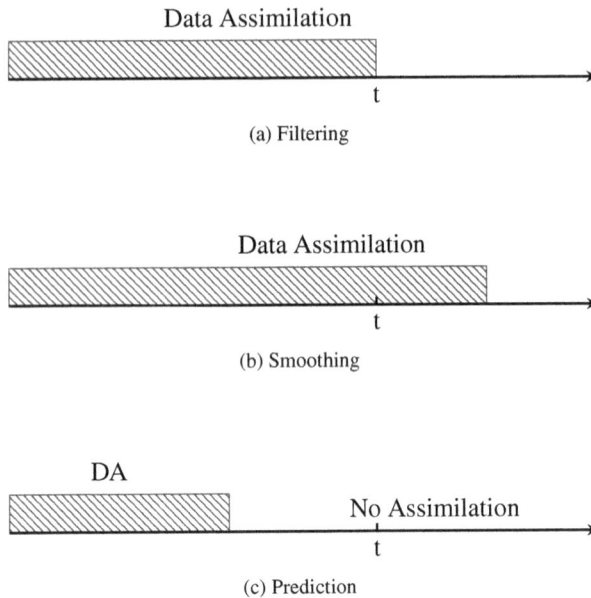

Figure 11.5: Types of estimation problems in simulation and prediction with data assimilation. Adapted and modified from Gelb (1974).

(ii) Smoothing – When some or all of the information up to a time after the time of estimation is used by the AMS.
(iii) Prediction – When the time of estimate is some time after the last available observation.

It is easy to draw a parallel to these three to what we practice today. Filtering is synonymous with "assimilation at initialization," where all of the data up to the present time is fed into the data assimilation system where the dynamical model adjusts the three-dimensional field of variables to all the observations to create an initialization field for further prediction or simulation. Smoothing is used for creating "reanalysis" fields. Large-scale numerical models are often run for creating a long-term adjusted and consistent set of fields for further analysis and understanding of past phenomenon and behavior of modeling system parameters as well as to understand the sensitivity of numerical algorithms to certain forcings, set ups, and other factors. One then borrows as much information within an influence window from past and future to feed into the AMS for creating a well-adjusted four-dimensional evolution of field variables.

The prediction problem is interesting. One can use the data assimilative framework in various ways to arrive at a number of different solutions (predictions) for the time of estimation. Let us illustrate this with one common

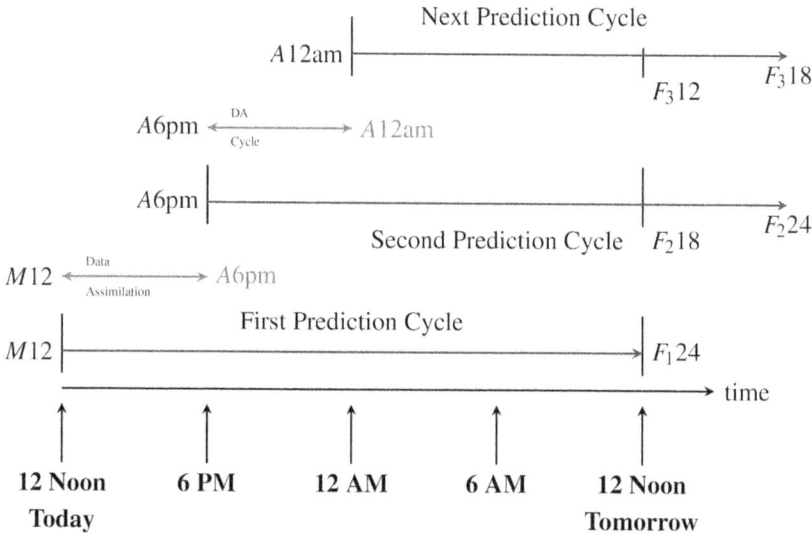

Figure 11.6: An example of daily weather forecast cycle with assimilation. Start from the bottom row and move upward in time with a 6-hourly interval.

example of weather forecasting every 6 hours. The weather model (very similar to the ocean models) is initialized with all the available climatology and data for say, 12 noon today. Then we issue a forecast for tomorrow's noontime by running the model for 24 hours in the prediction mode. Let's call the field at today's noontime as M12 and call that 24-hour prediction as F24. Now, you have new observations coming in every so often (in real-time or near real-time) from all different kinds of instruments like on-land weather stations, weather balloons, and satellites and buoys. So, by 6 o'clock today, you have consider-able information from different sensors that your model prediction F24 (which started from M12) has not seen or adjusted to. So, why not use the new data, new information to update the model prediction that you did six hours ago and do another prediction for the next eighteen hours? That is, why not adjust the model estimate for six pm (from M12 run) with the data for the time-interval between 12 noon and six pm and create a new adjusted initial condition which is a better first guess than the previously estimated fields for 6 pm. And then run that estimate (A6pm) for another 18 hours to arrive at an updated forecast for tomorrow's noontime. And if you are the weather forecaster in the evening news cycle of your local channel, you have something new and exciting to talk about for tomorrow's noon forecast!

Mathematically, this is shown in Figure 11.6. Looking from the bottom of the plot toward the top in time,

For the first prediction, $M12 \longrightarrow 24$ hour forecast $\longrightarrow F_1 24$.

For the second, $M12 \longrightarrow$ Data for 6 hours $\longrightarrow$ assimilation $\longrightarrow$ A6pm $\longrightarrow$ 18 hour forecast $\longrightarrow F_2 18$ and 24 hour forecast $F_2 24$.

Then for the third cycle, A6pm $\longrightarrow$ Data for six hours $\longrightarrow$ assimilation $\longrightarrow$ A12am $\longrightarrow$ 12 hour forecast $\longrightarrow F_3 12$ and 18 hour forecast $F_3 18$ and 24 hour forecast $F_3 24$, and so on.

This is called sequential assimilation and is generally followed for weather forecasting in the regional weather centers. See Figure 11.6. Note that the forecast made at 12 midnight for Noontime tomorrow ($F_3 12$ should be better (more informed with more data and their assimilation and adjustment of the model at the A12am level) than those obtained previously at M12 and A6am levels. There are a number of different assimilation methodologies. Without going into the details of the elegant mathematical exposition of each (which are discussed in many wonderful and advanced texts on DA) we highlight some of the physical assumptions for the basis of these methodological developments.

- Direct Insertion and Nudging – Direct insertion is just that. Replace the predicted value with observation and let the model adjust it later. Nudging uses an adjustment time scale to bring in the observation to replace the estimate. You can think about this as a slow insertion.
- Kalman Filter – It is the original (Kalman, 1963) idea of expressing the estimate with a gain matrix of the observation. Given in matrix form,

$$X_a = X_f + K(X_o - X_f) \tag{11.10}$$

where X_a is the analysis (assimilation), X_f is the forecast, and X_o is the observational variable. K is the Kalman Gain Matrix. The original Kalman Filter is mathematically and arguably one of the best possible assimilation techniques; however, this is so computationally demanding that it is unrealistic and almost impossible to invert a huge matrix every time-step to get the time-dependent Gain Matrix ($K(t)$). Generally, one computes the "K" once and uses it throughout the assimilation run.
- Variational Data Assimilation – These are optimal estimation methods that seek to minimize some measure of squared error or "variance." They define a functional (integral of functions and their derivatives) and use the principles of variational calculus to minimize the measure of variance subject to certain constraints. The functional could be the sum of squares of the differences between the observed and model values of any variable (T, S, U, V, etc.). The variational problem of minimization can be solved by using Lagrange multipliers (see Daley, 1991 for details). Variational algorithms were first developed

for operational application in data assimilation in weather forecasting (Lorenc et al., 2000) and then used in ocean forecasting for the last two decades. A nice overview of variational data assimilation for global ocean is provided by Cummings and Smedstad, 2013. The first successful operational 4D-Var system was implemented at the European Centre for Medium-Range Weather Forecasts (ECMWF) using an incremental formulation (Courtier et al., 1994; Rabier et al., 2000). Mathematically, the 4D-Var system minimizes the following cost function, J composed of three parts.

$$J = J_b + J_o + J_s \qquad (11.11)$$

These include the quadratic measure of distance to the background, observation, and the balanced solution.

The background cost function term is determined from the covariance of the climatology and some prior estimate or analysis or forecast. The observation term is a distance measure based on covariances of analysis and observations; and the balanced solution cost function is determined from a distance between the analysis and the balanced state. For details of the complex mathematical construct, please see the excellent paper by Huang et al. (2009) and many references therein for the evolution of this methodology.

Both 3D-VAR and 4D-VAR schemes are available on the ROMS website. A large group of researchers is working to improve these schemes with more efficient algorithms.

· Adjoint model – It is a formal way of minimizing a quadratic cost function subject to the constraints of the governing equations in a time-dependent model forecast. The cost function is defined as the misfit between the model forecasts and observations. It is an iterative process of running the dynamical model forward and the adjoint model backward till one minimizes this cost function. So, the initial guess of control variables is run by the dynamical model forward in time and compared with observations to find the misfit. The adjoint model is then run backward in time to find the gradients of the cost function with respect to the control variables. From these gradients, an improved estimate of the model variables is made and a new misfit from another forward run. This iterative process goes on till the cost function is minimized. For details on mathematical formulation and applications, please see Bennett (1992) and ROMS website. As indicated in Chapter 10, ROMS has tangent linear and newer versions of Adjoint models for advanced data assimilation with ROMS.

11.6   EXAMPLE SIMULATIONS

Here we present two examples of basin-scale simulations. One is for the North Atlantic using a high-resolution ROMS model (Chaudhuri et al., 2011a, 2011b). The other is for the Bay of Bengal using again the ROMS model (Jana et al., 2018). The purpose of the two simulations was different and it is worth thinking about such scientific motivations behind a model simulation to be performed. We will also show how both of these simulations used a common performance evaluation metric represented by a single diagram, Taylor Diagram (Taylor, 2001) to evaluate their simulation results.

11.6.1   THE NORTH ATLANTIC SIMULATION

In this first example, we discuss a North Atlantic model simulation. The purpose of carrying out the simulations was to understand the variability of the North Atlantic basin to two different wind conditions related to the two phases of the North Atlantic Oscillation (NAO). Remember Sir Gilbert Walker's contribution to describe the strengths of storm systems over the Atlantic and relate it to the difference in sea level pressures on the Azores High and the Icelandic Low. The idea was briefly as follows. During the high NAO phase, the stronger westerlies would result in a more northward path of the Gulf Stream; and the low NAO phase would result in more southward excursion of the Gulf Stream after it leaves the coast at Cape Hatteras. How do you do this in a model? Well, a twin-experiment simulation was thought out by Chaudhuri et al. (2011b) in which the same model set up (a ROMS model) was forced with two different climatological forcing created from two different time-periods – one for a low-NAO period and one for a high-NAO period. Let us look into this now.

The spatial domain of the ROMS model extends from $15°$S to $75°$N and $100°$W to $20°$E. The domain is implemented using a $1/6°$ horizontal and 50-level vertical resolution Mercator grid, with the bottom depth set to 5500 m. A $5°$ climatology sponge layer (a layer along the domain boundary within which the numerically simulated field is slowly relaxed to climatology values within a fixed time period) is prescribed towards the Arctic and the South Atlantic boundaries. Vertical mixing is determined by the Generic Length Scale (GLS) (Umlauf & Burchard, 2005) scheme. The bottom boundary conditions are parameterized by quadratic bottom drag law for the momentum equations and zero flux for tracer equations. Horizontal diffusion and dissipation are parameterized using Laplacian operators. The grid bathymetry is derived from a combination of finer-resolution ETOPO2 topography database (Marks & Smith, 2006) for the GS region and a coarser-resolution ETOPO5 (Edwards, 1989) for the rest of the North Atlantic basin. This was needed to keep the bathymetric

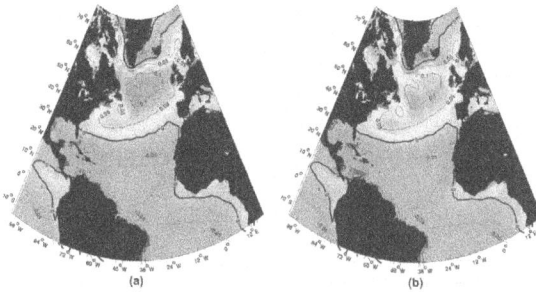

Figure 11.7: The North Atlantic ROMS model domain. Wind stress: NCEP reanalysis data derived mean zonal wind stress (Nm^{-2}) for high (a) and low NAO (b) phases. Zero contour line is shown in bold, whereas positive and negative contours are shown in solid and dashed lines, respectively. Republished with permission of Elsevier Science and Technology Journals, from "Response of the Gulf Stream transport to characteristic high and low phases of the North Atlantic Oscillation," Chaudhuri et al., in Ocean Modelling, 39 (3–4), 2011; permission conveyed through Copyright Clearance Center, Inc.

contours in the GS region from crossing the domain side walls, while keeping the shallow regions intact.

For the high/Low NAO wind and other forcing conditions, the National Centers for Climate Prediction (NCEP) reanalysis fields (heat flux, shortwave radiation, meridional and zonal wind stress components) from 1980 to 1993 and from 1958 to 1971 are selected because these periods witness sustained high and low NAO phases respectively. See Figure 11.7 for clear differences in the wind stress fields between the low and high NAO climatologies.

The initial condition for temperature and salinity were derived from Levitus and Boyer (1994) $1°$ climatology. Two simulations (high and low NAO forcing) were run for 14 years, six of which were needed for the initial state of rest to reach dynamical equilibrium or a steady state of kinetic energy. The 14th year of the simulations was saved at every 3-day intervals to compute the statistics of the parameters of interests.

An interesting aspect of these two simulations was the validation of the model skill using the so-called Taylor Diagram shown in Figure 11.8. Taylor Diagram (see Taylor, 2001 for even more details) provides a statistical summary of how well model simulations and observation patterns match each other in terms of their correlation, their root-mean-square difference, and the ratio of their standard deviations. The radial distances from the origin represent the standard deviations, whereas the azimuthal positions show correlation coefficients. These two parameters measure model-observation differences in phase. The distances between observation and simulations in Figure 11.8 represent

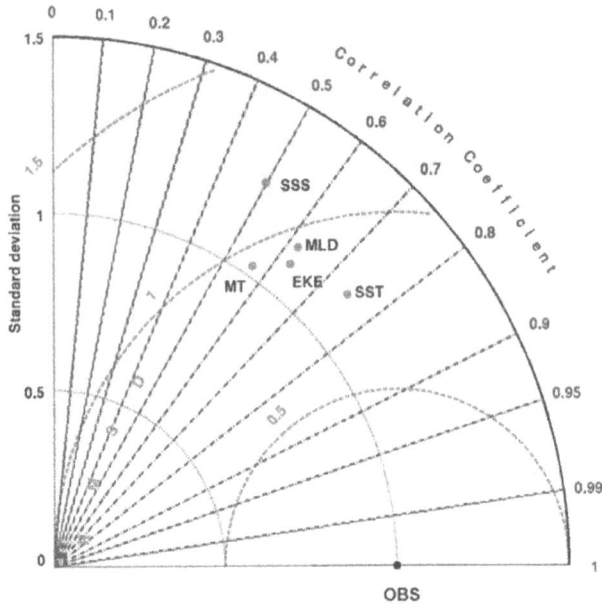

Figure 11.8: Taylor diagram: Taylor diagram representing model skill. The x- and y-axes are normalized standard deviation and the outer arc axis is the correlation coefficient. Green dashed lines show root-mean-squared differences. The red dots show model-derived representations of normalized sea-surface temperature (SST), sea-surface salinity (SSS), mean transport (MT), eddy kinetic energy (EKE), and mixed-layer depth (MLD), respectively. The black dot is the normalized representation of all observations. Relative distance between black and red dots quantifies model skill, i.e. shorter distance signifies better skill. Republished with permission of Elsevier Science and Technology Journals, from "Response of the Gulf Stream transport to characteristic high and low phases of the North Atlantic Oscillation," Chaudhuri et al., in Ocean Modelling, 39 (3–4), 2011; permission conveyed through Copyright Clearance Center, Inc.

model-observation root-mean-square errors that measure differences in amplitude. Simulation features that match well with observations in both amplitude and phase appear closest to the observed point in the diagram (e.g. SST in Figure 11.8). Five parameters (SST, SSS, Mixed Layer Depth, eddy kinetic energy, and mean transport of the Gulf Stream) were chosen for the skill test.

   The simulation results are described in detail by Chaudhuri et al. (2011b) for the high/low NAO simulation related to the Gulf Stream. Here we highlight two aspects which was intriguing from the model simulations and which illustrate

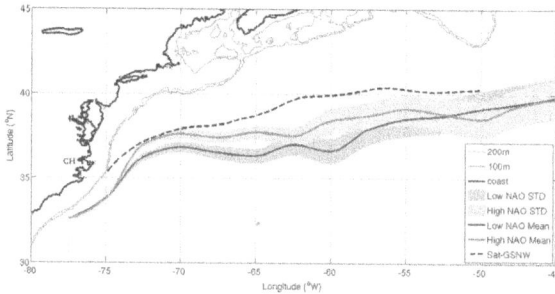

Figure 11.9: High and low NAO Gulf Stream path: Annual mean Gulf Stream path estimated by upper-layer (50–400 m) integration for high (red) and low (blue) NAO simulations. The standard deviations associated with the estimates are shown with shading. For interpretation of the references to color in this figure legend, the reader is referred to the web version of this article. Republished with permission of Elsevier Science and Technology Journals, from "Response of the Gulf Stream transport to characteristic high and low phases of the North Atlantic Oscillation," Chaudhuri et al., in Ocean Modelling, 39 (3–4), 2011; permission conveyed through Copyright Clearance Center, Inc.

the power of simulations in understanding our large-scale force-response system in the North Atlantic. First of all, the Gulf Stream path was found to follow a more northward (southward) path during high (low) NAO simulation. The GS path was defined as the location of the 15°C at 200 m depth. This is shown in Figure 11.9 from Chaudhuri et al. (2011).

Second and more importantly, this study addressed an ambiguity in the understanding of the interrelationship between transport, path, and NAO forcing. While some investigators reported high GS transport during high NAO (with northward path preference) and vice versa, some have found lower GS transport during high NAO period (with northward path preference). To address this ambiguity, the transport of the GS was calculated from the two simulations at several places between Florida Straits and 55 W by integrating the model velocities with depth as appropriate. These were then compared with observations available from different sources for the GS transport at different times in the past. The simulation results clearly indicated that the GS transport along the western boundary (from Florida Straits to Cape Hatteras) was higher during low-NAO period than in high-NAO period. However, diminished transport was found downstream of Cape Hatteras during the same low-NAO period due to weakened recirculating cells of the Southern recirculation gyre, northern recirculation, and deep western boundary current. The opposite setup is true for the high-NAO period when the Florida Current is relatively weak, and the GS

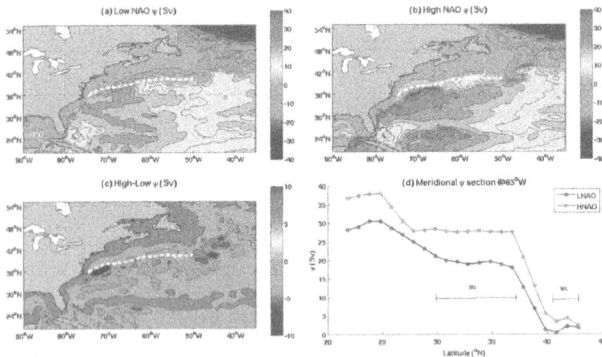

Figure 11.10: Barotropic Streamfunction: Barotropic streamfunction for the GS region during (a) low NAO and (b) high NAO phases. Stream contours signify 4 Sv increments. (c) Difference between high-NAO and low-NAO barotropic streamfunctions. White dashed line represents satellite-derived long-term GS north wall position (Drinkwater et al., 1994), which is provided as a reference for the reader. (d) Meridional Section across 65 W from (a) and (b) showing the barotropic transport streamfunction contributions from the southern recirculation gyre (SRG) and northern recirculation gyre (NRG). Republished with permission of Elsevier Science and Technology Journals, from "Response of the Gulf Stream transport to characteristic high and low phases of the North Atlantic Oscillation," Chaudhuri et al., in Ocean Modelling, 39 (3–4), 2011; permission conveyed through Copyright Clearance Center, Inc.

downstream of Hatteras is much stronger with a northward displaced path. This asymmetric impact of the basin-wide wind system is shown in Figure 11.10 in terms of barotropic streamfunction (proxy for wind stress curl forcing) from the model. This model simulation was also used to investigate the difference between the responses of the Eastern versus Western North Atlantic to the high and low NAO winds. The response parameters were different from the previous study described above (but the same model simulations). Here, the low-salinity intrusion through the Labrador Current into the western North Atlantic was found to be concurrent with high-salinity intrusion of the Mediterranean Out-flow water in the eastern North Atlantic during the extremely low-NAO phase of 1997–1998. This event was related to a concurrent westward shift of the forcing field, the sub-polar front. That is another story that you can look up in Chaudhuri et al. (2011a).

There are a number of excellent simulations for the North Atlantic as mentioned in some of the earlier chapters. The reader is highly encouraged to look at those, some of which are also mentioned in the references of these two recent studies briefly described here.

11.6.2   A BAY OF BENGAL SIMULATION

Now, let us move to the other side of the world, the Bay of Bengal (BoB). While one-fourth of the world's population lives around this basin (BoB), the economies of the surrounding countries have limited resources to survey the ocean. In addition, the ocean-atmospheric setup of the summer monsoon and the reversing seasonal wind patterns (southwesterly during summer/fall and northeasterly during winter/spring) have made understanding the details of the circulation of the Bay a bit challenging. So, with the help of satellite observations and limited oceanographic observations, the numerical modeling simulations are a big help to progress in our understanding of the behavior of the changing climate in the Bay of Bengal. We thus discuss a couple of recent simulations (again with ROMS) as examples of how the simulations are helping to characterize and quantify some aspects of the circulation of the Bay.

The Bay of Bengal is known for its salinity contrast from Northern Bay to Central and Southern Bay, due to the huge freshwater discharge in springtime from Monsoonal rain as well as from a number of rivers in the Indo-Gangetic plane. In fact, the Bay of Bengal was chosen as one of the primary calibration sites for ground-truthing NASA's Aquarius Salinity mission. To understand the impact of seasonal river input on the freshwater dispersion in the Bay, a modeling system was set up by Jana et al. (2015). The key element of this set of simulations was to include the river runoffs in the BoB region. The ROMS model domain (see Figure 11.11) extended from 76°E to 100°E, and 4°N to 24°N with an average horizontal resolution of $1/12°$ (≈ 9 km) and 32 terrain following vertical levels. For better simulation of the upper ocean dynamics, the stretching parameters were chosen as $\theta = 7.0$; $b = 0.1$ (see Section 10.6). Such a choice of θ allowed for 12 vertical levels in the upper 100 m at the position of maximum depth in the Bay. The bathymetry field was extracted from the ETOPO2 (Smith & Sandwell, 1997) 2-min topography data.

The model domain has oceanic boundaries at the eastern, southern, and western edges where the climatological tracer and velocity fields were imposed from the 0.25° WOA01 monthly climatology (Boyer et al., 2005). Initial conditions were derived from the January values of WOA01. The model was started from rest and forced by the climatological monthly forcing fields. Forcing fields containing wind stress, air temperature, air density, relative humidity, and specific humidity were taken from the 0.5° monthly climatology of the Comprehensive Ocean Atmosphere Data Set (COADS) (da Silva et al., 1994) supplemented by evaporation and precipitation from Coordinated Ocean Research Experiments version 2 (CORE2) air–sea fluxes (W. Large & Yeager, 2009), and surface heat fluxes from the Objectively Analyzed Air–sea Fluxes (OAFlux) Project (Yu et al., 2008; Yu & Weller, 2007).

To study the impact of seasonal river input, a twin experiment was designed – one with river runoffs and one without the river runoffs. A total of

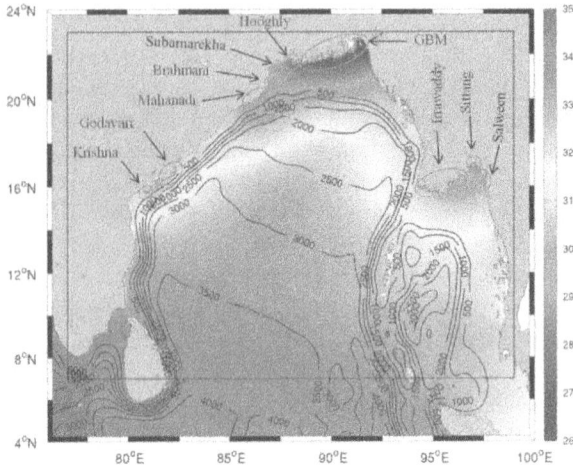

Figure 11.11: The model domain with climatological annual salinity (shaded background). Contours represent the model bathymetry values. The dotted black line represents the boundary of the domain of analysis. Magenta points along the boundary are the locations of the point sources for the ten rivers shown. Republished with permission of Elsevier Science and Technology Journals, from "Impact of seasonal river input on the Bay of Bengal simulation," Jana et al., in Continental Shelf Research, 104, 2015; permission conveyed through Copyright Clearance Center, Inc.

ten rivers (see Figure 11.11) were included in the river runoff simulation. Both of the simulations were carried out with exactly the same initial, boundary, and surface forcing conditions.

The rivers were included in the model as point sources with two key attributes (salinity and flow). A non-zero salinity at the river mouth is more like specifying the estuarine salinity at the coastal node of the model domain. The flow discharge of the river at the mouth was distributed vertically at the point sources according to a pre-determined monotonically decreasing function of depth in the upper 50 m. The climatological monthly mean discharge data were derived from the Global River Discharge Database (RivDIS v1.1) (Vörösmarty et al., 1998). However, discharge data for the Hooghly, Meghna, and Salween were obtained from observational studies, as they were unavailable in the RivDIS data set. The average values of freshwater discharge for the Hooghly are 3000 m^3/s during the southwest monsoon and 1000 m^3/s during the dry season (Sadhuram et al., 2005). Data on the Salween discharge were obtained from Syvitski et al. (2005).

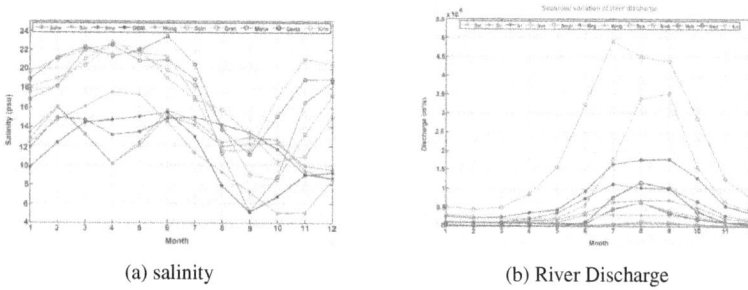

(a) salinity   (b) River Discharge

Figure 11.12: Left: Seasonal cycles of salinity are shown at the point sources corresponding to different rivers (shown in Domain Figure). Right: River discharge for ten rivers. Republished with permission of Elsevier Science and Technology Journals, from "Impact of seasonal river input on the Bay of Bengal simulation," Jana et al., in Continental Shelf Research, 104, 2015; permission conveyed through Copyright Clearance Center, Inc.

Due to scarcity of the observational salinity data at the river mouth and the overestimation of salinity in the climatological values at the coastal nodes (due to OA, which influences the coastal salinities to be biased by the open ocean high salinities), a seasonal cycle of salinity was constructed for each river mouth from past synoptic observations. The detailed procedure involves a bias correction from climatology (see Jana et al., 2015 - equation 2). The resulting seasonal cycles are shown in Figure 11.12. This is a very important development as this seasonal cycle can now be used for any other modeling simulation study in the future with river set up in the Bay of Bengal. The other option would be to couple an estuarine model to the ocean model and use a dynamic coupling to force the ocean model with both salinity and flow at the river point sources for the ocean domain.

The twin experiments were run for 15 years each. The initial adjustment from the state of rest to a dynamic equilibrium took about 3 years for temperature and 7 years for salinity. Note that the Bay of Bengal is much smaller than the North Atlantic and thus takes less time to adjust its kinetic energy (2–3 years); however, the salinity adjustment is a function of the initial bias between the river-input and climatological values nearby, which takes a longer time to redistribute over the whole volume.

The final year of model simulations was compared and the freshwater plume characteristics and dispersion within the Bay were quantified. Figure 11.13 shows the formation and dispersion of the plume water (identified by the 32.5 psu isohaline contour) as simulated by the RR simulation during the final year of the simulation.

It was possible for the first time to quantify the two large plumes in the Bay of Bengal in terms of their Kelvin Numbers. Kelvin Number is the ratio

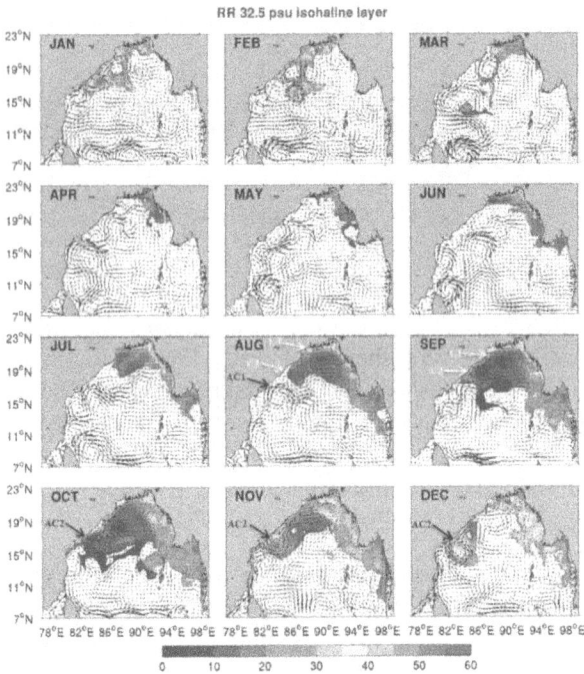

Figure 11.13: Depth of 32.5 isohaline layer showing the vertical extent of the freshwater plume. Spreading of the freshwater in the central part of the northern BoB is shallow and limited to upper 20 m. Arrows denote the surface circulation. Note that the horizontal spreading of the freshwater plume is the largest during October and the freshwater flows out of the Bay in December. The plume water dispersion is related to the advection of eddies as marked by the arrows. Republished with permission of Elsevier Science and Technology Journals, from "Impact of seasonal river input on the Bay of Bengal simulation," Jana et al., in Continental Shelf Research, 104, 2015; permission conveyed through Copyright Clearance Center, Inc.

of the horizontal spreading scale of the plume to the baroclinic Rossby Radius (Garvine, 1995). Considering the depth of the simulated GBM plume at 22°N as 35 m, cross-shelf spreading length scale of 200 km, and a baroclinic Rossby Radius (with $\triangle\rho = 5$) of 25 km, the Kelvin Number for this large-scale plume was determined to be 8. For the ISS plume at 17°N, which is also surface advected with a shallower depth of about 25 m and a baroclinic Rossby Radius ($\triangle\rho = 2$) of 16 km, a spreading scale of about 50 km, its Kelvin Number was about 3. These calculations put the ISS plume to be comparable dynamically to the plumes in Rhines and off of Delaware Bay, while the GBM plume competes with the most large-scale coastal current plumes such as the Scottish Coastal Current plume (see Table 4 of Garvine, 1995). Two large-scale freshwater plume formations follow the river discharge cycle reasonably well. First, the GBM plume forms in April (see Figure 11.13). Then the ISS plume appears in May, when parts of the GBM plume move southward and get advected by underlying eddies to the east. By June, the two plumes from GBM and ISS merge into a large plume in the north-northeastern part of the Bay. Further movement and dispersion of this freshwater plume system from the head of the Bay is primarily dominated by eddies and the cyclonic gyre in the northern Bay during June through September. Parts of the GBM plume move southward during post-monsoon along with the EICC and flows out of the Bay in December.

Comparison of the RR simulation to the Non-RR simulation clearly highlighted the realism that the river-input can bring to our understanding of the circulation. It was clear from these simulations that the river input is necessary for addition to precipitation-minus-evaporation in setting up the near-surface stratification and for the formation of the plumes and the barrier layer in the BoB. One needs to appreciate that while satellite observations and targeted observations can tell us about synoptic local dynamics of a region, the model simulations adjusts such fields and can provide a more in-depth, four-dimensional view of the processes working in ever-transient but seasonally repeating situations like that exist in the Bay of Bengal.

A follow-up simulation with the same setup was carried out by Jana et al. (2018) to investigate the sensitivity of the upper ocean to different winds and river-input condition. Sensitivity experiments are a very effective way to understand the behavior of a non-linear system like the ocean to differing strengths of a certain forcing parameter. The NAO simulations shown in Section 10.4.1 can also be thought as sensitivity simulations of the North Atlantic basin to differing wind strengths from high and low-NAO conditions. Many existing simulations are essentially sensitivity experiments with varying forcing and internal parameters (e.g., viscosity coefficients, mixing parameterization, diffusion coefficients etc.) as well as horizontal and vertical resolutions and different initial and boundary conditions.

A set of four different experimental simulations using two strong/weak winds and two estuarine/zero-salinity river input conditions were carried out by Jana et al. (2018). They used the same model configuration as Jana et al. (2015). The sensitivities were analyzed by comparing and contrasting the model simulations in terms of responses such as surface circulation, thermohaline structure, freshwater plume dispersion, and the coastal upwelling along the western boundary. The different impacts of winds (mixing) on the upper layer circulation which is dominated more by salinity (in terms of barrier layer) were identified. The salinity variations identified the possible extent and distribution of the freshwater plume in the extreme (zero-salinity at the coast) scenarios. The use of estuarine salinity resulted in more realistic simulations all over the domain. They concluded that the use of estuarine salinity and stronger winds (probably from satellites) would reduce existing model biases of high temperature.

There are a number of excellent simulations in the Bay of Bengal from many different groups over the last few decades (starting with Bryan, 1969a) and the reader is encouraged to look at those, which are mentioned in the references of these two recent studies briefly described here.

In recent years, thanks to the international collaborations between Indian scientists, agencies, and their counterparts in the US, Australia, Japan, Europe, Russia, and Africa, our knowledge about the circulation in the Bay of Bengal, Arabian Sea and the South Indian Ocean is increasing rapidly. Two examples are the application of HF Radar data for understanding the coastal circulation in multiple regions around the coast of India (Odisha coast, Gulf of Khambat, Andhra coast, and Andaman Seas), where Radar networks are available. See (Mandal, Sil, Gangopadhyay, Jena, & Venkatesan, 2020; Mandal, Sil, Gangopadhyay, Jena, Venkatesan, & Gawarkiewicz, 2021; Mandal et al., 2018) and the references therein for details. These studies focus on extracting high frequency (shallow water tidal) signals from HF Radar, understanding their variability due to interaction with bathymetry and gulf morphology (asymmetry) and identifying meso- and submesoscale features in shallow coastal regions allowed by the high resolution (6 km) HF Radar signals. Assimilation of such HF Radar data (surface velocity fields) for operational modeling would be the next step in the coming years.

11.6.3 THE GENESIS REGION OF THE BRAZIL CURRENT – AN OGCM APPLICATION

Finally, let us briefly explore another study that exploited a product of reanalysis done by an ocean General Circulation Model (OGCM). This OGCM is called OCCAM (Ocean Circulation and Climate Advanced Model), described by Webb et al. (1998). It is on a global domain encompassing the southern oceans, with a 1/12° horizontal resolution and 66 vertical levels. It reproduced the mean features of the general circulation along the Brazil coast very well (Cirano et al., 2006). This model results were used in conjunction with hydrographic data and ADCP-derived horizontal velocity fields to investigate the genesis region of the Brazil Current by Soutelino et al. (2011).

Concurrent observations clearly indicated that the region of genesis of the Brazil Current between 10°S and 20°S is inhibited by three anticyclonic eddies constrained by a set of topographic features. These eddies were so named: the Illheus eddy, the Royal-Charlotte eddy, and the Abvrolous eddy. The topographic features are the Illheus Bight in the north, the Royal-Charlott Bank, the Abrolhos Bank, and the Vittoria-Trinidad Ridge. See left panel of Figure 11.14 for the geographical setting and the eddies. The climatological mean

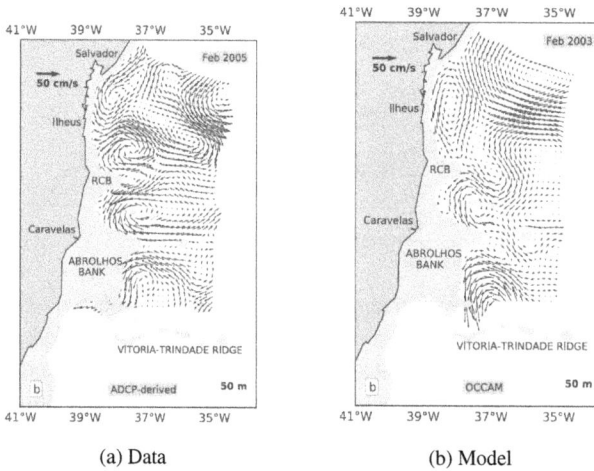

(a) Data   (b) Model

Figure 11.14: Left: Observed non-divergent velocities at 50 m calculated from OEII ADCP data. Right: OCCAM 2003 February mean velocity fields at 50 m. Republished with permission of John Wiley and Sons – Books from "Is the Brazil Current eddy-dominated to the north of 20°S?," Soutelino et al., in Geophysical Research Letters, 38(3), 2011; permission conveyed through Copyright Clearance Center, Inc.

circulation did not show such eddies. By comparing and contrasting the data and OGCM circulations fields, Soutelino et al. (2011) proposed that the Brazil Current generation region north of 20°S is dominated by these three eddies differently in different seasons. A comparison of observations and simulation fields from February is presented in Figure 11.14, which vindicated such a hypothesis.

11.7 PREDICTION

Let us talk about prediction now. Ocean prediction is an initial value problem as well as a time-scale dependent problem.

First of all, the better the initial condition, the better the prediction, or at least that is what you hope. This idea of dependency on the initial condition has been proven by the so-called "Butterfly Effect," elegantly demonstrated by Edward Lorenz of MIT in 1963 (Lorenz, 1963). This is also known as the "sensitivity to initial condition." Let us consider the simple non-linear system of atmospheric convection that is known today as Lorenz's equations.

They are

$$\frac{dx}{dt} = \sigma(y - x)$$

$$\frac{dy}{dt} = x(\rho - z) - y$$

$$\frac{dx}{dt} = xy - \beta z \tag{11.12}$$

Physically, think about a two-dimensional fluid layer being heated from below and cooled from above. See equation 11.12 for their mathematical dependencies on each other. Here, (x, y, z) are related to the rate of convection and temperature variations, and $(\sigma, \rho,$ and $\beta)$ are parameters related to some non-dimensional numbers of diffusivity and convection and physical dimension of the layer itself. Without going into the details, you can appreciate that these equations are nonlinear, time-dependent, three-dimensional, non-periodic, and deterministic. Lorenz assumed the values $\sigma = 10; \beta = 8/3$, and $\rho = 28$, and showed that the system exhibits a chaotic behavior for these values and if we just change these values by very little (to the inappreciable decimal places).

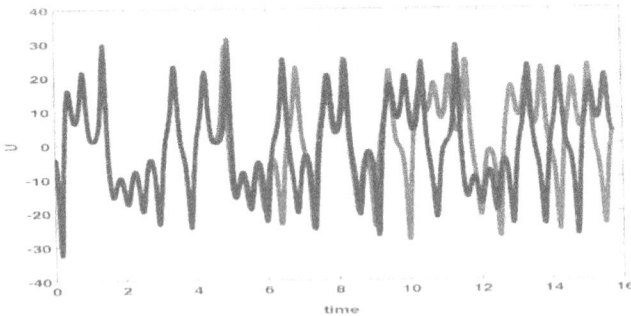

Figure 11.15: Lorenz solution showing sensitivity to the initial condition (from Wikipedia).

A simpler version of the solution is shown below in Figure 11.15. For a very minute difference in the initial value of a variable (from 0.832479 to 0.832, which is a round-off error in a calculation used by many), the two solutions diverge considerably.

Now, the solution is thus very sensitive to the initial condition. So, appreciating this kind of nonlinear set of equations describe both ocean and atmosphere (and for that matter climate as a whole), we can conclude that Ocean/atmosphere/climate Prediction is an initial value problem that is sensitive to the initial condition of choice and thus will differ in its final predicted state from one another if we change the assumed initial condition only a little. This can also lead to chaotic solutions at times even if our initial condition (which we can never specify perfectly – there will always be some gap in specification, even at the finest resolution level) is the best that we can come up with.

Second, ocean prediction is a time-scale problem too. Are we predicting the ocean for the Next six hours? Next 24 hours? Next couple of days out? A week ahead? Or 2 weeks from now? Do we want to know the state of the ocean a month or a season ahead in advance to plan for the next cruise? Do we want to predict the ecosystem and the habitat of Lobster or squid or Mackerel or Tuna stocks 1, 2, 3 years in advance? How about temperature and sea-level rise a decade from now or 50-years from now or for the twenty-second century?

There are two bearings on the above two scenarios.

(i) A particular initial condition suitable for one time-scale of prediction, might not be applicable (or need more information) for a separate time-scale of interest.

(ii) A perfect initial condition for one particular time-scale of prediction might work for a certain length of time within the period of prediction and then the model skill might be deteriorating.

The second one is related to the internal dynamics of the nonlinear system – how different dynamical processes might be important and dominant at different times of evolution during the prediction time window – it is related to the predictability of the system for a particular process. It is also somehow linked to the ocean's memory of its initial condition. How long does the system keep the memory of the initial condition? The memory might be dependent on both the model and the processes that we are trying to predict (and their representation in the model equations).

The first one is more like this – is one snapshot of the ocean sufficient for predicting the ocean for all time scales? Or if we could apply the same snapshot to different models representing different dynamical processes relevant for different time-scale evolution – will that always work?

These are important questions that a research forecasting exercise should think about and identify models that have skills to predict certain processes, phenomena and time scales. As a renowned forecaster said once in the early days of ocean prediction, "Unless you try to predict with what you have (models and data) you will never know whether your prediction is right or wrong. So, go ahead, predict and verify with new data. Predict, analyze, verify, understand the gaps and then improve the model parameters and try again." This is the game of prediction. The only thing certain about the future is its uncertainty. So, ocean prediction is an interesting problem of predicting the future ocean at different time-scale with uncertainties or errors. Always look for those error bounds of a forecast. Those are important.

Let us now look at some of the practical applications of predictions.

11.8   THE INTEGRATED OCEAN OBSERVATION SYSTEM (IOOS)

In 1998, during the Year of the Oceans, the United States Congress called for the establishment of an "Integrated Ocean Observing System" or **IOOS**. The purpose of IOOS is to routinely provide *data* and *information* required for more rapid detection and timely **prediction** of state changes in the oceans.

The overarching theme of the IOOS program was to have ONE system, addressing SEVEN societal goals. These are:

- Improve the safety and efficiency of marine operations
- Improve national/homeland security
- Improve forecasts of natural hazards and mitigate their effects more effectively
- Improve predictions of climate change and their effects
- Minimize public health risks
- Protect and restore healthy coastal marine and estuarine ecosystems more effectively
- Sustain living marine resources

Later in 2004, the US Commission on Ocean Policy released its report calling for the implementation of an IOOS to make effective use of existing resources and enhance operational capabilities over time to address the above seven societal goals. It was this report that emphasized the need for ecosystem-based management (EBM) and asked to strengthen the regional approach as a means of implementing the EBM.

Over the next few years, this regional implementation strategy evolved in the creation of eleven different (sometimes overlapping – as the ocean knows no boundaries and flows from one region to the other) regional associations made up of academic institutions, national and state agencies, and industries to set up eleven different Coastal Ocean Observation Systems or COOS-es. These are spread over around the US Coast and are shown in Figure 11.16. Specifically, they are starting from the northeast, Northeast Regional Association COOS (NERACOOS), Mid-Atlantic Regional Association COOS (MARACOOS), SouthEast (SACOORA), Gulf of Mexico (GCOOS), Caribbean (CARICOOS) in the Atlantic sector and Southern California (SCCOOS), Central California (CENCOOS), Northern US (NANOOS), Alaska (AOOS) and Hawaii-Pacific islands (PacIOOS) in the Pacific sector. Each of these entities is now thriving with their own data products and region-specific modeling systems that produce real-time monitoring information as well as short and long-term predictions as appropriate. The reader is encouraged to explore their websites for much more information on their excellent efforts. We will discuss one of these efforts (MARACOOS), specifically one modeling component in the next chapter as an example. In parallel, many other countries have established their own regional and national programs around their own coasts. Those national efforts have been

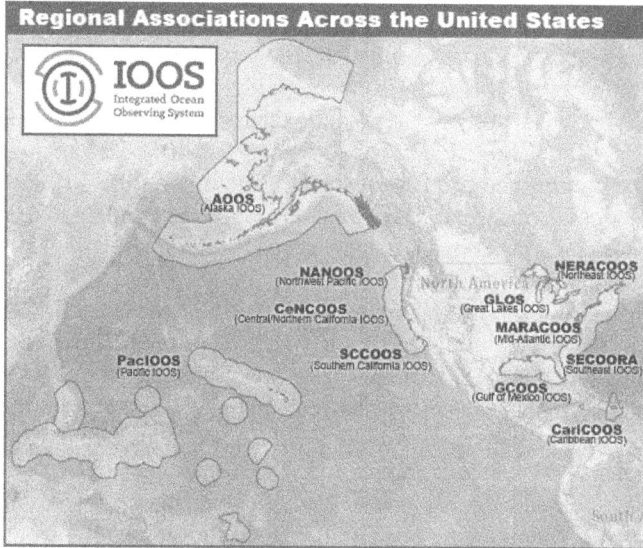

Figure 11.16: Regional Coverage of the US regional coastal ocean observation systems. Credit: NOAA IOOS.

very successful as well in monitoring and providing useful ocean predictions for the benefit of the society.

11.9 APPLICATIONS

Let us consider two applications now. The first is futuristic. How can we explore and exploit the emerging technologies such as machine learning and artificial intelligence for future prediction in a changing climate? This is where innovations are needed. The second is an example of a success story of the application of the ocean-atmospheric prediction system with internet and communication technology that did and can save lives in a natural disaster.

11.9.1 SIMULATION, PREDICTION, CLIMATE CHANGE, AND MACHINE LEARNING

As the climate is changing, the background state of the oceans is changing. The winds and the sea surface temperature are interacting at the air-sea interface and are responding and adjusting to such changes on a global scale. Thus the simulations presented in this chapter are to be treated as guides for future simulations and the results are relevant to our present and past climatic backgrounds. New results, new simulations, new predictions are to be sought in the face of climate change and evolving technologies such as machine learning and artificial intelligence. Application of machine learning for automatic identification of eddies and fronts (Gulf Stream, Kuroshio, other regions) and using those for initialization in real-time (using synoptic features, see Chapter 12) would enhance real-time forecasting in each IOOS activity. Feature-based assimilation using machine learning and artificial intelligence algorithms is not too far in the future. Such applications will enhance not only our understanding to adapt to forecasting in a changing climate; they will broaden our understanding of the pace of change in our current climate.

Learning based on past data might not be fully applicable for accurate predictions in a changing climate. This is where innovation is needed. How does one predict regime shifts that have been observed (Gangopadhyay et al., 2019; Silver et al., 2021a) in various physical, biological, and geological time series? Abrupt changes have happened and will happen in the future. How can we predict those? Responses at multiple time scales are captured by several observational data sets. Simulations and predictions should be able to capture these multiscale variabilities at each instance for any location. Physics-guided machine learning algorithms might be one approach where underlying dynamics is preserved by the physical constituents of the system's behavior; while the variations in the higher frequency can be better captured by short-term available data-learned statistics-based intelligent networks.

11.9.2   OBSERVATION, PREDICTION, TECHNOLOGY, AND SOCIETY

Let me draw your attention to a recent example of interlocking benefits of ocean and atmospheric prediction, communication and computer technology for mitigation of the impact of cyclone Fani (May 2019). When it comes to cyclones, there's a strong link between advanced communication systems, computer technology, and ocean technology. Together they have the ability to save lives. On May 3, 2019, Cyclone Fani struck Odisha, a state in India with a population of nearly 46 million individuals. The 157-mile-per-hour cyclone was classified as a category 4 or 5, mirroring the force of Puerto Rico's Hurricane Maria. About 2 million people live in the region where Fani made landfall; 11 districts along the state of Odisha were placed on red alert, and 900 shelters were set up to house evacuees. These shelters were really blessings for people to go and seek safety.

Unfortunately, 14 people lost their lives in Fani. But compare this to the year 1970 and 1999 when two other devastating cyclones hit the same region. About half of a million people lost their lives nearly 50 years ago when Cyclone Bhola hit the region (Emanuel, 2005; Longshore, 1998). We did not have satellites or computer technology with modeling capabilities to forecast cyclones then. In 1999, the Odisha Super Cyclone occurred and 10,000 individuals lost their lives. We could predict that the cyclone was coming, but could not communicate to about 6000 fishermen well in advance for them to return home safely. Twenty years later, Cyclone Fani, which occurred in the same location as the Odisha Super Cyclone, affected many fewer lives due to the availability of advanced ocean technology. Forecasts for Fani were issued five days ahead of the cyclone. Issuing the forecast just five days in advance resulted in fishermen being aware of their need to leave the ocean for safety. During Cyclone Fani, about 1 million lives were saved because of the efforts of academia, the government, NOAA, and Weather forecasting Centers in India including the Indian Met Department and all of the observation capabilities of satellites and the use of the internet to communicate. That is the power of science – satellites, modeling, and observations are helping to save lives in a way that could not be done in the 1970s and the 1990s. Just imagine. From half-a-million fatality in 1970 to ten thousand in 1999 and then to a handful of people lost in 2019. That is a somber measure of progress (or success) in our understanding of the ocean and atmosphere and their modeling as a predictive system. Science and communication together can really save lives. You cannot mitigate something if you don't know when and where it's coming. And you have to know it well in advance so that you can take the message out to the people who might be affected. That is where prediction helps society. Naturally, the Ocean knows no boundaries and the idea of the water planet as our only real home have brought the IOC to institute a Global effort in connecting all different IOOS from all over the world oceans. This has led to the formation of the Global Ocean Observing System (GOOS). And then, why only ocean? Think about the whole ocean-atmosphere-geosphere-biosphere system that we talked about at the beginning of this book in Chapter 1. This has led to the establishment of the Global Earth Ocean Observing System of Systems (GEOSS). There is a lot to learn from one another in this all-connected world ocean of ours.

CONCLUSION

In this chapter, we discussed the process of simulation and prediction in a general way. Two important aspects: (i) grids and domain set up and (ii) initialization schemes were discussed in some detail. A number of available observation data sets were discussed from various instruments and various data collection agencies from around the world. Multiple techniques of OI and the concept of data assimilation are briefly outlined, which are required for both simulation and prediction. Three example simulations for North Atlantic, Bay of Bengal, and Brazil Current System were presented.

We discussed the art of prediction and the challenge of predicting a nonlinear system such as an ocean basin or a coastal region in some detail. Examples of predictions using numerical models, by various agencies such as IOOS are being used for monitoring purposes and such applications were briefly mentioned at the end.

FURTHER READING

Bretherton, F. P., Davis, R. E., & Fandry, C. B. (1976, July). A technique for objective analysis and design of oceanographic experiments applied to MODE-73. In *Deep Sea Research and Oceanographic Abstracts, 23*(7), 559–582, Elsevier.

Carter, E. F., & Robinson, A. R. (1987b). Analysis models for the estimation of oceanic fields. *Journal of Atmospheric and Oceanic Technology, 4*(1), 49–74.

Emanuel, K. (2005). *Divine wind: the history and science of hurricanes.* Oxford University Press,

Gelb, A. (1974). *Applied optimal estimation,* 304 pp. MIT Press.

Lorenz, E. N. (1963). Deterministic nonperiodic flow. *Journal of Atmospheric Sciences, 20*(2), 130–141.

Swallow, J., & Worthington, L. (1961). An observation of a deep countercurrent in the western North Atlantic. *Deep Sea Research, 8,* 1–19.

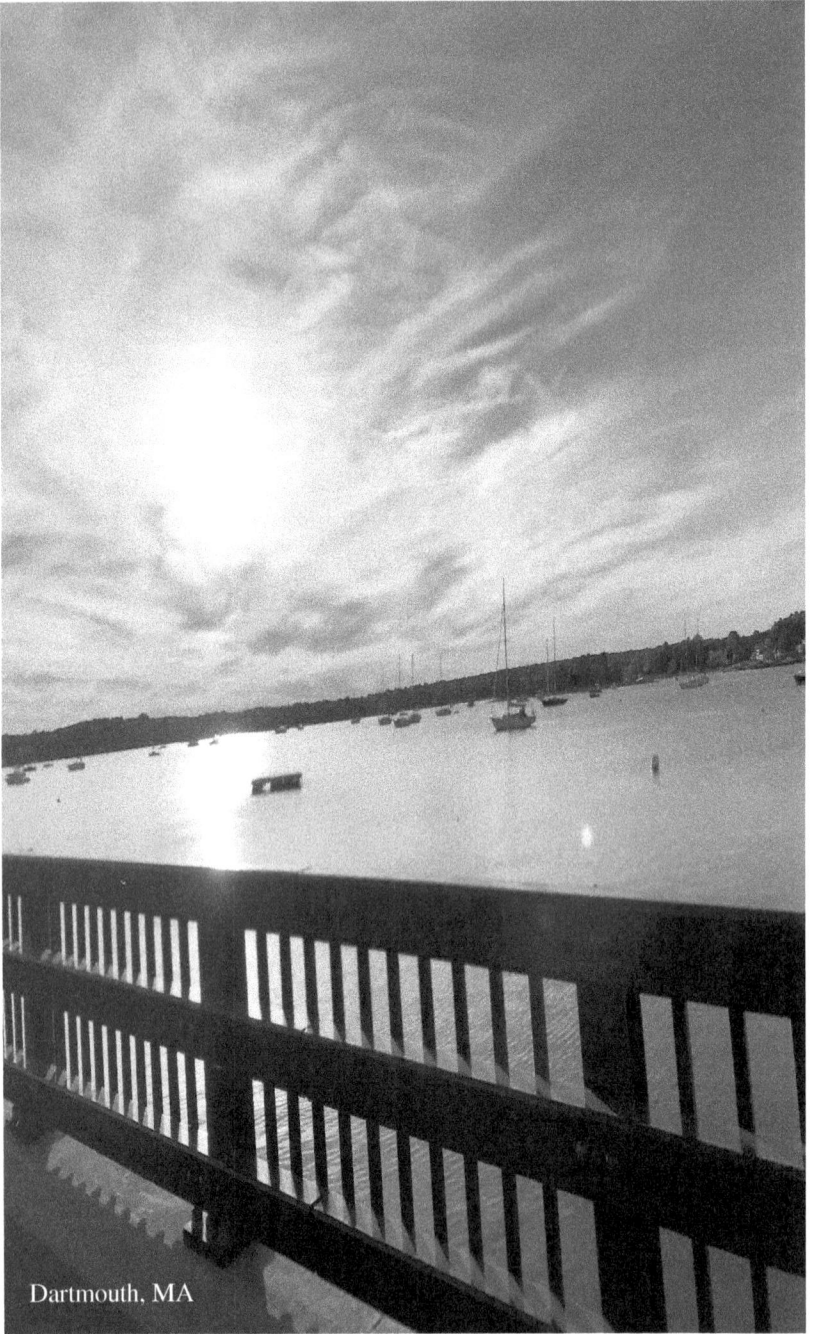

Dartmouth, MA

12 Synoptic Ocean Modeling

> It is at this moment in time that homo sapiens are making substantial progress in understanding the planet upon which they live
>
> ———————————————
> Allan R. Robinson (1932–2009)*

OVERVIEW

In the last chapter, we saw the important elements of simulating and predicting the ocean using a numerical model for studying and understanding various processes. We discussed the need for adequate and appropriate observations and creating the best initial fields for both simulations and predictions. We realized the usefulness of assimilating new data to adjust the model simulation toward reality. We emphasized and showed examples of current efforts to provide real-time forecasts for societal benefit via IOOS and GOOS efforts.

In this chapter, we focus on a particular way to generate the initial condition based on a synthesis of information from past observation of oceanic features and utilizing present-day satellite data. These initial conditions are particularly suitable for short (1–7 days) and synoptic (mesoscale) forecasting. These synoptic ocean predictions use a methodology called the *Feature-Oriented Regional Modeling System (FORMS)*. We describe the historical development of the FORMS methodology first. This is followed by the description of FORMS for the western North Atlantic(WNA), which includes the Gulf Stream and Ring region, the Coastal Gulf of Maine, and Georges Bank. An operational forecasting system for the WNA, which was run under MARACOOS, is described in detail. Finally we show the application of FORMS for process studies and for a few worldwide applications.

12.1 THE SYNOPTIC OCEAN

Ocean prediction is an initial value problem as well as a time-scale dependent problem. A simple question to ask is, "If we know or can guess today's ocean how far out can we predict?"

———————————————

*Credit/Source: Claire T. Carney Library Archives and Special Collections. UMass Dartmouth, Office of Chancellor Peter Cressy papers, URC 6.

DOI: 10.1201/9780429347221-12

If we look at the ocean from a satellite, we see a "synoptic" view of the ocean. The word "synoptic" means "seen together." It comes from the original Greek word "synoptikos." The "seen together" has this feeling of "togetherness" – multiple entities existing together as we see the ocean – at multiple scales – all at the same time – in a synoptic view.

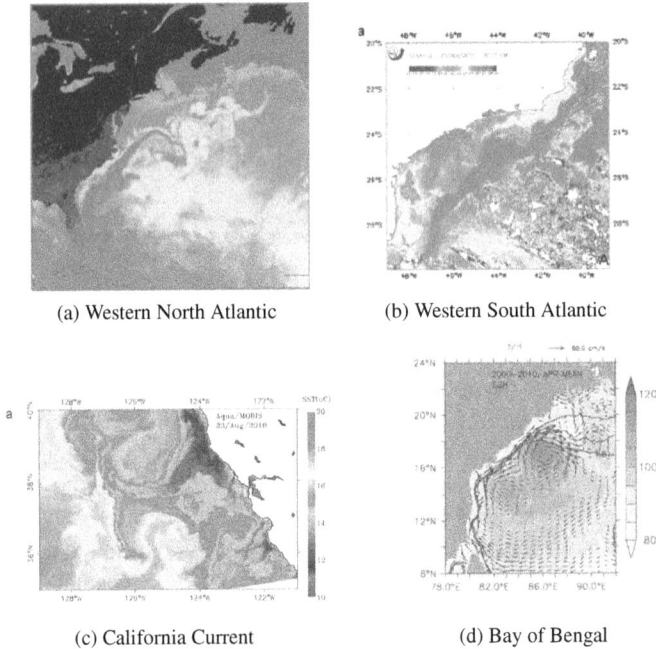

(a) Western North Atlantic   (b) Western South Atlantic

(c) California Current   (d) Bay of Bengal

Figure 12.1: Synoptic Views of four different oceans using Satellite images of temperature (North and South Atlantic and the California Current)and altimetry-derived sea surface velocity vectors for the Bay of Bengal. Credit: (a) https://commons.wikimedia.org/wiki/File:Golfstrom.jpg. (b) Courtesy: Ilson Silveira's Group at IOUSP. (c) Republished with permission of Elsevier Science and Technology Journals, from "The California Current System: A multiscale overview and the development of a feature-oriented regional modeling system (FORMS)," Gangopadhyay et al., Dyn. of Atmosphere and Oceans (52), 2011; permission conveyed through Copyright Clearance Center, Inc. (d) Republished with permission of John Wiley & Sons, from "On the nature of meandering of the springtime western boundary current in the Bay of Bengal," Gangopadhyay et al., Geophysical Research Letters, 40(10), 2013; permission conveyed through Copyright Clearance Center, Inc.

Look at Figure 12.1 for four different oceans. What do you see? You see fronts – the Gulf Stream, the Kuroshio, the Brazil Current, the Somali Current. You see eddies – the Gulf Stream Rings – warm core rings to the northern side and cold core rings to the southern side; similarly for Kuroshio. You see freshwater plumes and upwelling regions in the California Current, you see meandering, eddies, and upwelling regions in the Brazil Current region.

These are our oceans as we see them from satellites, a "synoptic" view. We need to predict this multiscale view for the future!

We can look for some help from Meteorology. The synoptic scale in meteorology (also known as large scale or cyclonic scale) is a horizontal length scale of the order of 100–1000 kilometers or more. This corresponds to a horizontal scale typical of mid-latitude depressions (e.g. extratropical cyclones). Most high- and low-pressure areas seen on weather maps (such as surface weather analyses) are synoptic-scale systems driven by the location of Rossby waves in their respective hemispheres.

The Rossby Radius for the deep ocean eddies is around 50–100 km; while that for the coastal ocean and shelf eddies could be as small as 5–10 km! There is your multiscale co-existence problem. A synoptic Ocean is a multiscale view of the ocean where the large-scale oceanic boundary currents (with length-scales of 100–200 km), eddies with (50–100 km) length scale, to smaller eddies with 5–10 km deformation radius to 10–50 km shelf-slope fronts, jets, filaments, upwelling, and freshwater plumes all exists together.

In the above sense, the "synoptic" scale of the ocean is a multiscale view, and how long do we want to predict this multiscale ocean? Again, let us look at meteorology. The migrating cyclones and anticyclones which control our daily weather have short lifespans, about 3–7 days. However, the oceanic eddies live much longer (weeks to months) and the ocean is known to have some longer memories than the atmosphere. Things are slowly evolving in the ocean than in the atmosphere. So, one would expect that 3–7 days of ocean prediction at a synoptic scale might be a worthwhile goal for synoptic ocean prediction. Things might be a little more complicated than that since the synoptic ocean is also multiscale with all its entities.

It was with that understanding that the idea of defining the synoptic state of the ocean in terms of a number of multiscale entities called "Features" evolved in the eighties. Examples of these oceanic features would be the Gulf Stream, the Rings, the Kuroshio, the shelf-slope fronts, the basin-scale and sub-basin scale gyres, etc.

Let us repeat this idea of synoptic ocean prediction using Features. What do we see from the images from satellites, which are now revealing synoptic views of several oceanographic parameters of the surface ocean such as the sea surface temperature, the sea surface salinity, the sea surface height and the sea surface chlorophyll? We see that the ocean is full of multiscale features such as the eddies of different scales and orientations, large-scale to shelf-scale fronts, gyres, upwelling regions, and finer-scale entities. These features span multiple scales from basin-scale to sub-basin-scale to shelf-scale and their characteristics are relevant for the physics, chemistry, biology, and sediment movement of the oceanic regions.

Now, how does the surface relate to the subsurface? We cannot see the subsurface from Satellites; however, we might have an idea of the synoptic character of the three-dimensional form of these features from past oceanographic surveys and might be able to construct data-based and knowledge-based expressions for the features linking the parameters. These analytical-empirical three-dimensional physical constructions of temperature, salinity, velocity profiles of the features are generally referred to as "Feature models."

If you have a synoptic view of the ocean, these feature models would allow you to construct a three-dimensional state of the ocean by using those feature models in a kinematic synthesis framework. Such a three-dimensional field set can be treated as the best-guess initial field to a numerical dynamical model such as those discussed in Chapter 10. The model would then allow for further prognostic simulation and prediction. The integrated fields would also allow us to understand processes and phenomena including evolution of these three-dimensional features themselves and their interactions with one another and with the background field. Such field simulations could then be used for societal benefits, for weather forecasts, for fisheries and ecosystem management, for studying climate change (comparative state simulations) and many other applications including search and rescue and naval applications, ship routing, oil spill, and offshore oil exploration and exploitation.

In this chapter, we will learn about this unique way of synoptic ocean prediction. A generic name for this approach has evolved over time and it is called the "Feature oriented Regional Modeling System" or FORMS. We will first give an overview of the FORMS approach in the next section. Our focus is on the WNA as the region of implementation. We then describe how to identify and develop a "feature model" for a western boundary current such as the Gulf Stream and an eddy. Then we will know how to merge multiple features together to form an initialization. Then we will look into a numerical modeling exercise with such initialization in an operational model setting for MARACOOS. Finally, we will outline the application of this technology for other regions in the world oceans with what exists today and where we could go in the future.

12.2 FEATURE ORIENTED REGIONAL MODELING SYSTEM (FORMS)

Let us think about typical oceanic regions for a moment. We know that the western side of the ocean basins is dominated by the western boundary currents – the large-scale meandering synoptic features. The eastern boundary current regions are more dominated by the transient eddies and upwelling regions, with a disorganized equatorward flow bringing colder polar and sub-polar waters to the tropical regions. Now, there are also subsurface thermohaline flows, and so the wind-driven features are interacting with the subsurface flows, and probably in layers and in opposing directions, as the case may be in particular regions. So, the consideration of the development of feature models and putting them together in a synthesized three-dimensional circulation is interesting and challenging at the same time.

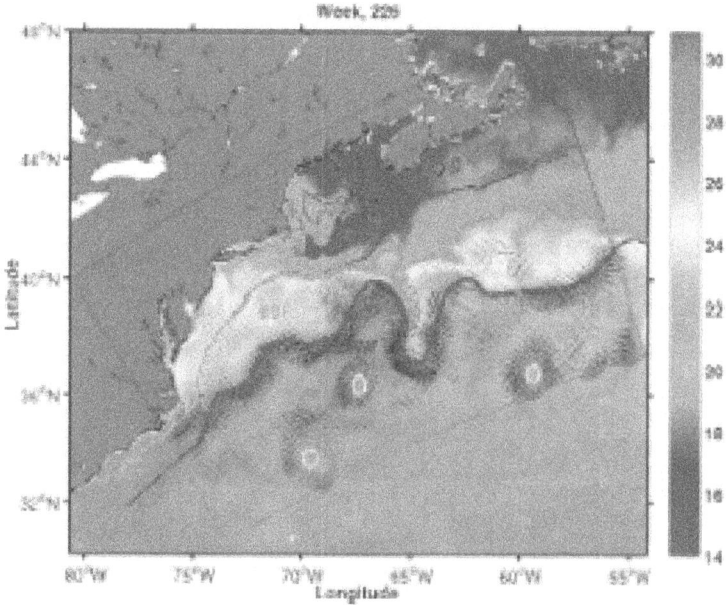

Figure 12.2: NOAA CoastWatch blended 5-day composite of sea surface temperature (SST) overlaid with absolute geostrophic velocities computed from sea surface height from AVISO, centered on 14 August 2006 at 1.1-km resolution. The front analysis shows the shelf-slope front (SSF) in green, the Gulf Stream north wall in blue and the rings in white. Republished with permission of Elsevier Science and Technology Journals, from "An operational ocean circulation prediction system for the western North Atlantic: Hindcasting during July–September of 2006," Andre Schmidt and Avijit Gangopadhyay, Continental Shelf Research (63), 2013; permission conveyed through Copyright Clearance Center, Inc.

The WNA is a great example to follow the idea of synoptic ocean modeling. Look at the WNA using a satellite image, as shown in Figure 12.2. You can see the Gulf Stream, the Rings, and large regions of interacting multiscale eddies and other features. Following the logic of sub-regional differences within the larger WNA, it is worth thinking of a multiscale multi-nested modeling system utilizing any of the modeling systems described in Chapter 10.

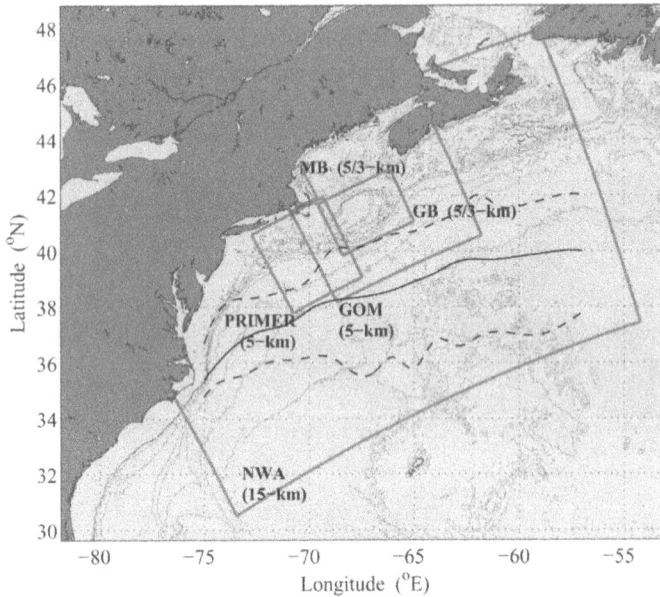

Figure 12.3: Multiscale nested modeling domains for the WNA. The Northwest Atlantic (NWA) domain has a horizontal resolution of 15 km with triple-nested domains in Gulf of Maine (5 km resolution), Georges Bank (5/3 km), and Massachusetts Bay (5/3 km). Feature-oriented initialization and updating helps in transferring information between such regions through multiscale circulation structures, which exist, evolve, and interact over these domains in a dynamical model. Republished with permission of Elsevier Science and Technology Journals, from "Feature-oriented regional modeling and simulations in the Gulf of Maine and Georges Bank," Gangopadhyay et al., Continental Shelf Research (23), 2003; permission conveyed through Copyright Clearance Center, Inc.

As a first example, we choose the HOPS modeling system with a suite of nested architecture to address different multiscale processes and applications in the WNA, as shown in Figure 12.3. Multiple domains are shown to resolve multiple specific features important for regional and sub-regional circulation. Let us consider two specific ones for the development of FORMS. First the deep region dominated by the Gulf Stream and its rings – called the Gulf Stream Meander and Ring (GSMR) region. Second is the shallow coastal regions encompassing both the sub-basins of the Gulf of Maine (GOM) and the shallow-to-deep circulation around the Georges Bank (GB). Together, we call this coastal region as the Gulf of Maine-Georges Bank (GOMGB) domain.

Once you have selected a region and identified the dominant features of circulation and thought about a problem of simulation (for process studies) or for prediction (maybe in real-time for some application in industry, academia or Navy or for some event); you are almost ready to initialize a model. You realize that from all the past data and new data, you can create new feature models, but you only have the feature models for parts of the region where the features exist, and the whole region is definitely not fully covered or occupied by these features. So, you need to fill up the whole region with some background. This is what FORMS does. It is a methodology to bring in synoptic feature models within the background of mean circulation (to fill the gaps between features) using a multiscale objective analysis approach. The multiscale objective analysis approach is built upon the OA methodology discussed in Section 11.4.3 before.

A little bit of history on the development of Feature Models for synoptic prediction and its evolution to FORMS is in order here. Utilizing data and synoptic structures of features for ocean modeling was first developed by A. Robinson et al. (1988) and extended later by A. Robinson et al. (1989) and Glenn and Robinson, 1995. They essentially developed the first generation of feature models for the Gulf Stream (with symmetric profiles), its Rings and used a Quasi-Geostrophic model and a primitive equation model for operational forecasting for the Navy. Later on, a three-part series of studies by Gangopadhyay et al., 1997 (GRA97), A. Robinson and Gangopadhyay, 1997 (RG97), and Gangopadhyay and Robinson, 1997 (GR97) expanded the feature model idea to sub-basin scale gyres, asymmetric Gulf Stream with transport increasing from Hatteras to 60°W, and the implementation of a deep western boundary current (DWBC) to cover the depth of the North Atlantic down to 4500 m. The overall objective of this series of studies was threefold: (1) to present a methodology for deriving a dynamically balanced regional climatology that maintains the synoptic structure of the permanent fronts embedded in a mean background circulation, (2) to present a methodology for using such a regional climatology for calibrating and validating dynamical models, and (3) to use similarly derived synoptic realizations as initialization and assimilation fields for mesoscale nowcasting and forecasting.

This methodology was further extended to include the Gulf of Maine/Georges Bank (GOMGB) region, developed as a *feature-oriented regional modeling system (FORMS)* and implemented in a data assimilative operational mode by (Brown, Gangopadhyay, Bub, et al., 2007; Brown, Gangopadhyay, & Yu, 2007) for the western North Atlantic. This operational system ran on a weekly frequency for the MARACOOS program from 2009 through December 2016. Some of the operational model results and their usefulness for validation with drifters, assimilation of gliders, and OSSEs have been elaborated by Schmidt and Gangopadhyay (2013), Gangopadhyay, Schmidt, et al. (2013), and by Schofield et al. (2010). A key component of the FORMS methodology is to provide the initial field for model initialization. In other words, it should provide you with initial fields for temperature and salinity at all depths of the model. Right? You need all seven primary variables, u, v, w, t, s, p, and ρ – at time $t = 0$. Now, note that the strong western boundary currents and their eddies are designed as velocity-based feature models; while, the coastal features are designed on the basis of their water-mass (t-s) structures. So, the velocity-based feature models will need to be converted to equivalent water-mass fields, and different techniques exist that primarily use the "thermal-wind" relationship discussed in Section 3.6. On the other hand, a simple geostrophic relationship (discussed in Section 3.3) can yield an equivalent velocity field for the water-mass feature models in the coastal regions. These are explained in detail in later sections of this chapter. In summary, the Feature-oriented methodology requires developing a synoptic circulation template for the region, called the basis template. This template is designed from a feature-based synthesis of the regional circulation patterns. From this basis template, a map of strategic sampling locations for placing feature model profiles is produced. These profiles provide synthetic synoptic expressions for fronts, jets, eddies, gyres, and other circulation structures and water masses at the initialization or updating phases. Features are represented by both analytical structures and by synoptic sections and profiles, which are developed by analyzing past synoptic high-resolution observations for each feature.

The FORMS methodology can be described in the following steps: (i) identify circulation and water mass features relevant to the process and within the domain; (ii) develop a regional synthesis schematic of features that represent these processes and spans within the domain; (iii) investigate past synoptic studies to identify synoptic data sets (in-situ and satellite) to get information of the structures of the features; (iv) develop feature models and verify them against synoptic data; (v) Seek regional climatology at sufficient resolution to provide for the background circulation; (vi) Use multiscale objective analysis to construct initialization with synoptic features melded with the background climatology; (vii) calibrate the modeling system with this initialization for simulation in a nowcasting/forecasting environment to properly tune the feature model parameters; (viii) Use data assimilation as necessary and as available with feature models to create the best-guess initial condition and carry out some verification studies. You will then be ready to use the FORMS for your own region for multiple process studies as well as operational modeling. Note that FORMS is essentially an initialization methodology irrespective of the numerical model system of your choice. So, it can be used for any of the models listed in the Table of Models in Chapter 10. We would encourage you to re-read this section of the FORMS overview after you complete this chapter.

In the next section, we will consider the FORMS approach for the western North Atlantic. This system has been developed over the last three decades to become operational for IOOS. Let us look at some of the critical elements of this feature-oriented system so that you can use it for your own region.

12.3   FORMS – WESTERN NORTH ATLANTIC

As mentioned earlier, from a modeling and process studies perspective, the WNA is a region that can be thought of as a combination of two sub-regions. These are the deeper Gulf Stream Meander and Ring (GSMR) region and the shallower coastal region of the Gulf of Maine and Georges Bank (GOMGB). Look at the satellite picture in Figure 12.2 and the schematics in Figures 12.4 and 12.9. The open ocean is dominated by the Gulf Stream and its Rings. The stream exchanges mass with the surrounding gyres to its north and south and interacts with the DWBC as it crosses over the thermohaline flow to the east of Cape Hatteras. The coastal region of the Gulf of Maine and Georges Bank is a shallow water system above 300m water depth whose circulation is seasonally dependent on local atmospheric fluxes, pressure gradients from inflows from Labrador sea, and constrained by topography. Clearly, the dominant features of circulation are different in these two regions, and building a realistic combined circulation system is challenging. The two systems (deep and coastal) exchange water-masses across and along the continental shelf break, where the warm core rings impinge upon the southwestward flowing shelf-slope front (created from the buoyancy difference between the slope and shelf).

Realize that many regions of the word ocean have a very similar setup like the WNA. There is a deep and a shallow region adjacent to and interacting with each other. So, the FORMS methodology that we describe here for the WNA can be applied to other regions very efficiently. While the methodology is portable and the Feature Model equations are general, each region has its own dynamics, which makes the parameters and their variations and ranges unique and interesting. At the same time, we can all learn from one region of a particular dynamics that might be similar to another region. So, modeling and process studies in one region with FORMS are very valuable to design and develop a new FORMS in a new region and carry out process studies and operational modeling in the new region.

Next, we will describe the feature models for the Gulf Stream region. We will also see how the individual velocity-based feature models were originally put together in a kinematically consistent system with multiple features inter-linked with each other in an interacting manner.

Then we discuss the water-mass feature models developed for the GOMGB region. Then we discuss merging the two (GSMR and GOMGB) for the WNA to develop an initialization field for the whole modeling domain shown in Figure 12.3, encompassing the two sub-regions of GSMR and GOMGB. A data-assimilative forecasting system for the WNA is then discussed with model verifications.

12.3.1 GSMR – DEEPER REGIONS

Let us again look at the western North Atlantic using a satellite image as shown in Figure 12.2. The very first thing that one needs to realize is that this synoptic view of this regional ocean shows a number of distinct dominant "features" of circulation. These are: (i) the large-scale Gulf Stream in its meandering state, which can be demarcated by the north wall location; (ii) the warm core rings to the north of the stream in the slope water; (iii) the cold core rings to the south in the Sargasso sea. Less obvious are the basin-scale (iv) baroclinic southern recirculation gyre (SRG) to the south; a (v) barotropic northern recirculation gyre (NRG) to the north and (vi) a baroclinic slope gyre to the north with a different configuration than the NRG. And then there is the thermohaline component (see Chapter 5) – the DWBC. So, close your eyes and think of a circulation cartoon with all these features interacting with one another and maintaining their own identity at the same time. A circulation schematic from Gangopadhyay and Robinson, 1997 is presented in Figure 12.4. This is the first step toward developing a feature model for a new feature in your own region. You need to identify the synoptic features of circulation.

12.3.2 FEATURE MODELS – GULF STREAM AND ITS RINGS

Once the identification of the features in the regional circulation is established (preferably through a schematic to visualize its synoptic state), then one needs to find relevant studies that previous researches have published to define the basic water property and flow structure of the feature in a three-dimensional framework. Remember your work on developing a feature model is possible because so many others before you have gone out there in the ocean and collected data and analyzed them to understand how the feature can be characterized and what its characteristic properties are. A list of relevant studies is a good thing to create, like the one that was done for the Gulf Stream Meander and Ring Region and shown in Table 12.1.

The velocity structure of the Gulf Stream was described in a feature model form by GRA97 as follows.

$$u(x,y,z) = \gamma(y)\left(\underbrace{[U^T(x) - U^B(x)]\phi(x,z)}_{\text{Baroclinic}} + \underbrace{U^B(x)}_{\text{Barotropic}} \right) \tag{12.1}$$

Here, $\gamma(y)$ is the non-dimensional form of the horizontal velocity of the Stream across it; and $\phi(z,x)$ is the non-dimensional vertical profile of the velocity at the axis and along the axis (x); U^T and U_B are the top and bottom velocity at the axis, which can vary in the along-stream direction(x). $\gamma(y)$ has

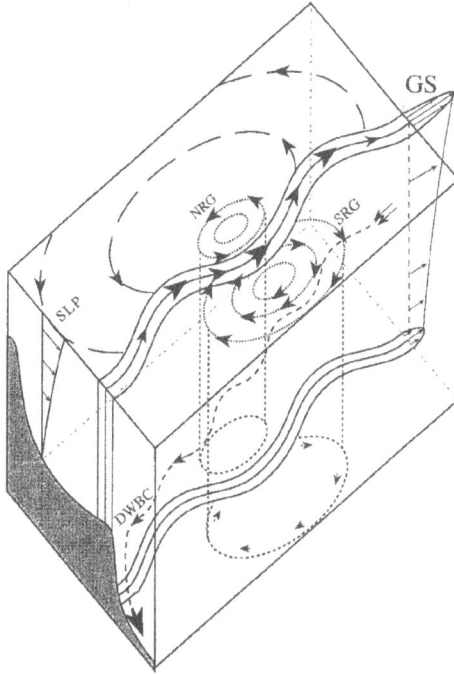

Figure 12.4: Schematic of the prevalent features in the Gulf Stream mean-der and ring (GSMR) region: Gulf Stream (GS), deep western boundary cur-rent (DWBC), northern recirculation gyre (NRG), south-ern recirculation gyre (SRG), and slope water gyre (SLP). Republished from "Gangopadhyay, A., A.R. Robinson, and H.G. Arango, 1997: Circulation and Dynamics of the West-ern North Atlantic, I: Multiscale Feature Models. Journal of Atmospheric and Oceanic Technology, 14(6), 1314–1332." American Meteorological Society. Used with permission.

a value of unity along the axis, and $\phi(z,x)$ is 1 at the surface and zero at the bottom, respectively; z is positive upward. Note how the first part of the RHS represents the baroclinic part of the velocity structure while the second part represents only the barotropic part. See Figure 12.5(a) for the assumed feature model structure for the GS. Note that this structure is sufficiently generalized so that it can be used for any other large-scale meandering current such as the Kuroshio and the Brazil Current.

So, where do you obtain the velocity structure form? This is where you need the benefit (and the blessings) of previous studies such as those listed in Table 12.1. GRA97 used the high-resolution cross-stream survey from Pegasus for the Gulf Stream; Calado et al., 2008 used similar observations for the Brazil

Gulf Stream	Niiler and Robinson (1967)
	Knauss (1969)
	Richardson and Knauss (1971)
	Robinson et al. (1974)
	Halkin and Rossby (1985)
	P. L. Richardson (1985)
	Hall and Bryden (1985)
	D. R. Watts (1985)
	A. Robinson et al. (1989)
	Spall and Robinson (1990)
	Hogg (1992)
Deep western boundary current (DWBC)	Pickart (1992b)
	Pickart (1992a)
	Pickart and Hogg (1989) and Pickart and Watts (1990)
Rings	Lai and Richardson (1977)
	McWilliams and Flierl (1979)
	Olson (1980)
	O. Brown et al. (1986)
Southern recirculation gyre (SRG)	Schmitz (1976)
	Worthington (1976)
	Hogg (1992)
Northern recirculation gyre (NRG)	Hogg (1983)
	Hogg and Stommel (1985)
	Hogg et al. (1986)
Slope (SLP)	Bisagni and Cornillon (1984)
	Cornillon et al. (1989)
	Churchill and Cornillon (1991)

Table 12.1

List of Relevant Studies for Developing Feature Models for the GSMR Region. Republished (Table 1) from "Gangopadhyay, A., A. R. Robinson, and H. G. Arango, 1997: Circulation and Dynamics of the Western North Atlantic, I: Multiscale Feature Models. Journal of Atmospheric and Oceanic Technology, 14(6), 1314–1332." ©American Meteorological Society. Used with permission.

Current System (BCS). Nowadays, new shipboard ADCPs should give you a better understanding of the representation of both the horizontal ($\gamma(y)$) and vertical ($\phi(z,x)$) structures.

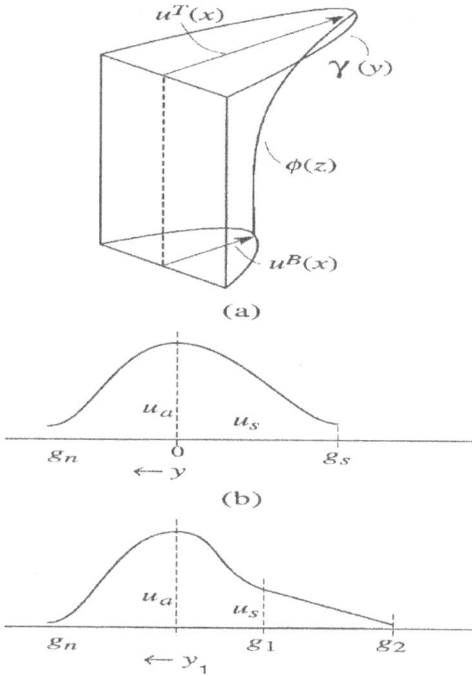

Figure 12.5: (a) A typical thin jet model and its parameters; (b) an asymmetric Gaussian velocity distribution; and (c) the piece-wise continuous horizontal velocity distribution for the Gulf Stream feature model. Republished from "Gangopadhyay, A., A. R. Robinson, and H. G. Arango, 1997: Circulation and Dynamics of the Western North Atlantic, I: Multiscale Feature Models. Journal of Atmospheric and Oceanic Technology, 14(6), 1314–1332." American Meteorological Society. Used with permission.

The non-dimensional feature model structures defined by equation 12.1 are further implemented by GRA97 with parametric variations for the Gulf Stream as follows.

$$
\begin{aligned}
&= e^{-y^2/g_n^2} \text{ for } 0 \le y \le g_n \\
\gamma_{GS}(y) &= ay + be^{y/l_1} + c \text{ for } -g_1 \le y \le 0 \qquad (12.2)\\
&= a_1(-y - g_2 + g_1) + b_1 e^{(y-g_1)/2} \text{ for } -g_2 \le y \le g_1
\end{aligned}
$$

$$\qquad (12.3)$$

Note that the northern side of the GS is represented by one single exponential form while the southern side of the GS is a combination of two piece-wise linear segments modulated by exponential to match the slope and continuity of derivatives. See Figure 12.5b for this analytical structure. The free parameters (a,b,c) are given in the Appendix of GRA97.

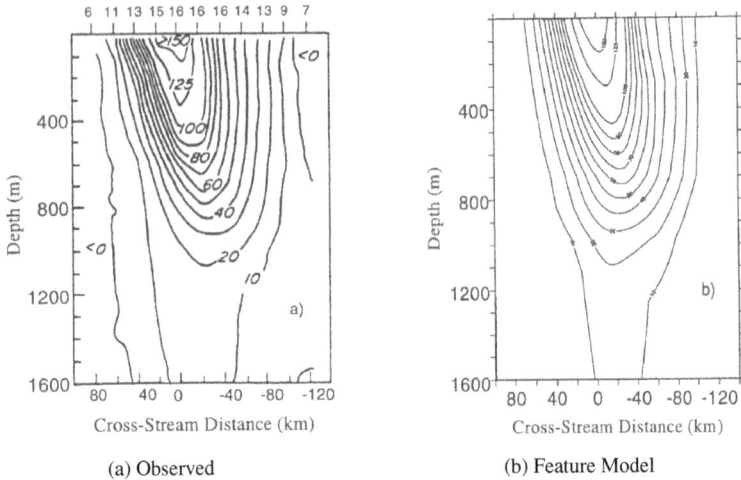

(a) Observed   (b) Feature Model

Figure 12.6: Gulf Stream velocity sections (a) from Pegasus average synoptic profile and (b) from multiscale feature model. Republished from "Gangopad-hyay, A., A. R. Robinson, and H. G. Arango, 1997: Circulation and Dynamics of the Western North Atlantic, I: Multiscale Feature Models. Journal of Atmospheric and Oceanic Technology, 14(6), 1314–1332." American Meteorological Society. Used with permission.

The non-dimensional vertical structure ($\phi(z,x)$) of the cross-stream velocity at Cape Hatteras is obtained from the Pegasus section observations by normalizing the axis velocity with respect to its surface value. With such parameter choices, one can reconstruct the 'Feature model' velocity section across the Gulf Stream by choosing just five parameter values, the top and bottom velocities of the Gulf Stream at Cape Hatteras, and the three width parameters (g_n, g_1, and g_2). This feature model section is shown along with the Pegasus section in Figure 12.6.

Once the feature model is constructed, it needs to be verified against its data-derived realizations for a number of properties such as the potential vorticity and the barotropic and baroclinic components of the total along-stream transport. The general form of the feature model as given by equation 12.1 was chosen to identify both the barotropic and baroclinic components as indicated by its velocity components.

For every new feature model construction of a velocity structure, this verification process is critical, as the transport matching is related to the realistic reproduction of baroclinic and barotropic instability and growth in the ensuing integration. The parameters of choices need to be tuned at this stage to match the transport components as well as the potential vorticity of the current to make the stream barotropically and baroclinically unstable and grow meanders with typical wavelengths and growth time-scales to lead to form warm and cold core rings in a realistic manner.

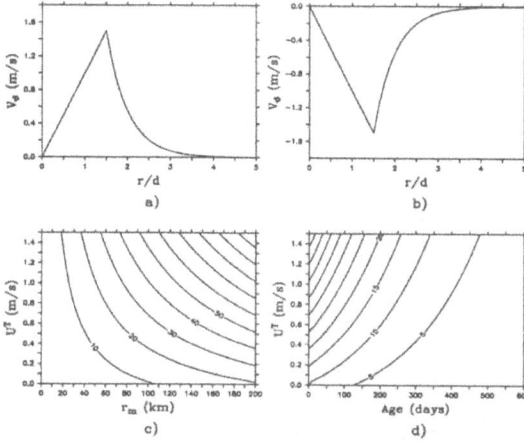

Figure 12.7: The non-dimensional horizontal velocity distribution for (a) cold core and (b) warm core rings. Transport variation for a typical synoptic ring as a function of (c) maximum speed at the surface and non-dimensional axial distance r/d and of (d) U^T and ring age. Republished from "Gangopadhyay, A., A. R. Robinson, and H. G. Arango, 1997: Circulation and Dynamics of the Western North Atlantic, I: Multiscale Feature Models. Journal of Atmospheric and Oceanic Technology, 14(6), 1314–1332." American Meteorological Society. Used with permission.

Similarly, several observational studies in the list in Table 12.1 were used to develop the feature model velocity distributions of the warm or cold core rings generated by the Gulf Stream. The rings are assumed to be either circular or elliptical and modeled in the horizontal with a linear velocity profile increasing to a maximum value v_m followed by an exponential decay. This is shown in Figure 12.7 for a cyclonic and anticyclonic ring. For the circular eddies or warm and cold core rings, the azimuthal velocity components can be given as,

$$v_\theta(r,z) = \frac{r}{r_0} v_m \phi_R(z) e^{-t_0/t_a} \text{ for } 0 \le r \le r_0$$

$$= v_m e^{\alpha(1-r/r_0)} \phi_R(z) e^{-t_0/t_a} \text{ for } r > r_0 \qquad (12.4)$$

a)

b)

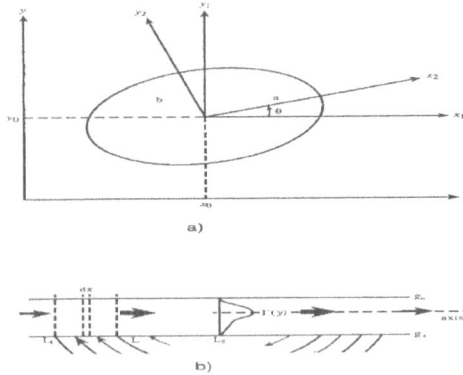

Figure 12.8: (a) Schematic of an elliptic gyre. (b) A synoptic perspective of the kinematic synthesis. Republished from "Gangopadhyay, A., A. R. Robinson, and H. G. Arango, 1997: Circulation and Dynamics of the Western North Atlantic, I: Multiscale Feature Models. Journal of Atmospheric and Oceanic Technology, 14(6), 1314–1332." American Meteorological Society. Used with permission.

The elliptical eddies or the gyres of the WNA and other regions can be expressed in terms of their velocity distribution as follows.

$$u(r,z) = u_p(r,z)\cos\theta - v_p\sin\theta \tag{12.5}$$
$$v(r,z) = v_p(r,z)\cos\theta - u_p\sin\theta \tag{12.6}$$

where

$$u_p(r,z) = \gamma(r)[(U^T - U^B)\phi(z) + U^B] \tag{12.7}$$
$$v_p(r,z) = \gamma(r)[(V^T - V^B)\phi(z) + V^B] \tag{12.8}$$

where U^T is the zonal component of the maximum top velocity (tangential) at a distance $r = \sqrt{a^2\cos^2\alpha + b^2\sin^2\alpha}$ from the center of the ellipse with major and minor axes of (a,b). Please see GRA97 for more details. A representative configuration of the elliptical sub-basin scale gyres is given in Figure 12.8.

We will discuss other types of feature models for temperature and salinity fronts a little later in Section 12.3.3. A list of other types of Feature Models and their representative equations for further application, adaptation, and extension is provided later.

The age and size parameters of the Gulf Stream rings can now be modified based on the new 40-year studies in this region by Gangopadhyay et al., 2020, Silva et al., 2020 and Silver et al., 2021a.

12.3.2.1   A Kinematic Synthesis

The structure of the synoptic and the mean circulation can be quite different. Compare the satellite picture in Figure 12.2 with any climatology of the WNA. The synoptic structures that you see in this figure are missing in any climatology. Synoptic structures exist and evolve on many scales, and a characterization of the mean and the synoptic state of a region of the ocean can be represented by a linked set of multiscale features. The synoptic state of an oceanic region is generally characterized by interacting free currents and vortices embedded in a background circulation.

A generalized kinematical constraint that links the multiscale structures is derived in terms of their interaction scales. Interacting scales include large-scale, subbasin-scale, mesoscale, cross-axis jet-scale, and submesoscale.

Along the axis of the stream, the recirculation gyres inject and extract mass and momentum both barotropically and baroclinically, the mechanism of which has been kinematically modeled in terms of multiscale interaction parameters. The geographical distribution of the structures can then provide a set of consistency conditions among the feature model parameters. Such kinematically mass balanced conditions are also consistent with values of individual feature model parameters in the realistic observational range. Let us explore this idea for the GS-SRG system.

The fundamental concept here is to interpret the mass exchanges between two features as a two-component exchange made up of the barotropic (depth-averaged) and baroclinic (depth-dependent) flows. Consider the Gulf Stream (GS) and Southern Recirculation Gyre (SRG) interaction. As the stream flows downstream from Cape Hatteras, it is joined by the SRG, which is anticyclonic and thus adds transport by adding both baroclinic shear flow and barotropic velocity (bottom) (see Fig. 12.8b). Due to continuity, the total mass flux entering normal to the southern edge of the GS between L_1 and L would increase the transport of the stream over this length. Clearly, the baroclinic flux "input" I_{bc} from the SRG into the stream from L_1 to L is given by

$$
\begin{aligned}
I_{bc} &= \int_{-H}^{0} \int_{L_1}^{L} \gamma_{SRG}(r)(U_{SRG}^{T} - U_{SRG}^{B})\phi_{SRG}(z)\,dr\,dz \\
&= \Gamma_{SRG}\Phi_{SRG}(U_{SRG}^{T} - U_{SRG}^{B})
\end{aligned}
\tag{12.9}
$$

where

$$
\Gamma_{SRG} = \int_{L_1}^{L} \gamma_{SRG}(r)\,dr
$$

$$
\Phi_{SRG} = \int_{-H}^{0} \phi_{SRG}(z)\,dz
$$

where γ and ϕ are non-dimensional representations of the horizontal and vertical structure of individual features with appropriate subscripts.

Now, we assume that this baroclinic influx into the stream gets distributed in the baroclinic structure of the stream and does not affect the stream's barotropic flow. We also assume that the stream's original velocity distribution in the horizontal ($\gamma(y)$) is not affected by this influx (or the vertical shear structure of the SRG (ϕ_{SRG}). The only thing that the SRG influx can change is the amplitude of relative vertical shear of the horizontal axis velocity or ($U^T - U^B$). This influx can then be formally represented by a small but appreciable increase in the stream's top and bottom velocities in an integral sense over the whole water column and over the interacting distance. This increase of Stream's baroclinic transport can be written as,

$$T_{bc} = \int_{-H}^{0} \int_{-g_2}^{g_n} \gamma_{GS}(y)(U_i^T - U_i^B)\phi_{SRG}(z)dydz$$
$$= \Gamma_{GS}\Phi_{SRG}(U_i^T - U_i^B) \qquad (12.10)$$

where

$$\Gamma_{GS} = \int_{-g_2}^{g_n} \gamma_{GS}(y)dy$$

Equating the influx from SRG, I_{bc} from Equation 12.9 to the increase of transport in the GS, (T_{bc} from Equation 12.10, we get the following simple ratio relationship,

$$\frac{U_{SRG}^T - U_{SRG}^B}{U_i^T - U_i^B} = \frac{\Gamma_{GS}}{\Gamma_{SRG}} \qquad (12.11)$$

In a similar manner, we can equate the barotropic influx from the SRG to the barotropic increase of Stream transport in an integral sense. Mathematically, this is given by

$$\Gamma_{SRG}HU_{SRG}^B = \Gamma_{GS}HU_{GS}^B \qquad (12.12)$$

Using Equation 12.12 with Equation 12.11, we get the fundamental interaction relationship between the two features (GS and SRG in this case), which is,

$$\frac{U_i^T}{U_{SRG}^T} = \frac{U_i^B}{U_{SRG}^B} = \frac{\Gamma_{SRG}}{\Gamma_{GS}} \qquad (12.13)$$

What does this equation tell us? It simply states that top velocity increase at the GS axis due to influx from the SRG is proportional to the ratio of the interaction lengths of SRG to that of the GS and the top velocity of the SRG.

Similar relationship holds for the bottom velocity increase. Clearly, the increase in the top and bottom velocities of the GS due to influx from other features such as the SLP and the NRG would be related similarly to their interaction lengths. Note that the interaction length Γ has the units of length as $\gamma(y)$ is dimensionless.

Extending these to get the individual increases (between L_1 and L) at the top and at the bottom,

$$U_i^T = U_{SRG}^T \frac{\Gamma_{SRG}}{\Gamma_{GS}} \tag{12.14}$$

$$U_i^B = U_{SRG}^B \frac{\Gamma_{SRG}}{\Gamma_{GS}} \tag{12.15}$$

This equation can be used for numerical implementation for the kinematic synthesis and will allow for transport increase consistent with the interactions between two features in a general sense.

Now, for the WNA, this influx of mass from SRG to the GS happens in reality up to about 60°W from Cape Hatteras and then the stream reduces its flow by recirculating parts of its flow into the Southern recirculation gyre (see the right side of the bottom panel (b) of Figure 12.8). So, accounting for all three sub-basin scale gyres injecting mass into the stream west of 60°W, the total velocities at top and bottom of the GS at 60°W can be given by,

$$U_{60W}^T = U_{CH}^T + \frac{\Gamma_{SRG}}{\Gamma_{GS}} U_{SRG}^T + \frac{\Gamma_{SLP}}{\Gamma_{GS}} U_{SLP}^T + \frac{\Gamma_{NRG}}{\Gamma_{GS}} U_{NRG}^T \tag{12.16}$$

$$U_{60W}^B = U_{CH}^B + \frac{\Gamma_{SRG}}{\Gamma_{GS}} U_{SRG}^B + \frac{\Gamma_{SLP}}{\Gamma_{GS}} U_{SLP}^B + \frac{\Gamma_{NRG}}{\Gamma_{GS}} U_{NRG}^B \tag{12.17}$$

From observations listed in Table 12.2, we can consider that the NRG is purely barotropic, i.e. $U_{NRG}^T = U_{NRG}^B = U_{NRG}$; and that $U_{SLP}^B = 0$. These observations then lead us to six free parameters, $U_{CH}^T, U_{CH}^B, U_{SRG}^T, U_{SRG}^B, U_{NRG}$, and U_{SLP}^T for the velocity-based kinematically synthesized circulation model for the GSMR region.

The GS transport at 60°W can be obtained by integrating the velocities across and through the water column. Using equation 12.17 and the above considerations for the free parameters, it is easy to obtain the following,

$$T_{60W} = T_{CH} + \Gamma_{SRG} \left(\Phi_{SRG} (U_{SRG}^T - U_{SRG}^B) + H U_{SRG}^B \right)$$
$$+ \Gamma_{SLP} \Phi_{SRG} U_{SLP}^T + \Gamma_{NRG} H U_{NRG} \tag{12.18}$$

This transport constraint binds the free parameters of the feature model set for the regional circulation of the GSMR region. Note that a fixed amount of increase $(T_{60W} - T_{CH})$ can be achieved in multiple combination of shear and bottom velocities of the features or the free parameters of the system. That is the power of the feature model system for initializing a model with different conditions to verify and calibrate the model for realistic performance and predictability.

In this integrated framework, we have seen how a set of free parameters can lead to kinematically constrain the transport of the system through linked interactions between features. The barotropic constraints would define a range of bottom velocity variations for realistic transport increase between two points of the stream, the baroclinic constraints would do the same for the vertical shear profile variations of the stream within reasonable limits. While this method of prescribing the alongsteam transport variation for a WBC in a numerical modeling set up is relatively easy (but not trivial) to implement following the equations 12.11 to 12.13 above and Figure 12.8b); if you have considerable data from a cruise along-stream and across-stream a current to appropriately resolve its along-stream variation, a quicker (but approximate) alternative would be to prescribe observed variation of the $t - s$ properties and thermocline slope across the stream. Such alternate methodology was applied by Calado et al. (2008) and followed by Soutelino et al. (2013) for the Brazil Current System. However, it is worth to keep in mind that the transport of the Brazil Current is about 10 Sv, about an order of magnitude smaller compared to that of the Gulf Stream, which is in the range of 100–150 Sv. Clearly, this system allows different ways of imposing baroclinicity at the initial state. A comprehensive dynamical sensitivity study was carried out by RG97 to investigate the dispersion behavior of the meandering Gulf Stream with a focus on wave growth and phase speed of the meandering between Cape Hatteras and 65°W. This multiscale multi-parameter system allows for different initial conditions by varying parameters associated with GS, SLP, SRG, NRG and DWBC (see Table 12.2, adapted from GRA97). For example, the shear at Cape Hatteras, the shear of SRG, strengths of SLP and NRG will be varied with and without a DWBC for the central set to understand the impact of different feature on wave-growth and phase-speed. The parameter choices (from Table 12.2) will enable any user to calibrate each simulation to provide the best initial condition for the synoptic state of the Gulf Stream/DWBC system.

Two primary characteristics of the Gulf Stream meandering behavior, namely, the wave-growth characteristics and the ring formation and absorption statistics were studied via both quasi-geostrophic and primitive equation dynamics in the second part of the three-part series of studies mentioned earlier (GRA97, RG97, GR97). In achieving realistic dispersion characteristics, a comprehensive methodology for dynamical model tuning and validation in this limited region ocean was developed. The realistic regimes of parameter variation were identified on the basis of observational growth rate and phase speed of the Gulf Stream meanders.

Parameter	Value	Range	Comment
g_n	45 km		GS width in the slope side
g_1	40 km		G1 width in the Sargasso side
g_2	110 km		G2 width in the Sargasso side
U_H^T	1.30 m s^{-1}	1.0–1.5 m s^{-1}	GS top velocity at Cape Hatteras
U_H^B	0.0 m s^{-1}	0.0–0.5 m s^{-1}	GS bottom velocity at Cape Hatteras
Z_{lnm}	4600 m		Level of no motion
S_{tilt}	30 km	15–30 km	GS axis tilt
Z_{tilt}	800 km	700–1000 m	Depth of maximum tilt
g_W^n	50 km		DWBC northern width
g_w^s	50 km		DWBC southern width
Z_{DWBC}^T	800 m		DWBC shallowest depth
Z_{DWBC}^B	4600 m		DWBC bottom depth
U_{DWBC}^T	0.0 cm s^{-1}		DWBC velocity of Z_{DWBC}^T
U_{DWBC}^B	20 cm s^{-1}		DWBC velocity of Z_{DWBC}^B
C_{SRG}^{lon}	60°W		Longitude for SRG center
C_{SRG}^{lat}	38°W		Latitude for SRG center
A_{SRG}	700 km		SRG major axis
B_{SRG}	500 km		SRG minor axis
U_{SRG}^T	7 cm s^{-1}	6–10 cm s^{-1}	SRG velocity at Z_{SRG}^T
U_{SRG}^B	2 cm s^{-1}	2–5 cm s^{-1}	SRG velocity at Z_{SRG}^B
C_{NRG}^{lon}	56°W		Longitude for NRG center
C_{NRG}^{lat}	42.4°N		Latitude for NRG center
A_{NRG}	50 km		NRG major axis
B_{NRG}	100 km		NRG minor axis
U_{NRG}	3.5 cm s^{-1}	3–5 cm s^{-1}	NRG barotropic velocity
C_{SLP}^{lon}	62°W		Longitude for SLP center
C_{SLP}^{lat}	40°N		Latitude for SLP center
A_{SLP}	1800 km		SLP major axis
B_{SLP}	400 km		SLP minor axis
U_{SLP}	3 cm s^{-1}	3–6 cm s^{-1}	SLP top velocity

Table 12.2

Central Set of Multiscale Circulation Model Parameters. Republished from "Gangopadhyay, A., A. R. Robinson, and H. G. Arango, 1997: Circulation and Dynamics of the Western North Atlantic, I: Multiscale Feature Models. Journal of Atmospheric and Oceanic Technology, 14(6), 1314–1332." ©American Meteorological Society. Used with permission.

Long-term simulations within these realistic regimes provide statistics of ring production and interaction behavior. The observed range of transport variability of the Gulf Stream system further constrains the parameter selection. This exercise of calibrating a modeling system is extremely valuable to understand the capabilities and limits of a numerical modeling system. Please see RG97 for details of such procedure. Note that while RG97 is applicable for the Gulf Stream region, other regions will have their own intrinsic dynamical behavior which need to be identified and quantified for the modeling system to be calibrated for realistic simulations.

12.3.3   GOMGB – COASTAL FORMS

The multiscale synoptic circulation system in the Gulf of Maine and Georges Bank (GOMGB) region can also be similarly presented using a feature-oriented approach. Prevalent synoptic circulation structures, or "features," are identified from previous observational studies. See the schematic developed for the GOMGB region in Figure 12.9. These features include the buoyancy-driven Maine Coastal Current, the Georges Bank anticyclonic frontal circulation system, the basin-scale cyclonic gyres (Jordan, Georges, and Wilkinson), the deep inflow through the Northeast Channel (NEC), the shallow outflow via the Great South Channel (GSC), and the shelf-slope front (SSF). Their synoptic watermass $(t - s)$ structures are characterized and parameterized in a generalized formulation to develop temperature-salinity frontal feature models. The multiscale circulation in any coastal region evolves as a response to winds, tides and buoyancy forcing. The temperature and salinity structures of the typical re-occurring Coastal fronts are results of typical buoyancy forcing including heating/stirring, freshwater influx from rivers, dense water inflow/outflow, and inter-basin exchanges. These effects create variations of temperature and salinity along and across a coastal front (or feature), which is different than the way a western boundary current (such as the Gulf Stream) gathers mass as it meanders downstream and interacts with its adjacent gyres.

12.3.3.1   Water-Mass Front FM

Considering the above differences between the coastal water-mass fronts and the WBC velocity fronts, the three-dimensional synoptic structures of the temperature and salinity of a water-mass front are represented by a combination of a normalized empirical shear in the vertical and an empirical-analytical horizontal distribution across a water-mass current/front/flow feature. This functional expression from Gangopadhyay, Robinson, Haley, Leslie, et al. (2003) is given below:

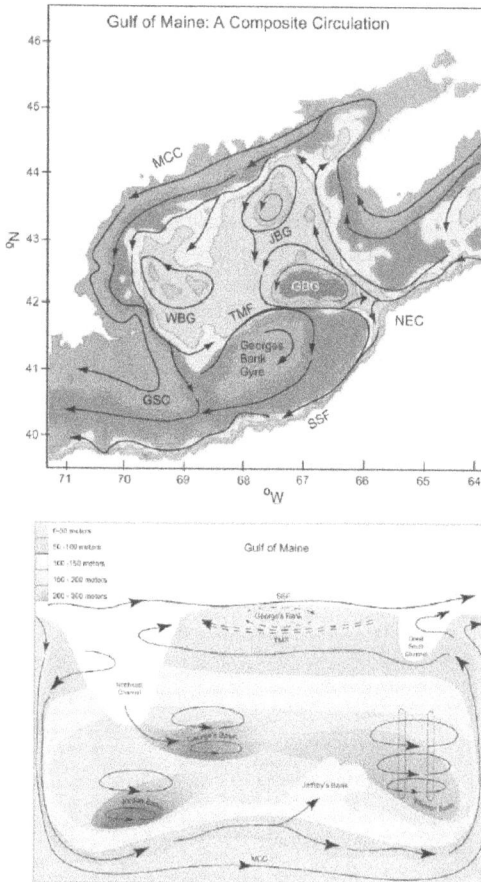

Figure 12.9: (a) A schematic of circulation features in the Gulf of Maine. GSC, Great South Channel; SSF, Shelf/Slope Front; NEC, Northeast Channel; GBG, Georges Basin Gyre; TMF, Tidal Mixing Front; WBG, Wilkinson Basin Gyre; MCC, Maine Coastal Current; JBG, Jordan Basin Gyre. (b) A three-dimensional bathymetric perspective of the regional circulation features. The basins are the three deep regions in the interior Gulf. The vertical-mixing region is predominantly in the Wilkinson Basin. Republished with permission of Elsevier Science and Technology Journals, from "Feature-oriented regional modeling and simulations in the Gulf of Maine and Georges Bank," Gangopadhyay et al., Continental Shelf Research (23), 2003; permission conveyed through Copyright Clearance Center, Inc.

$$T(x,y,z) = T_a(x,z) + \alpha(x,z)\gamma(y) \qquad (12.19)$$

where

$$T_a(x,z) = [T_0(x) - T_b(x)]\,\phi(x,z) + T_b(x) \qquad (12.20)$$

Here $T_a(x,z)$ is the along-stream tracer amplitude variation, $\gamma(y)$ is the cross-frontal distribution of the tracer gradient (α); while x is the dimensional along-stream coordinate, y the cross-stream coordinate with origin at the center of the current, z the dimensional vertical coordinate (positive upward) and $\phi(x,z)$ is the non-dimensional temperature/salinity profile.

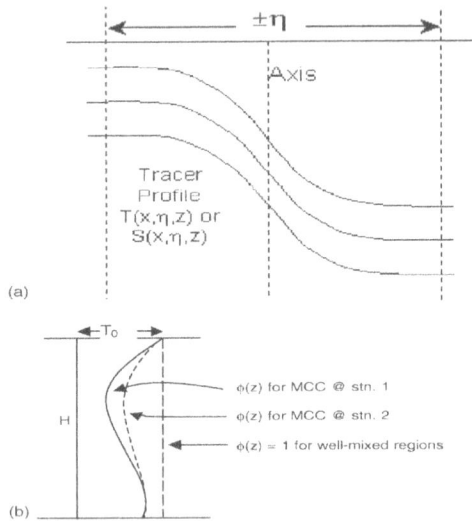

Figure 12.10: A schematic representation of a coastal current/front TSFM. It is defined by horizontal gradients of the tracer (temperature/salinity) as shown in (a) and a vertical structure $\phi(x,z)$ which varies with along-stream distance (x) as shown in (b). Republished with permission of Elsevier Science and Technology Journals, from "Feature-oriented regional modeling and simulations in the Gulf of Maine and Georges Bank," Gangopadhyay et al., Continental Shelf Research (23), 2003; permission conveyed through Copyright Clearance Center, Inc.

A few important things distinguish this formulation from the velocity-based feature models discussed earlier in section 12.3.2. First of all, this feature model construct has an axis profile $T_a(x,z)$ and a depth-dependent cross-stream temperature or salinity gradient distribution ($\alpha(x,z)\gamma(y)$), which varies in the along-stream direction. Second, the along-stream variations of surface and bottom temperature/salinity representation ($\alpha(x,z)$) allow for the effects of river runoff, buoyancy flow influx, and other anomalous water masses. Third, the horizontal distribution function $\gamma(y)$, if chosen as a step-function, would allow for various configurations such as plumes or pools within a buoyancy front. Finally, the along-stream variation of $\phi(x,z)$ allows for representing the effects of vertical mixing, branching of currents, and joining of frontal systems along the path of the front.

An example schematic of a $t-s$ feature model for the Maine Coastal Current (MCC) is shown in Figure 12.10. The upper panel shows a typical cross-section of the feature model while the lower panel shows a different depiction of a choice of $\phi(x,z)$ for the MCC. Note the advantage of choosing $\phi(x,z)=1$ for a winter-mixed scenario. A detailed description of the MCC feature model, with its along-stream variations of parameters for four different seasons, and similar characteristics of all other six different feature models are described in detail by Gangopadhyay, Robinson, Haley, Leslie, et al., 2003. We present the relevant studies that were used for developing the detailed feature models in Table 12.3.

12.3.3.2   Shelf-Slope Front (and Upwelling) FM

In the coastal oceans, a shelf-slope front (SSF) is a typical feature that is at the boundary of the shallow, colder, and fresher shelf water and the deeper, relatively warmer, and saltier slope water. The relative t-s characteristics might be different depending on the season and region and topographic setup. Essentially, the SSF can be thought of as a melding region of two different water masses, one from the shelf and the other from the slope.

A schematic of the feature model representation of an SSF is shown in Figure 12.11, which has the following characteristics, (i) the mean location at the surface from satellite observations (or following a particular isobath as the case may be); (ii) the isotherm and isohalines that represent the mean location; (iii) the slope (tilt); (iv) the width; and (v) the melding function. So, given a temperature or salinity profile for the shelf water ($T^{sh}(z)$) and that for the slope water ($T^{sl}(z)$), the SSF temperature and salinity distribution across the front at a location (x, alongstream) can be modeled by:

$$T(y,z) = T^{sh}(z) + [T^{sl}(z) - T^{sh}(z)]m(y,z) \qquad (12.21)$$

where the melding function $m(y,z)$ is given by

$$m(y,z) = \frac{1}{2} + \frac{1}{2}\tanh\left[\frac{y - \theta \times z}{\gamma}\right]$$

where θ is the slope ($\tan\alpha$) in Figure 12.11, y is the distance in the cross-frontal direction, z is the depth, and γ is the e-folding half-width of the front ($\eta/2$).

The parameters for the SSF feature model for the GOMGB region were derived from the studies listed in Gangopadhyay, Robinson, Haley, Leslie, et al., 2003, specifically utilizing the comprehensive climatology of the shelf-break

Features	Selected Studies
Maine Coastal Current (including South Channel Outflow)	Beardsley et al. (1997), Bisagni et al. (1996), Brooks (1985, 1987, 1990), Brooks and Townsend (1989), Chapman and Beardsley (1989), Holboke and Lynch (1995), D. Lynch (1999), D. Lynch et al. (1996), D. Lynch et al. (1992), Mavor and Huq (1996), Mountain and Manning (1994), Naimie (1995, 1996), and Naimie et al. (1994), P. C. Smith (1989)
Georges Bank Anticyclonic circulation, Tidal Fronts	Bisagni et al. (1996), Butman and Beardsley (1987), Butman et al. (1987), Flagg (1987), Houghton et al. (1982), and Loder et al. (1992)
Jordan Basin Gyre	Beardsley et al. (1997), Brooks (1985), Pettigrew et al. (1998), and D. Wright et al. (1986)
Wilkinson Basin Gyre	W. Brown (1998), W. Brown and Beardsley (1978), W. Brown and Irish (1992, 1993), and Mountain and Jessen (1987)
Georges Basin Gyre	Beardsley et al. (1997), Brooks (1985), Pettigrew et al. (1998), D. Wright et al. (1986), and Xue et al. (2000)
Northeast Channel inflow	Bisagni and Smith (1998), Brooks (1987), and Ramp et al. (1985)
Shelf-slope front	Allen et al. (1983), Gawarkiewicz et al. (2001), Sloan (1996), and W. Wright (1976)

Table 12.3

List of GOMGB Features and Selected Studies. Republished with permission of Elsevier Science and Technology Journals, from "Feature-oriented regional modeling and simulations in the Gulf of Maine and Georges Bank," Gangopadhyay et al., Continental Shelf Research (23), 2003; permission conveyed through Copyright Clearance Center, Inc.

front south of Georges Bank and Nantucket Shoals developed by Linder and Gawarkiewicz, 1998.

The above formulation for the SSF is sufficiently general for its application for melding any two different water masses with an appropriate melding function with a choice of different e-folding width criteria. Thus, this formulation given in Equation 12.21 has been used for Upwelling feature models later for the Arabian Sea (Shaji & Gangopadhyay, 2007) and for the California Current

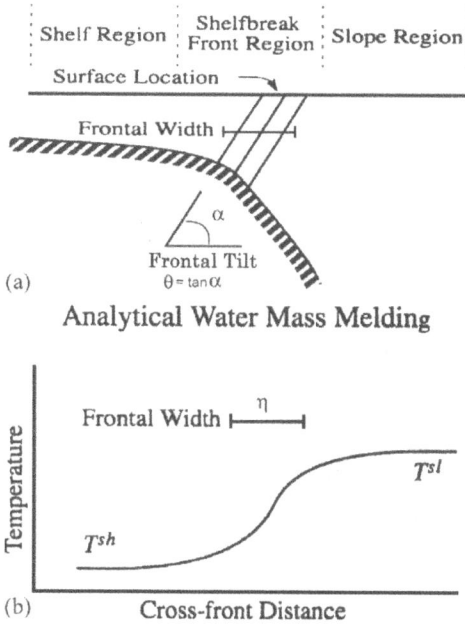

(a)

Analytical Water Mass Melding

(b)

Figure 12.11: Schematic representation of (a) shelf-break feature model front and (b) analytical water mass (T-S) melding from shallow to deep regions. Republished with permission of Elsevier Science and Technology Journals, from "Feature-oriented regional modeling of oceanic fronts," Gangopadhyay, A. and Robinson, Allan R., Dynamics of Atmosphere and Oceans, (36), 2002; permission conveyed through Copyright Clearance Center, Inc.

System (Gangopadhyay et al., 2011). In upwelling situations, the shelf and slope water masses are usually replaced by the appropriate upwelled and offshore water and the formulation is modified slightly to match the mathematical melding with realistic temperature and salinity changes across the upwelling region. This formulation was also cast in a polar coordinate system for non-frontal systems such as a plume (Gangopadhyay et al., 2005) or a pool of anomalous water or an asymmetric eddy off of a meander (Calado et al., 2006).

12.3.3.3   Coastal Eddy and Gyre t/s Feature Models

Next, let us look at a temperature and salinity feature model for an eddy and a gyre feature in the coastal ocean. A simple model for the temperature and salinity distribution across a circular eddy like the Gulf Stream's warm or cold core rings can be developed by a combination of contribution from the core tracer amplitude to the ambient water's condition just outside of ring or the "background" water-mass. Following observational studies (see Table 12.1) on GS rings, an exponential combination is chosen and is given below,

$$T(r,z) = T_b(z)\left[1 - exp(-\frac{r}{R})\right] + T_c(z)exp(-\frac{r}{R}) \qquad (12.22)$$

where T_b and T_c are the background and core tracer values, respectively; the decay scale R is generally taken as 3–5 times the Rossby Radius. This formulation has two very interesting characteristics: (i) the core temperature fades out at a distance r, which is much larger than the Rossby Radius, preserving the mesoscale nature of the eddy core; (ii) the background temperature starts fading in at this distance and continue beyond the mesoscale radius of deformation.

The temperature and salinity distribution for an elliptical gyre feature model can be given in the polar coordinate system as,

$$T(r,\theta,z) = T_c(z)\alpha(\theta)\gamma(r) \qquad (12.23)$$

where $T_c(z)$ is the vertical core tracer profile, $\alpha(\theta)$ is generally a delta or step function, and $\gamma(r)$ is a linear exponential or a combined function depending on the gyre characteristics.

The set of feature models for the GOMGB region are developed here on the basis of their water-mass properties. See Figure 12.10. These temperature and salinity feature models will first be geostrophically adjusted to derive a consistent baroclinic velocity component. Second, additional barotropic flow fields will be supplied, either externally or by adjusting the level of no motion in short-term (1–2 day long) simulations. This ensures consistency between water masses and circulation fields. This method of providing a consistent set of u,v,t,s fields at initialization starting with the water-mass feature models is called the "forward approach." Contrast it to the "backward approach" followed for GSMR velocity-based feature models, where the "u,v" field will be used with the thermal wind equation to get to the equivalent "t,s" fields.

Let us now look at developing a synthesized initial field for the whole coastal region of GOMGB, like the domain at 5 km resolution in Figure 12.3. First, a set of strategic locations based on the above dynamical considerations for all the features is identified and called the "synoptic circulation template" for the region. The template for the GOMGB is shown in Figure 12.12. These are effectively the locations representative of the synoptic features in the domain. In addition to the MCC, the TMF, the SSF front, and the inflow/outflow system, the topographically controlled gyres are strategically sampled for initialization as well. The gyres' mean locations and geographical extents are preset from observational studies as default and listed in Table 4 of Gangopadhyay, Robinson, Haley, Leslie, et al. (2003). The eddy or gyre structures are best resolved with a cross-hair sampling as shown in Fig. 12.12. The major advantage in using such a template is the ability to resolve synoptic structures on the basis of past oceanographic knowledge of the region, even when there is a lack of observations for the synoptic features. Note that only a handful (270 for GOMGB)

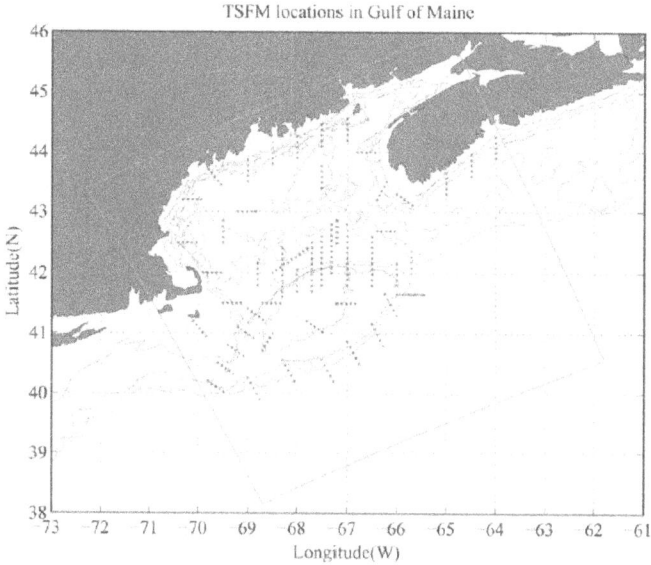

Figure 12.12: A synoptic feature-oriented circulation template for the GOMGB region used for initialization and updating of numerical models is presented here. The red marks are pseudo-CTD stations where feature model t-s profiles are generated. The profile locations are strategically selected along and across circulation features such as the Maine Coastal Current, the Tidal mixing front, the shelf-slope front, the Northeast Channel inflow, the Great South Channel outflow, and the three basins, Wilkinson, Jordan, and Georges. Republished with permission of Elsevier Science and Technology Journals, from "Feature-oriented regional modeling and simulations in the Gulf of Maine and Georges Bank," Gangopadhyay et al., Continental Shelf Research (23), 2003; permission conveyed through Copyright Clearance Center, Inc.

of stations are needed to describe the synoptic behavior of this regional ocean. The "default" template should be used as a guide to place the synoptic feature model profiles and can be adjusted according to available satellite or in situ observations for a specific nowcasting, forecasting, hindcasting, or updating exercise.

A six-step procedure is followed to generate the three-dimensional initialization fields for modeling in the GOM or GB domains. These activities are: (i) placing the individual synoptic TSFMs in a collective feature-oriented circulation template; (ii) creating shelf objective analysis or OA T-S fields; (iii) creating the deep OA fields; (iv) melding the shelf and slope T-S OA fields through the SSF; (v) adjusting the shelf and slope dynamic height OA fields to

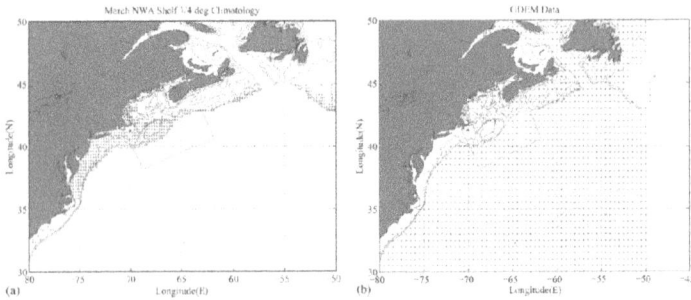

Figure 12.13: (a) Distribution of profiles, which were used as background climatology for melding with feature model profiles in Figure 12.12. The NWA shelf climatology for RTDOC of AFMIS during March–April 2000 was derived from NODC and NMFS data sets. (b) A typical coarse ($1°$) resolution climatological data sets available from GDEM is shown here. Republished with permission of Elsevier Science and Technology Journals, from "Feature-oriented regional modeling and simulations in the Gulf of Maine and Georges Bank," Gangopadhyay et al., Continental Shelf Research (23), 2003; permission conveyed through Copyright Clearance Center, Inc.

bottom topography; and (vi) generating the equivalent barotropic streamfunction for the numerical dynamical model.

First, the temperature and salinity feature-model fields are selectively sampled along and across the important features, as shown in Figure 12.12. The red dots in Figure 12.12 are the locations chosen for sampling the five important features of circulation for this region. This strategic feature-oriented arrangement (feature model profiles in the template locations) ensures the presence of synoptic structures in the initialization and updating fields for nowcasting and forecasting. Temperature and salinity profiles for the individual TSFMs are obtained using the equations in Section 12.3.3 and placed at their appropriate locations in Figure 12.12.

Second, this collection of temperature and salinity profiles are treated as pseudo-CTD observations and objectively analyzed (Carter & Robinson, 1987a; P. F. Lermusiaux & Robinson, 1999; D. Watts et al., 1989) with appropriate background seasonal climatology in the coastal shelf domain. Different climatologies for different regions can be mixed at this stage for the background (See Figure 12.13). The resulting shelf OA field for temperature and salinity thus combines the synoptic structures in a background of available climatology, appropriate for nowcasting and forecasting. The multiscale OA is performed in two stages (P. F. Lermusiaux & Robinson, 1999). In the first stage, the largest dynamical scales are resolved at each level, using estimated large-scale e-folding spatial decays, zero crossings, and temporal decay. The background field for this first stage is the horizontal average of the climatology data. In the second stage, the synoptic dynamics of interest (mesoscale and sub-basin-scale) are resolved using its estimated space-time decays. The background for this second stage OA is the first-stage OA. The primary assumption made in this two-scale OA is that the errors in the climatology (first-stage) and synoptic (second-stage) dynamical scales are statistically independent. This procedure effectively and smoothly melds the synoptic feature model or observational profiles and climatology for the shelf region. For a choice of typical multiscale OA parameters see Table 5 of Gangopadhyay, Robinson, Haley, Leslie, et al. (2003).

Such a procedure can result in initialization fields for different regions at different resolutions. Medium-range (10-day long) simulations (with multiple forcing conditions) from a gulf-scale (5 km resolution) GOM domain and a finer-scale (5/3 km resolution) GB domain during the spring and summer seasons have been presented to show the validity of this approach (Gangopadhyay, Robinson, Haley, Leslie, et al., 2003). Brown, Gangopadhyay, Bub, et al. (2007) implemented this system in an operational setting for a combined GS-GOMGB region in 2007–2008, which is discussed next.

12.3.4   INITIALIZATION WITH FORMS – WNA

It is important to realize that the GSMR feature models (GRA97) discussed earlier (Section 12.3.2) were developed starting from a velocity-based formulation which is more appropriate for strong currents such as the Gulf Stream and its rings. In contrast, the GOMGB feature models (Gangopadhyay, Robinson, Haley, Leslie, et al., 2003) discussed in Section 12.3.3 were developed on a T-S based formulation, more applicable for buoyancy-dominated flow fields in the shallow coastal regions. In the coastal regions, the circulation is defined and dominated by changes in the water-mass properties and their subsequent interactions with bathymetrically constrained flow around banks and within the sub-basin-scale gyres.

The FORMS methodology (and any other initialization process) demands prescribing the tracer (T and S) as well as the momentum (u and v) fields in a consistent manner as the initial condition. As mentioned earlier, the forward approach (using geostrophy and providing some known or first-guess barotropic component) can be used to derive the velocity fields from the water-mass feature models for any region and a first-guess initial condition can be constructed for further simulation and/or prediction. Similarly, the backward approach (using thermal wind to invert velocity to density and then density to t-s using either observation of t or s or climatological salinity (e.g., Soutelino et al., 2013) can be used for a purely velocity-based feature model to construct the initial field. However, many of the global oceanic regions have both deep and coastal regions

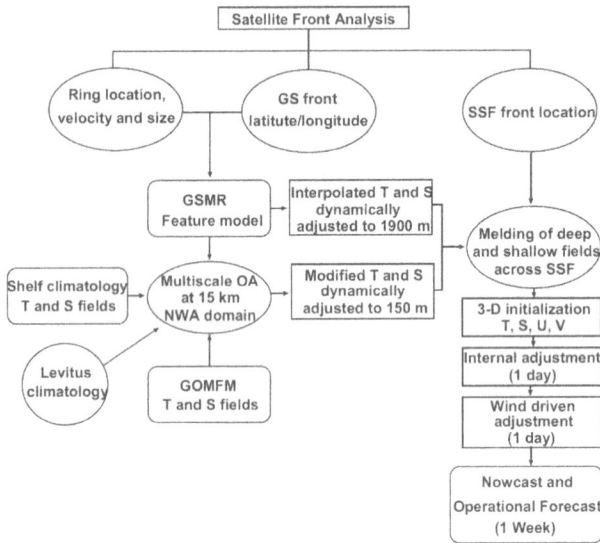

Figure 12.14: Flow diagram of the circulation model initialization and forecasting protocol starting with the analysis of satellite SST image. Republished with permission of Elsevier Science and Technology Journals, from "An operational ocean circulation prediction system for the western North Atlantic: Hindcasting during July–September of 2006," Andre Schmidt and Avijit Gangopadhyay, Continental Shelf Research (63), 2013; permission conveyed through Copyright Clearance Center, Inc.

are adjacent to one another along a shelf-break (where bathymetry changes sharply), exchanging mass and interacting with each other at multiple scales with their own features and across the shelf-break. Most of the WBCs, including the Gulf Stream in WNA and the Kuroshio in Western North Pacific, are great examples of such situations of complex deep-littoral interactions. So, in more generality, the FORMS approach was developed to account for such challenging situations as in the western North Atlantic, which we describe here.

The initial velocity field is generally obtained by using geostrophic balance to a T-S field for the GOMGB region using a prescribed level of no motion of about 150 m. For the GSMR region, the tracer fields are inverted from the initial velocity field by using a quasi-geostrophic formulation with a prescribed level of no motion of about 1900 m. So, the overall challenge for the whole WNA region is to meld/merge these two seemingly different sets of feature models with climatology into one consistent three-dimensional T-S-u-v field set, which preserves the original distribution of the sub-regional T-S-u-v fields computed with the different levels of no motion for the respective sub-regions. This robust methodology for the WNA domain is presented in Fig. 12.14 as a flow diagram of different steps. For details of each step see Schmidt et al., 2012. These are briefly described below.

The critical element for this deep-shallow merger is the design of a feature model for the shelf-slope front (SSF). Specifically, the melding across the shelf-slope front is performed with the GSMR feature model derived WNA domain T-S fields seaward of the observed shelf-slope front with the GOMGB domain T-S fields that are landward of the front, according to a set of prescribed frontal structure, width (15 km), and depth (100–300 m) parameters. See Figure 12.11 for the frontal structure of the SSF. The FORMS methodological steps are:

(i) Analyze satellite-infrared imagery in terms of the GS, the GS rings, and the shelf-slope front.

(ii) Use the velocity-based GSMR feature model to derive u, v fields in WNA.

(iii) Generate T-S fields for the 15-km WNA domain using the velocity-based feature models in #2 above. The WNA FM dynamic height has to be adjusted to the bathymetry of the continental slope, assuming a level of no motion of 1900 m.

(iv) A separate GOMGB FORMS is used to yield coastal T-S fields in the GOMGB region.

(v) Generate T-S fields for the combined shallow GOMGB and mid-Atlantic Shelf (MAS) regions using the appropriate FM and a background climatological hydrography.

(vi) The GSMR T-S fields from step 3, the water-mass-based feature model profiles for the GOMGB region from step 5 and the Levitus Climatology for the WNA region are combined using a multi-scale objective analysis into the WNA 15 km domain for the upper 300 m in the water column.

(vii) The GOMGB FM profiles and the MAS Levitus climatological profiles are treated as synoptic data in the background of the GSRFM T-S fields from step 6. The produced GOMGB-WNA dynamic heights have been adjusted to the bathymetry, assuming a level of no motion of 150 m.

(viii) Meld the dynamic height-adjusted WNA FM field (step 3) with the GOMGB-MAS T-S adjusted fields (on the shelf) from step 6 across the shelf-slope front. Specifically, meld (i.e., across the shelf-slope front) the GSRFM WNA domain T-S fields seaward of the observed shelf-slope front with the GOMGB-MAS T-S fields that are landward of the front, according to a set of prescribed frontal structure, width (15 km), and depth (100–300 m) parameters. See Figure 12.11.

(ix) Adjust the dynamic height field of the melded 15 km resolution WNA domain T-S field to the local bathymetry. This is a four-step process: 1) the geostrophic surface velocity is computed in both shelf and deep domains, 2) the two surface fields are melded, 3) the corresponding surface dynamic height field is obtained by solving the Poisson equation, and 4) the subsurface dynamic height field is obtained by integrating the density anomaly associated with the melded TS fields from the surface downward. Any nonzero bottom velocity results after the fourth step is further compensated by the barotropic streamfunction computation in step 10 below.

(x) Generate the initial barotropic streamfunction field for the WNA domain.

The FORMS initialization methodology described here can be applied to any oceanic region with deep, coastal, and shelf-slope subregions. For the WNA region, it was implemented with a choice of level of no motion of 1900 m for the deep and 150 m for the coastal GOMGB and MAS. The melding was done across a 300-m deep shelf-slope front with a maximum slope of 0.008. The MARACOOS operational implementation (2009–2016) was typically run for 7–10 model days, during which the first available SST image was assimilated to update the model SST field. This initialization protocol, along with the adjustment and forecasting protocols, is shown in Figure 12.14. Specifically, the geostrophic computation of dynamic height field from the FORMS-derived T-S fields yields the baroclinic velocity components. The velocity field is then vertically integrated to determine the barotropic velocity (Brown, Gangopadhyay, Bub, et al., 2007). The curl of the barotropic velocity field generates the barotropic vorticity, which in turn yields the transport streamfunction via the solution of a Poisson equation (Gangopadhyay & Robinson, 2002). The resulting fields are then used as the initial conditions or updating fields for primitive equation model simulations.

One of the approaches for regional modeling in the world oceans is the use of these knowledge-based feature models. This approach is distinctly different from the basin-scale approach that requires models to develop the inertia fields in a so-called "spin-up" period, which could require multiple years (3–10) of numerical integration prior to resolving synoptic mesoscale fields. This feature-oriented approach takes a few days of adjustment (because of its synopticity to begin with in the feature models) and has been used for regional simulations and operational forecasting for the past two decades. Ocean forecasting and realistic data-driven simulations require initialization and assimilation fields, and an efficient representation of synoptic realizations is valuable in this regard.

12.3.5 A FORECASTING SYSTEM FOR WNA

The development of FORMS for the WNA was carried out over a number of decades with the goal to apply it for operational forecasting. The first successful implementation of the combined GS-GOMGB region was carried out in the 2000–2001 period and was described by Brown, Gangopadhyay, Bub, et al. (2007). The assimilation methodology at initialization utilizes all data available up to the initial model day including SST, bottom temperature, and other available CTD casts. This data-assimilative system has been discussed in detail by Brown, Gangopadhyay, and Yu (2007). This system was operational on a weekly basis from 2009 through 2016 and is now available on an as-needed basis. All of the weekly forecasts are available online (https://www.smast.umassd.edu/modeling/RTF).

We want to remind our readers about a particular challenge of forecasting with models at increasingly higher resolutions and newer and more efficient algorithms in the future. Quoting RG97, "We distinguish here among validation, calibration, and verification. By model validation we mean simply that the model physics is appropriate to represent the regional phenomena and processes of interest. Model calibration is achieved via determination of necessary regional and computational parameters. Finally, the dynamical model set needs to be verified against two particular aspects of the regional circulation. First, statistical verification is established when the model results agree with past data of synoptic realizations and their evolution in a statistical sense. Second, synoptic verification is achieved when detailed agreement with actual synoptic realizations and evolutions is materialized."

Two example forecast verifications are briefly presented here, which utilized the FORMS-WNA development, used the (Brown, Gangopadhyay, Bub, et al., 2007; Brown, Gangopadhyay, & Yu, 2007) modeling system and was described in detail by Schmidt et al. (2012). First is the frontal forecast. Modifications were carried out using the FM parameters to further calibrate the initialization, which led to a better growth rate and better ring formation forecast.

Figure 12.15 shows the modified front forecast against analyses for the 2 week-long forecasts (yearday 226 and 233) and the 7-day predicted temperature-velocity field for the ring formation (model day 7 starting on yearday 233). The AVISO-generated surface velocity observation for August 28 is also shown for comparison. The skill scores for these two examples of the modified forecasts were computed using Equation 2 of Schmidt et al. (2012).

This large-scale predictability assessment has been augmented by mesoscale drifter trajectory comparison over a 1–4 day period for this operational system. About 22–28 drifters out of a possible 33 drifters showed varying degrees of skill in the TD analysis presented here. This analysis indicates the need for increasing the operational horizontal resolution of this model system. The simulated drifter comparison (Fig. 12.16) indicates that the model simulation performs better after an initial adjustment period of 6–12 hours. Such an initial adjustment period could be critical for implementing model-based forecasts in a real-time search and rescue operation.

Many of today's operational models have been well validated and well calibrated. However, it is the space and time-scales of the instabilities and the eddies – which are difficult to capture in a statistical sense and the synoptic verification becomes challenging. Statistical verification is generally limited to comparing simulated EKE with altimeter-derived EKE over a larger region or sub-region than the eddy-scale (Rossby radius). Synoptic verification means getting the timing and location of an eddy correct in a prediction. Can we predict

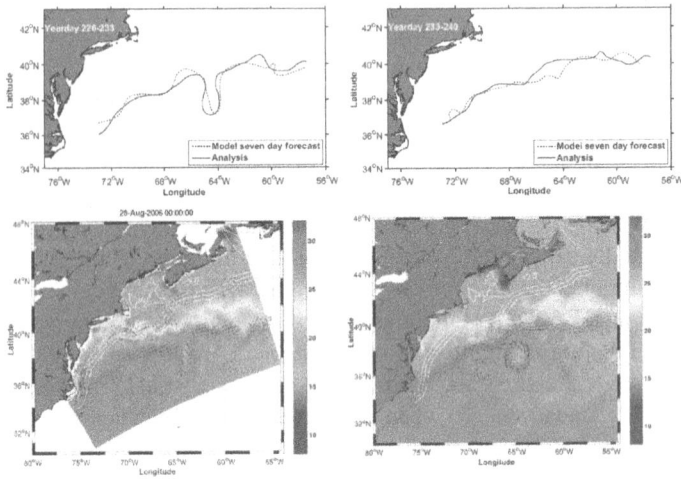

Figure 12.15: (a) Upper panel: Comparisons of the 15 1C isotherm at 200 m from model (prediction on day 7, dashed line) to the SST/AVISO analysis (solid line) for yearday 226–233 (left) and yearday 233–240 (right). (b) Lower panel: Model 7-day forecast of yearday 233 (left) and AVISO-generated surface velocity observation for August 28 (right). Republished with permission of Elsevier Science and Technology Journals, from "An operational ocean circulation prediction system for the western North Atlantic: Hindcasting during July–September of 2006," Andre Schmidt and Avijit Gangopadhyay, Continental Shelf Research (63), 2013; permission conveyed through Copyright Clearance Center, Inc.

an eddy in a general circulation to occur at the right time and at the right location, as seen later in the observations? This is a current area of research. Assimilation techniques are improving and we are looking more into the surface-subsurface linkages. Surface Data is sparse (even with satellite), extrapolating surface information into the water column is possible during initialization; however, getting the correct fields of features during integration is an ongoing challenge.

12.4   PROCESS STUDIES WITH FORMS

The FORMS approach has been used for multiple process studies as it is very effective in simulating the oceanic conditions for various conceptualized situations. Such conceptual simulations then provide a fully three-dimensional model-adjusted description of the ocean and its evolution in time. All of the fields and their gradients and energy terms can be determined quantitatively from such simulated fields for a number of different sets of parameters in the

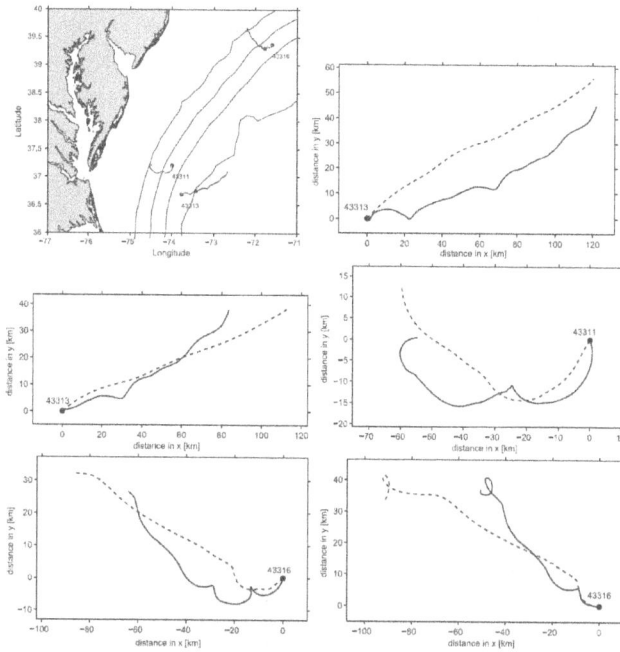

Figure 12.16: Drifter 43311, 43313, and 43316 trajectories, model (dash line), and data (solid line) for yearday 240. The left upper panel shows the drifters, geographical locations. Simulation starting on day 29 (right upper panel), 30 (left middle panel), 31 (right middle panel), and 31 (left lower panel) August with 3, 2, 2 and 3 days of model simulation, respectively, and 1 September of 2006 at 0 h with 3 days simulation (right lower panel). Republished with permission of Elsevier Science and Technology Journals, from "An operational ocean circulation prediction system for the western North Atlantic: Hindcasting during July–September of 2006," Andre Schmidt and Avijit Gangopadhyay, Continental Shelf Research (63), 2013; permission conveyed through Copyright Clearance Center, Inc.

Feature models. Thus, comparing and contrasting the resulting simulations one can understand the impact of the changes in the different attributes or features like baroclinicity, transport, bottom temperature, thermocline slope, different shapes of vertical structure (stratification), etc. This leads to better understanding of the processes involved. Additional data-assimilation exercises might also lead to an understanding of errors and error propagation or error minimization through feature assimilation in the future.

We briefly discuss a few examples of such process studies here.

12.4.1   THE STRAIT OF SICILY

One of the early applications of FORMS to a process study was demonstrated from the data obtained during a real-time exercise on-board R/V
Enterprise in the Strait of Sicily during Aug-Sept 1996. After the survey, the
data from this cruise was utilized to develop 'feature models' for the three
major circulation elements: the Atlantic Ionian Stream (AIS), the Adventure
Bank Vortex (ABV), and the Ionian Shelf Vortex (ISV). See Figure 12.17 for
the location, extent, and setup of the region and its features. The feature models
were not available during the real-time exercise and were built afterward from
about 200 strategic stations (along and across the features) from a total of 1000+

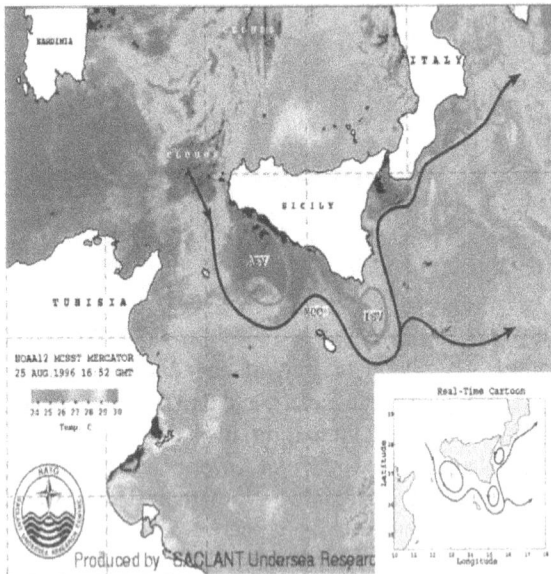

Figure 12.17: (a) Schematic of AIS from Lermusiaux and Robinson (2001).
Cartoon of the summer surface circulation features identified by Robinson et
al. (1998) for the strait of Sicily region (De Agostini, 1998). It is superimposed
on the satellite SST distribution for 25 August 1996 (Saclantcen). The path of
the AIS is shown with its bifurcation region. Also shown are other features
of circulation: ABV, MCC, and ISV. Republished with permission of Elsevier
Science and Technology Journals, from "Feature-oriented regional modeling
of oceanic fronts," Gangopadhyay, A. and Robinson, Allan R., Dynamics of
Atmosphere and Oceans, (36), 2002; permission conveyed through Copyright
Clearance Center, Inc.

stations available from the study of P. Lermusiaux et al. (2001). The usual process of developing a FORMS initialization was then followed and simulations were carried out with the HOPS modeling system described in Chapter 10. See Gangopadhyay and Robinson (2002) for the details of the process study simulations and feature model set up.

Two main dynamical conclusions resulted from this process study in this region of the Strait of Sicily: (i) the underlying bathymetry and the coastal boundary play a major role for both generation and subsequent maintenance of the ABV than for the ISV; (ii) the dynamical evolution of the ISV is more governed by the baroclinic structure and variability of the frontal configuration of the IAS and its bifurcation near the Ionian Shelf. Such inferences are very useful for future NATO surveys in reducing operational sampling and time for routine exercises.

12.4.2 A MEANDER EDDY UPWELLING SYSTEM IN BRAZIL CURRENT

The eddy feature models are very useful to understand processes related to eddy-current interactions. A particularly interesting question is what happens in cases along various coasts, when even if there is not enough upwelling-favorable wind, there is upwelling. In such situations, sometimes, an eddy is observed, and it is hypothesized that there is eddy-induced upwelling (within the eddy). There could also be upwelling due to the interaction of the eddy with a meander of a boundary current such as the Brazil Current or the EICC in the Bay of Bengal.

The regional ocean off the coast of southeast Brazil ($20° - 28°S$) is a known region of the current-eddy-upwelling region. The upwelling is very visible in satellite images (see Figure 12.18a). The study by Calado et al., 2010 applied the FORMS methodology with previously developed feature models for the Brazil Current and eddies (Calado et al., 2006; Calado et al., 2008) in a modeling set up using the Princeton Ocean Model (POM).

The model was set up with the FORMS initialization as shown in Figure 12.18b with the feature models for the Brazil Current and the Cabo Frio eddy. A month-long regional simulation was carried out starting from the eddy-current configuration without any upwelling. The simulation was intentionally not forced with any wind stress to eliminate the effect of wind-driven coastal upwelling.

The initial surface fields of temperature and velocity and their evolution after 30 days is shown in the upper panels of Figure 12.19. Clear signature of the development of the upwelling is visible in the surface cooling along the coast and across the shelf by day 30. The three-dimensional evolution of the upwelled thermocline is seen in the lower panels of Figure 12.19. The results confirmed the near Cape Sao Tome, the coastal eddy and the Brazil current interact and that the cyclonic meander of the BC can cause and/or enhance coastal upwelling.

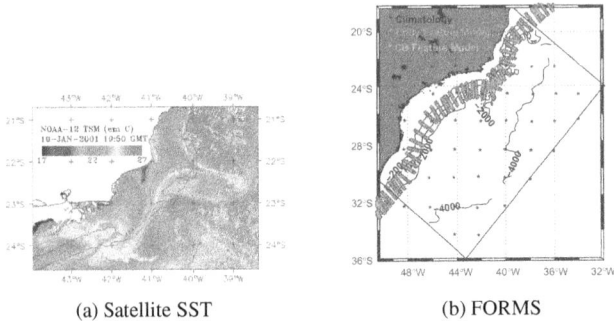

(a) Satellite SST   (b) FORMS

Figure 12.18: (a) An AVHRR image from 10 January 2001 exemplifying the recurrent CST and CF cyclonic meanders and the coastal upwelling in the northern portion of the SEBRA region. The blueish-yellow colors are associated with the cooler and fresher Coastal Water (CW) on the shelf, and the reddish colors mark the presence of the warmer and saltier Tropical Water (TW). Image courtesy of W. Lins e Mello, Brazilian Navy Ocean Research Institute, IEAPM. (b) The regional template of the feature models. The black stars denote the grid point locations of the Boyer et al., 2005 climatology. The feature model "data" locations corresponding to the BC front and CF eddy are represented by red and green stars, respectively. Republished with permission of Elsevier Science and Technology Journals, from "Eddy-induced upwelling off Cape São Tomé' (22°S, Brazil)" Calado L. et al, Continental Shelf Research (30), 2010; permission conveyed through Copyright Clearance Center, Inc.

See Calado et al. (2010) for more detailed dynamical discussions. Two other related recent examples of the application of the FORMS technique for process studies are to identify the roles of topography and shear for (i) the formation of the Brazil Current (Soutelino et al., 2013) and (ii) the generation of the Potiguar Eddy (Krelling et al., 2020). Both of these studies utilized the recently developed feature models for the BC and North Brazil Undercurrent (BC-NBUC) in a vertically-coupled transport-varying configuration.

Three important things to keep in mind for designing a process study or developing operational modeling with FORMS are below.

- The forward approach calls for developing t-s feature models and uses geostrophy to get the velocity structure, which is all required for initialization of a numerical model like HOPS, POM or ROMS or MOM, or any other model.
- The backward model uses the thermal wind relationship to get the density from the velocity-based feature models and then uses a known

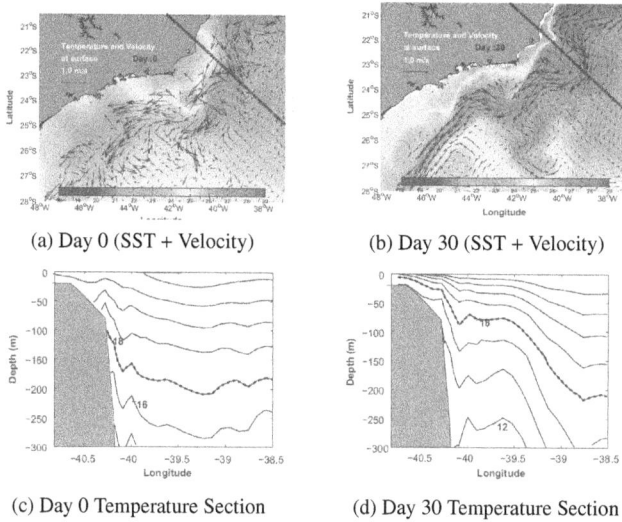

(a) Day 0 (SST + Velocity)

(b) Day 30 (SST + Velocity)

(c) Day 0 Temperature Section

(d) Day 30 Temperature Section

Figure 12.19: Upper panels: Model results of initial field day 0 and day 30 of the simulation prognostic phase: surface temperature and velocity horizontal fields. The dark straight lines represent the positions of the corresponding temperature vertical sections presented in lower panels. Model results of initial field day 0 and day 30 of the simulation prognostic phase: temperature vertical sections off Cape São Tomé. Republished with permission of Elsevier Science and Technology Journals, from "Eddy-induced upwelling off Cape São Tomé (22°S, Brazil)" Calado L. et al., Continental Shelf Research (30), 2010; permission conveyed through Copyright Clearance Center, Inc.

t-s relationship to get t-s from density. An excellent example of this procedure is provided in Soutelino et al. (2013).

- The procedures discussed here in this chapter for melding feature models for two different sub-regions (coastal and deep) developed by these two different approaches may be applicable to any complicated region in the world, of course, with appropriate modifications.

12.5 A WORLD OF FORMS

It is evident that both synoptic Ocean Prediction and Process Studies using the feature-oriented regional modeling approach are generic and can be applied to many different parts of the world's oceans. Every region is dynamically unique and requires modification of parameters and characteristics of the presented feature models to be applicable to the particular region. There are also many other types of multiscale features that are probably not yet formulated. There

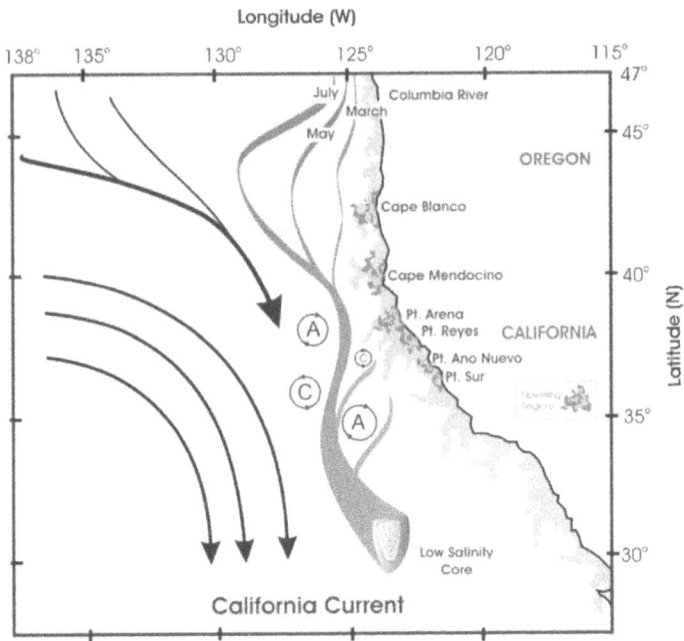

Figure 12.20: California Current System. This schematic shows the complexity in the meander-eddy-upwelling system of the California Current off Oregon and California in the spring/summer. The coastal jet coming from the north moves offshore as the season progresses (denoted by the March, May, and July lines) and contributes downstream to the low-salinity core of the California Current. This core delineates the higher variability region nearshore from the less active regions offshore. The offshore region consists of the mean southeastward flow of the wind-driven subtropical gyre. The inshore region is populated by upwelling centers concentrated near the capes. Eddies generated by the meandering jet are found in both regions and may exhibit either cyclonic or anticyclonic rotation. Temporally transient eddies of various sizes also occur. Republished with permission of Elsevier Science and Technology Journals, from "The California Current System: A multiscale overview and the development of a feature-oriented regional modeling system (FORMS)" Gangopadhyay et al., Dynamics of Atmosphere and Oceans, (52), 2011; permission conveyed through Copyright Clearance Center, Inc.

are also opportunities to develop multidisciplinary and climate change related feature models in the future, including human and ecosystem habitat interactions and socioeconomic dynamics. It is interesting to highlight a number of FORMS in other regions of the world ocean. One example is shown in Figure 12.20.

The FORMS for the California Current System is a great example of what is needed for a typical Eastern Boundary Current region. Its features included: (i) a low salinity pool for the California Current whose core was modeled as an elliptical pool with Fm parameters varying seasonally; (ii) a broad background flow accompanying the CC on its offshore edge; (iii) multiple upwelling regions around the capes; (iv) both cyclonic and anticyclonic coastal eddies in the transition zone between the CC and the coast; (iv) the California undercurrent flowing poleward at depths of 750–1000 m. See Gangopadhyay et al. (2011) for details on its setup with ROMS, validation, calibration, and verification with mesoscale simulations.

A number of FORMS-related studies have been successfully carried out in the Southwest Atlantic for the Brazil Current System. Process studies using these FORMS are described earlier in Section 12.4.2. As mentioned in Chapter 5, this is a classic region with the surface wind-driven flow interacting with the subsurface thermohaline flow and resulting in a multi-layer system of opposing flows and different water masses. The first development by Calado et al. (2008) was for the Brazil Current only with a water-mass formulation. Soutelino et al. (2013) developed a two-layer velocity-based feature model system for the Brazil Current and the Intermediate Western Boundary Current (IWBC) with a unique transport constraint.

One of the key elements of the FORMS initialization technique is to make sure that the initial field conserves mass over the whole domain. This has to be done before the model is started. This is usually achieved by computing the transport balance. Whatever flow comes in must equate to the flow going out. An iterative process of adjusting the parameters of the feature model structures of different features is generally invoked to make sure that a transport balance is reached within 1% of the total inward or outward transport. Soutelino et al. (2013) were very successful in meeting this transport constraint in a two-layer varying transport model.

To summarize, in this chapter, we have followed the evolution of synoptic ocean modeling following the first development of feature models from A. Robinson et al. (1988) through the nineties and 2000 and to its development for application to operational oceanography in the equatorial margins by Krelling et al. (2020). Table 12.4 lists a number of regional FORMS which have been developed so far. See the relevant references for some of these in Table 12.4.

We also provide a list of regions where new FORMS can be developed in the future for process studies and operational modeling at a regional scale (Table 12.5). It is also interesting to think about additional regions in the face of our new understanding and changing climate. FORMS developed for regions such as the Arctic, Antarctic, and the Global Conveyor Belt would be helpful for a better understanding of the Global Climate system. A good starting point to develop a systematic circulation schematic and look for past synoptic data would be the regional synthesis for each region in the book by K. H. Brink

Ocean	Sub-basin	Reference
Western North Atlantic	GSMR	Gangopadhyay et al. (1997)
WNA	GOMGB	Gangopadhyay et al. (2003)
SEBRA	Brazil Current System	Calado et al. (2008)
SEBRA	BC-NBUC System	Soutelino et al. (2013)
WNA Forecasting System	GSMR and GOMGB	Brown et al. (2008); Schmidt and Gangopadhyay (2012)
Eastern North Pacific	California Current System	Gangopadhyay et al. (2012)
Equatorial South Atlantic	Potiguar Basin	Krelling et al. (2020)
Arabian Sea	Western India Coastal Upwelling	Shaji and Gangopadhyay (2007)
Equatorial North Atlantic	TNBC – NBC Rings	Schmidt et al. (2012)

Table 12.4

List of Regions Where FORMS Have Been Developed.

and Robinson (2005) and Belkin et al. (2009) on fronts for 64 Large Marine Ecosystems.

Finally, it is worth mentioning that with the advances in pattern recognition, image processing, artificial intelligence (AI), and machine learning (ML), massive amounts of data can be analyzed to yield new feature models (Krasnopolsky, 2007). Learned parameters of these new intelligent feature models can help in realizing more realistic and synoptic (and hopefully more accurate) initialization, assimilation, and prediction in real-time. Multidisciplinary Feature models using simple Sverdrup model or temperature growth models for biology have been tried for process studies. One can think about a temperature-food-dependent fish model for species-specific short-term forecasting. Synoptic idealized process studies to understand the changing behavior of the Gulf Stream ring formation over the last four decades (Silver et al., 2021a) and marine heat waves (Gawarkiewicz et al., 2019) would be very valuable from a multiple equilibrium perspective. FORMS approach is also appealing for constructing climate-conditioned physical fields projections using climate system model outputs and then performing regional ocean simulations (downscaling physics from CMIP simulations to force ocean models with initialization mapped with warm and salty conditions; See Chapter 14). The idea

Region	large-Scale Fronts	Coastal-scale (water mass fronts, upwelling, plumes, etc.)
Western North Pacific	Kuroshio-Oyashio system	Yellow Sea CC, Korean CC, EC-SCC, SCSSS, Taiwan warm current, Tsushima Current
N Pacific	North Pacific Current	Alaska CC, Alaskan Stream, PWS circulation
S Pacific	EAC; Humboldt Current	Upwelling fronts in Central Chile
E North Atlantic	Azores Current, Canary Current, Guinea Current	Upwelling fronts (Permanent and seasonal)
Southwest Atlantic	Brazil-Malvinas Current	Patagonia Shelf
Southeast Atlantic	Agulhas retroflection	Angola-Benguela Front, Benguela upwelling system
NW European Shelf, Bering Sea	Greenland Current, NAC	Norwegian Current, Tidal Mixing Fronts, Celtic shelfbreak front
Equatorial Pacific	Equatorial Current System	Upwelling fromts, NEC, NECC, SEC, SECC, EUC
Southern Ocean	Antarctic Circumpolar current	Wedell fronts, Antarctic circumpolar shelf front
Indian Ocean	Somali Current, Agulhas current, WIUC	East African coastal Current
Bay of Bengal	Springtime WBC, Autumn EICC	GBM Seasonal Plume, ISS Plume, Seasonal Eddies
Indo-Pacific	S. Australian Current, Indonesian Throughflow	Plumes, Upwelling jets, warm pool
Medit. Sea	IAS, Leventine Inflow,	Meddies, EMed, WMed
G of Mexico	Loop Current	Eddies

Table 12.5

Example Regions Where FORMS Can Been Developed in the Future Around the World Oceans. This List Not Exclusive.

of dynamical FORMS with instability parameters at initialization might lead to enhanced (and more realistic) wave growth and eddy formation, which can be investigated in the near future.

CONCLUSION

In this chapter, we focused on a particular way to provide the initial condition based on past observation of oceanic features and present-day satellite data. These initial conditions are particularly suitable for short (1–7 days) and synoptic (mesoscale) forecasting. These synoptic ocean predictions use a methodology called the *FORMS*. We described the historical development of the FORMS methodology first. This is followed by a description of FORMS for WNA region, which includes the Gulf Stream and Ring region, the Coastal Gulf of Maine, and Georges Bank. An operational forecasting system for the WNA, which was run under MARACOOS is described in detail. Finally we show the application of FORMS for process studies and for a few worldwide applications. Extending FORMS to other regions of the world ocean awaits younger minds to follow through in their own regions of interest.

FURTHER READING

Brink, K. H., & Robinson, A. R. (2005). *The global coastal ocean-regional studies and syntheses* (Vol. 11). Harvard University Press.

Calado, L., Gangopadhyay, A., & Da Silveira, I. (2008). Feature-oriented regional modeling and simulations (forms) for the western South Atlantic: Southeastern Brazil Region. *Ocean Modelling*, *25*(1–2), 48–64.

Gangopadhyay, A., Robinson, A. R., Haley, P. J., Leslie, W. J., Lozano, C. J., Bisagni, J. J., & Yu, Z. (2003). Feature Oriented Regional Modeling and Simulation (FORMS) in the Gulf of Maine and Georges Bank. *Continental Shelf Research*, *23*, 317–353.

Gangopadhyay, A., & Robinson, A. (1997). Circulation and dynamics of the western North Atlantic. Part III: Forecasting the meanders and rings. *Journal of Atmospheric and Oceanic Technology*, *14*, 1352–1365.

Jana, S., Gangopadhyay, A., Lermusiaux, P. F., Chakraborty, A., Sil, S., & Haley Jr, P. J. (2018). Sensitivity of the Bay of Bengal upper ocean to different winds and river input conditions. *Journal of Marine Systems*, *187*, 206–222.

Lermusiaux, P. (1999). Data assimilation via error sub- space statistical estimation, Part II, Mid-Atlantic Bight Shelfbreak front simulations and ESSE validation. *Monthly Weather Review*, *1278*, 1408–1432.

Robinson, A. R., McCarthy, J. J., & Rothschild, B. J. (2005). *Biological-physical interactions in the sea* (Vol. 12). Harvard University Press.

Robinson, A. R., & Brink, K. H. (2005). *The global coastal ocean: Multiscale interdisciplinary processes* (Vol. 13). Harvard University Press.

Robinson, A., & Gangopadhyay, A. (1997). Circulation and dynamics of the western north Atlantic, II, Dynamics of meanders and rings. *Journal of Atmospheric and Oceanic Technology*, *146*, 1333–1351.

Kailua-Kona, Hawai'i

13 Interdisciplinary Modeling

Life comes from Life

Francesco Redi (1626– 1697)

OVERVIEW

So far we have learned in Chapters 8 through 12 about the art of modeling the physical ocean (u, v, w, t, s, ρ, p) and its applications at different temporal and spatial scales.

In this chapter, we venture into the domain of multidisciplinary multiscale modeling. We first provide a brief introduction on the interconnectedness of biology, chemistry, and geology to physics. Then we discuss simple biogeo-chemical and coupled physics-biogeochemical modeling frameworks. Simple biogeochemical processes such as ocean acidification (OA), hypoxia, harmful algal bloom (HAB), and denitrification are discussed. This chapter ends with a basic exposure to modeling fishery with simple stock-recruitment relationships.

13.1 INTRODUCTION

The ocean circulation can be described by the physical equations of motion. The circulation can also be simulated, understood, and predicted (with some uncertainties) by numerical models. But how about the biological processes in the ocean? How about the chemical gases that are dissolved in the seawater? How about the geology and sedimentation that the seawater is encountering all the time? Or in other words, how do biology, chemistry, and geology work in the ocean? How do we model the biogeochemistry of the ocean?

Our home, the earth, is much more than just a physical system called the hydrosphere containing the oceans. As described in Chapter 1, it has the lithosphere (part of the geosphere), the gases (atmosphere) on top of the oceans (hydrosphere) and the biotic compartments in the oceans (parts of the greater biosphere). The thermodynamic nature of chemical components and tracers in addition to their geochemical mobility connect the spheres. So, the whole Earth system is a big interdisciplinary system that works simultaneously with its physics, chemistry, biology, and geology at a wide range of spatial and temporal scales. In this chapter, we will first look at some important concepts of this

DOI: 10.1201/9780429347221-13

interdisciplinary system, specifically at the biogeochemical subsystem, and then discuss some approaches to modeling this complex system. In the next chapter, we will connect this interdisciplinary system to humans and social aspects to look at some approaches to "Climate Modeling" and "Climate Projections."

The physics of ocean circulation was set up in a force-response framework with Newton's laws of motion (especially, the second law, $F = ma$) which is the starting point of describing the motion of any particle on the rotating earth. That is the kernel of describing the mechanistic workings of a physical system. The starting point in understanding the Chemistry of any system is the "Periodic Table". It is a Table of Elements that can then be used to describe chemical properties, reactions, and changes in a system. But it is not a single equation like Newton's second law, $F = ma$. The situation in Biology is similarly complex to that in Chemistry. There is no single formulation like $F = ma$ that governs all or some of the biological systems. Non-linear and exponential growths are commonplace in biology. Different species act differently to the environment that they are in.

The geological time-scales are too large for our interest at this time; however, it is good to point out that at those longer time scales, the physical laws of motion can be helpful.

While the chemistry and biology in the ocean and the atmosphere are non-linear and complex, they are connected to each other and in a global sense to the underlying physical parameters such as the velocity components, the temperature, salinity, density, and pressure (and their gradients) of the fluid as appropriate. In other words, you can think of the biological or chemical variables as being driven by the physics or the physical flow system (u, v, w, t, s, p, ρ). This makes things easier (but not trivial) to include the biology and chemistry within a larger (physics–chemistry–biology) Newtonian ($F = ma$) framework with some additional terms as sources and sinks (or growth and mortality). Thus, the usual formulation of a biological or chemical variable is cast like a scalar tracer (e.g., temperature or salt equations) with advection, dispersion and diffusion terms similar to the primitive equations and then additional terms for sources and sinks of that variable which represents the biological or chemical variable of interest.

So, now we have a pathway toward interdisciplinary understanding and then modeling. We will thus start looking at the common threads of the overall Ocean-atmosphere-biosphere system – their compositions and their pathways within the overall system. It is best to start where it all began. The sun. The energy from the sun controls the biogeochemistry on the earth in some form or fashion. Once we understand that process, we can dive into modeling our global and regional interdisciplinary systems. In the first part of this chapter, we build from this understanding and look at some of the major components and their role in maintaining the overall balance (or imbalance) or equilibrium that exists today for many chemical compounds and biological species on Earth.

13.2   THE BASICS

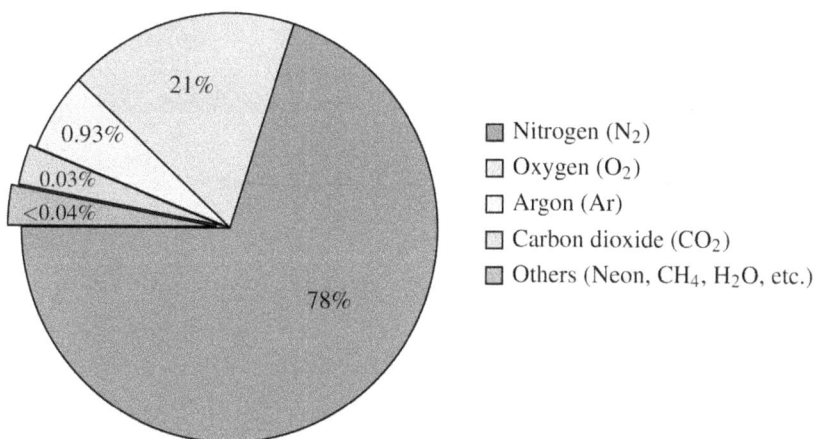

Figure 13.1: Composition of atmospheric composition (the percentages are not to exact scale).

Let us take a closer look at the constituents of gases in our atmosphere and the oceans. See Figure 13.1. The four most abundant gases in the atmosphere are also the most abundant gases dissolved in the ocean. In the atmosphere, their percent compositions are: nitrogen: 78.09%, oxygen 20.95%, argon 0.93%, and carbon dioxide 0.033%. Compare these to their composition in the ocean: nitrogen 62.6%, oxygen 34.3%, argon 1.6%, and carbon dioxide 1.4%. Note that the gas Argon is inert to other substances. This is why Argon is used to create an inert atmosphere, like in an incandescent or fluorescent bulb, to restrict the oxygen from corroding the hot filament, or in welding and in semiconductor processes where you need shielding from other atmospheric gases.

Another important thing to realize is that 97.2% of the total water on Earth is in the oceans; less than one percent is fresh, and a tiny fraction exists in the atmosphere as vapor. Water is a "universal solvent" and the ocean water contains many dissolved substances, mostly (>90%) chloride ions, such as sodium, magnesium, and calcium chlorides. Additionally, there are sulfur and other dissolved minerals and gases, which are nutrients necessary for the functioning of the marine ecosystems.

The four gases of importance from the global circulation perspective are nitrogen, oxygen, carbon dioxide, and water vapor. Together, they represent the

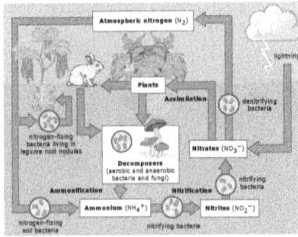

(a) Nitrogen cycle

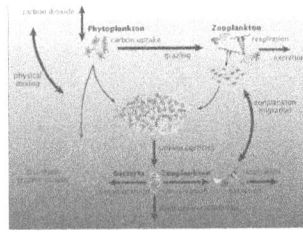

(b) Marine carbon cycle

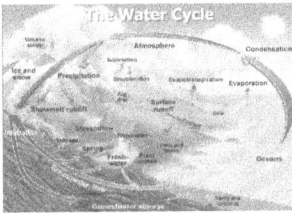

(c) Water cycle

(d) Global carbon cycle

Figure 13.2: Biogeochemical Cycles. (a) Nitrogen Cycle (Source: Mid-Atlantic integrated Assessment, 2003, US EPA http://www.epa.gov/maia/html/nitrogen.html). (b) Marine Carbon Cycle (Source: NASA Earth Observatory; Original work of NASA). (c) Water Cycle (Source: http://ga.water.usgs.gov/edu/watercycle.html; Author: John Evans and howard Periman, USGS). (d) General representation of the global carbon cycle. The broad translucent white arrows illustrate the general motion of the carbon cycle, depicting production and burial of carbon, and its uplift erosion and weathering resulting in the return of this carbon to the atmosphere. The orange arrows highlight loci of carbon (CO_2) exchange and transfer, such as via photosynthesis and respiration, including examples of human influences on carbon cycle processes. A hypothetical national boundary and exclusive economic zone (EEZ) are indicated as yellow dashed lines. Source: doi:10.1186/s13021-017-0077-x. Authors: Silvania Avelar, Tessa S. van der Voort & Timothy I. Eglinton. Credit to Jack Cook, Woods Hole Oceanographic Institution; CC-BY-SA-4.0.

biogeochemical cycle. One can also think and understand them one at a time or together. It is useful to think about the gases and their cycles around the ocean, atmosphere and land system. These different pathways are captured usually by four different cycles shown in Figure 13.2: (a) the nitrogen (nutrients) cycle; (b) the marine carbon cycle; (c) the hydrological (water) cycle; and (d) the global carbon cycle. There are variations of these cycles as well. Let us think about the ocean from a biogeochemical perspective now. See Figure 13.2d. Start from the rivers on the left. The rivers supply the ocean with most of its dissolved ions from weathering of rocks on land. Then look at the sun. The incoming shortwave radiation from the sun warms up the upper layers of the ocean and stratification begins in the water column. A thermocline is created and then continuously maintained by this process through daily solar heating. The thermocline separates the colder deeper water from the warmer upper layer waters. That was all physics. Now think biology and chemistry.

Biology first. See Figure 13.2 (a–b). The sunlight (specifically, the visible photosynthetically active radiation of 400–700 wavelength, PAR) is used by phytoplankton in the euphotic zone to convert the carbon dioxide and nutrients into organic compounds. The plants grow and then are used as food for small marine animals (herbivorous, e.g. copepods, euphausiids), which in turn become food for the larger animals (carnivorous). This process of "big eats the little" is also called the "food web." The whole process is a conversion and flow of energy (from the sun's visible radiation) and matter (dissolved nutrients from the rocks) through the biological systems.

Now think chemistry (Figures 13.2a,b,c). When the marine organisms die, they sink down below the main thermocline, taking with them the nutrients (nitrogen and phosphorous compounds), which were critical for photosynthesis. At some depth and deep in the water column, the bacteria decompose this lifeless (detritic particulate and dissolved organic) matter back into nutrients, making the water nutrient-rich. Vertical mixing and upwelling bring these nutrient-rich waters back up to the surface layers (euphotic zones), where photosynthesis converts these nutrients to grow plants again. These particulate living matters then fuel the lower trophic levels of the pelagic food web with more energy.

Now think geology (Figure 13.2a,b,c). Some of the sinking biomass particulate biogenic matter reaches the seafloor and decays there and gets decomposed by bacteria. Some of these get buried in the oceanic crust, and over time, becomes lithified into rocks and can be raised to form mountains. These processes change their chemical composition due to pressures between rocks and forces from colliding plates over long geological time-scales. Chemical weathering and erosion over time releases chemicals back to the rivers, which transport them back to the ocean.

Different gases have important relevance to different biogeochemical processes. For example, CO_2 and O_2 are relevant for photosynthesis and respiration. CO_2, H_2O, CH_4 are important to monitor for climate regulation and understanding climate change. CO_2 is critical for ocean acidification and building of calcium carbonate shells.

One can also set up a regional cycle with a few components from each discipline and test a hypothesis. Looking for dominant mechanisms at play for a complex interdisciplinary problem needs careful consideration from a system perspective. So, different interdisciplinary models have evolved over time to address such complex processes. We will discuss a few of them briefly later in this chapter.

13.3 THE BIOGEOCHEMICAL CYCLES

It is useful to look at the broader perspective of our global climate system with both inanimate mass and matter and life as we know it that exploits the matter to breathe and go through the life cycle. This is best done through the carbon cycle.

The notion of biosphere was first coined by the Russian scientist Vladimir Vernadsky in his 1926 book, "The Biosphere." He proposed the connectivity of life with inorganic matter within the geological cycle of the earth. The editors of his translated book (Vernadsky, 1998) explained one of the three empirical generalizations of Vernadsky's concept as follows: "Life makes geology. Life is not merely a geological force, it is *the* geological force. Virtually all geological features at Earth's surface are bio-influenced, and are part of Vernadsky's biosphere." They continued to explain further by stating, "Life, as he (Vernadsky) viewed it, was a cosmic phenomenon which was to be understood by the same universal laws that applied to such constants as gravity and the speed of light." Vernadsky makes a case for there to be "a new Physics" – a physics that would take into account the ideas of Newtonian physics, ideas of relativity (mass changing over to energy at the speed of light) and the basics of biology, chemistry, and geology and will be able to explain different evolution and anomalies of life and human life.

This idea that living matter is part of the geological force-response system is also finding some evidence in the modern concept of Anthropocene (Crutzen, 2002).

Let us follow this line of reasoning and look at the biogeochemistry of the carbon cycle. In this setup, the most important chemical reaction is the photosynthesis of cyanobacteria, algae, and plants that uses the energy from the sunlight to break down carbon dioxide and water to produce sugar and oxygen.

$$6CO_2 + 6H_2O \longrightarrow C_6H_{12}O_6 + 6O_2 \qquad (13.1)$$

This is a fundamental equation that uses CO_2 and water to produce sugar and oxygen in photosynthesis. The biochemical reversion of photosynthesis is *respiration*, taking place in all living organisms of the biosphere, in which the

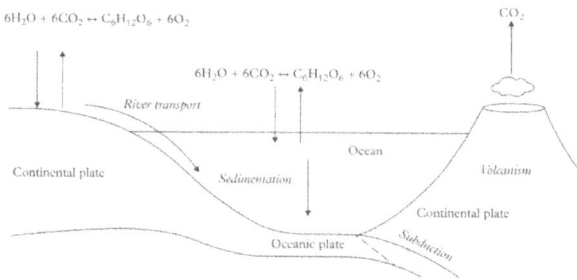

Figure 13.3: The organic carbon cycle based on the conversion of carbon dioxide and water into oxygen and sugars (biomass). The speed at which the sediments are buried and subsequently returned to the atmosphere through subducting plates that emit volcanic carbon dioxide links the organic carbon cycle to the tectonic cycle. From Dolman (2019). Reprinted with permission from Oxford University Press.

sugar produced by the photosynthetic primary producers is oxidized ("burned") by the oxygen in an exothermic process releasing energy and producing CO_2.

These fundamental cycles of carbon and oxygen are intercepted by the geological processes such as the minerals in the earth's crust (e.g. containing calcium, sulfur, silicon, iron, phosphorous), and the carbon cycle becomes an aide to the other cycles.

Equation 13.1 represents the processes in the organic carbon cycle shown in Figure 13.3. It allows organic matters to work and interact with the carbon cycle. On the other hand, imagine the long geological time-scales over which the minerals (mainly, calcium and silicon) work with the water falling on Earth's land surfaces and the rocks and soil react chemically as given by Langmuir and Broecker (2012),

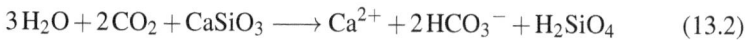

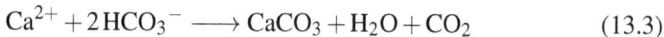

$$3\,H_2O + 2\,CO_2 + CaSiO_3 \longrightarrow Ca^{2+} + 2\,HCO_3^- + H_2SiO_4 \qquad (13.2)$$

$$Ca^{2+} + 2\,HCO_3^- \longrightarrow CaCO_3 + H_2O + CO_2 \qquad (13.3)$$

The first equation (13.2) shows how the water and carbon dioxide dissolves the calcium silicate to produce calcium and carbonate ions; and the second equation (13.3) shows how those calcium and carbonate ions can then react to produce water and carbon dioxide again, with one molecule of calcium carbonate, or shells (which are made from calcium carbonate), which gets fixed in the calcareous scales of planktonic coccolithophorids, which ultimately sink down to the ocean floor forming the calcareous ooze. See Figure 13.4.

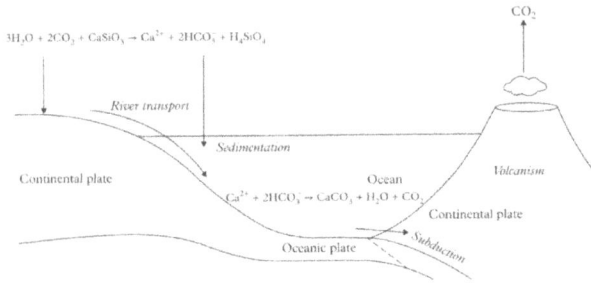

Figure 13.4: The inorganic carbon cycle containing the geological thermostat. The speed at which the sediments are buried and subsequently returned to the atmosphere through subducting plates that emit volcanic carbon dioxide then determines the net long-term balance of carbon dioxide in the atmosphere. From Dolman, 2019. Reprinted with permission from Oxford University Press.

Note that you start with two molecules of CO_2 in the first equation (13.2), and you end up with one molecule of CO_2 after forming the shells (equation 13.3). So, in this process of inorganic carbon cycling, the ocean water has stored one molecule of CO_2, which might have come as an excess input from the atmosphere due to the greenhouse effect. This is why it is said that the ocean is a huge reservoir of carbon and it has the capacity to hold some of the excess CO_2 that might have been produced in the last century from anthropogenic sources. Carbon dioxide, water vapor, and methane are three of the major constituents of the atmosphere, which, when increased in their composition, will raise the porosity of the atmosphere and will trap more heat than necessary and warm our planet even more. This is in short what happens in the greenhouse effect and these gases, which help the atmosphere become more susceptible to trap heat like the greenhouse are called the greenhouse gases.

Similarly, other elements like phosphorous, iron, and sulfur affect the carbon cycle and they each play their own role in the complex biogeochemical cycle of the Earth system. Furthermore, other greenhouse gases like methane (CH_4) and nitrous oxide (N_2O) play their role in the carbon and nitrogen cycle along with oxygen. All of these cycles and their interactive behavior is the fascinating study of the field of biogeochemistry of the climate system. To get a glimpse of these different cycles please see the recent advanced text by Dolman (2019), from which the above materials are synthesized. Much more details are provided in Dolman's book for understanding the complications and intricacies of the above processes.

We will now digress to modeling and processes of such interdisciplinary systems in a rudimentary framework.

13.4 VERTICAL DISTRIBUTION OF DISSOLVED GASES IN THE OCEAN

Let us look at the vertical distribution of some of these gases in the major oceans. First of all, realize that the surface layer of the ocean is usually saturated with atmospheric gases due to the direct exchange with the atmosphere. Below the surface layer, the concentration of a particular gas depends on multiple factors such as respiration, photosynthesiss, and decay rates and gases being released from hydrothermal, volcanic vents, or other sources at the sea bottom.

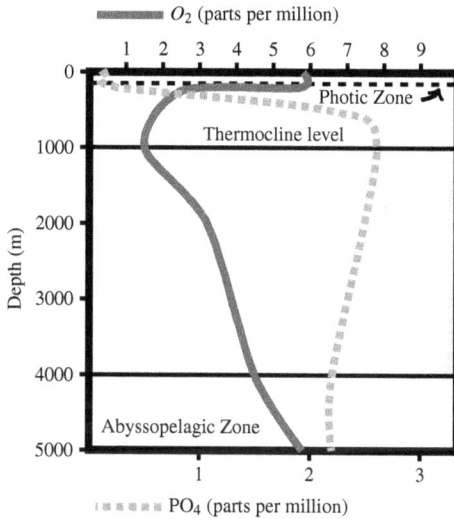

Figure 13.5: Example vertical profiles for oxygen and phosphate.

Due to such factors, the vertical profiles of different dissolved gases and elements are different. Example profiles for oxygen (O_2) and phosphate (PO_4) are shown in Figure 13.5. Oxygen tends to be abundant in the surface layers and deep bottom layer, with an *Oxygen Minimum Zone* (OMZ) in the pycnocline level. Why? The surface layer is rich in oxygen because of direct contact with the atmosphere as well as photosynthesis. The deep layer is rich in oxygen because of the thermohaline circulation (Chapter 5), which brings cold and oxygen-rich surface polar waters from the Arctic and Antarctic oceans. The bottom waters are getting replenished often by the sinking of polar waters in the three sinking regions (see Figure 5.2). There are also fewer organisms and less decay at depth, consuming less oxygen.

What happens in the depths of about 150 m (below the wind-driven mixed layer) down to the pycnocline layers is very interesting. Sinking food particles settle in these layers and become suspended in place because the water below the pycnocline is denser. Hence settling organic particles are subjected to microbial degradation and nutrients regeneration. This is why the nutrients are high at this level (see phosphate profile in Figure 13.2). A large number of organisms get drawn to this level for feeding and consume oxygen while respiring. Decay of unconsumed food also consumes some oxygen. The density difference between layers prevents mixing from below the oxygen-rich surface waters and from above to the deep layer (see Figure 13.2).

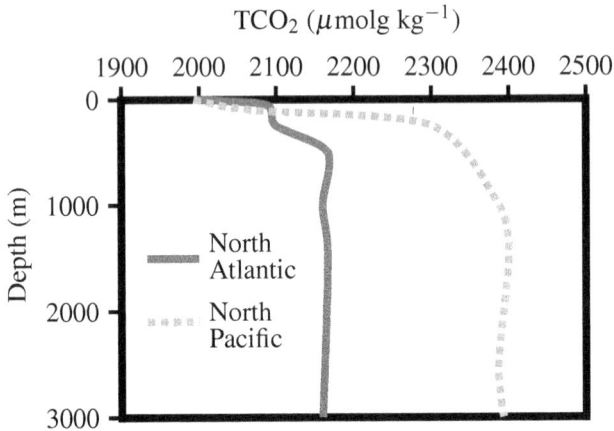

Figure 13.6: Example vertical profiles for total CO_2 in North Atlantic and North Pacific.

Typical vertical distribution of the total CO_2 (TCO_2) for the North Atlantic and North Pacific are shown in Figure 13.6. The observed reduction of TCO_2 in the surface layers is due to several reasons. First, the phytoplankton uses carbon dioxide for photosynthesis. Second, calcification (production of $CaCO_3$) by the photosynthetic planktonic coccolithophorids uses some of the CO_2 in the surface ocean. Third, sinking of particulate organic matter produced in the euphotic zone by both photosynthesis and calcification to deeper layers reduce the CO_2 surface concentration. At the same time, the remineralization at depths increases the TCO_2 to a considerable level for deeper water. Essentially, both the biological and physical processes of carbon transport vertically reduce the TCO_2 near the surface and increase it at deeper levels. Three main processes that make up the marine carbon cycle bring atmospheric CO_2 into the interior of the ocean and distribute it. These three processes are called the three "pumps": (i) the biological pump; (ii) the carbonate pump; and (iii) the solubility pump. The first two processes described above are often referred to as the "biological pump" of carbon transport. It is generally described as a three-phase process: (i) photosynthesis and its multiple effects, (ii) sinking of organic matter and associated processes, and (iii) gravitational settling in the bottom, transport, and mixing. The carbonate pump is the "hard tissue" component of the biological pump which describes how marine organisms like Coccolithophores produce hard shells out of calcium carbonate by fixing bicarbonate. The "solubility pump" of carbon transport is more of a physio-chemical process and can be thought of as a combination of two effects. First, the solubility of carbon dioxide depends on temperature inversely, i.e., the solubility is higher for cooler waters (polar regions). Second, the thermohaline circulation is driven by density differences (Chapter 5) – and thus, more CO_2 is carried by the sinking cooler waters from the polar regions to the deeper region of the subtropics and interior of the oceans.

13.5 LARGE-SCALE SPATIAL DISTRIBUTION OF DISSOLVED GASES IN THE OCEAN

Let us now look at the large-scale distribution of nutrients, oxygen and CO_2 in the global oceans. There are interesting aspects of the large-scale distribution of the biogeochemical properties that are important to understand how the overall climate system works in an interdisciplinary cycle, in a holistic manner with physics, chemistry, biology, and geology all in sync with the global ocean-atmosphere-land-ice system.

A great resource to explore this large-scale distribution is available to the curious mind, thanks to years of effort by many researchers all over the world participating in the World Ocean Circulation Experiment (WOCE). An Electronic Atlas of WOCE (https://www.ewoce.org) is available for the community to explore more than 350 tracer distributions along the WOCE Hydrographic Profile (WHP) lines. These plots are available for direct viewing in the WOCE gallery at http://www.ewoce.org/gallery. We encourage the students to go to this site, choose their basin of interest, and pick the WHP line identifier. And then you can choose a property from the drop-down menu. See Figure 13.7 for available sections for the four different oceans (Atlantic, Pacific, Indian, and the Southern Ocean). We shall discuss a couple of sections next to understand the spatial distribution of nutrients, oxygen, and dissolved inorganic carbon (as a proxy for CO_2).

Let us investigate the main features of nitrate-nitrogen, oxygen, and TCO_2 along section P15, which runs from 60°N (Bering Sea) to below 60°S (Antarctica) meridionally across the Pacific Ocean. See Figure 13.8 for nitrate (top), dissolved oxygen (middle), and TCO_2 (bottom). Let us start with the surface nitrate concentrations in the mid-latitudes, where it is low (pink area near the top) in both hemispheres. It is because the nutrient nitrate is being utilized by the plants (phytoplankton) for photosynthesis in the presence of light and by organisms – it helps them grow and build their cellular membranes and DNA. So, from the Equator to mid-latitudes, nitrates are consumed by organisms and their concentrations drop down to near-zero at the surface. Now, why doesn't

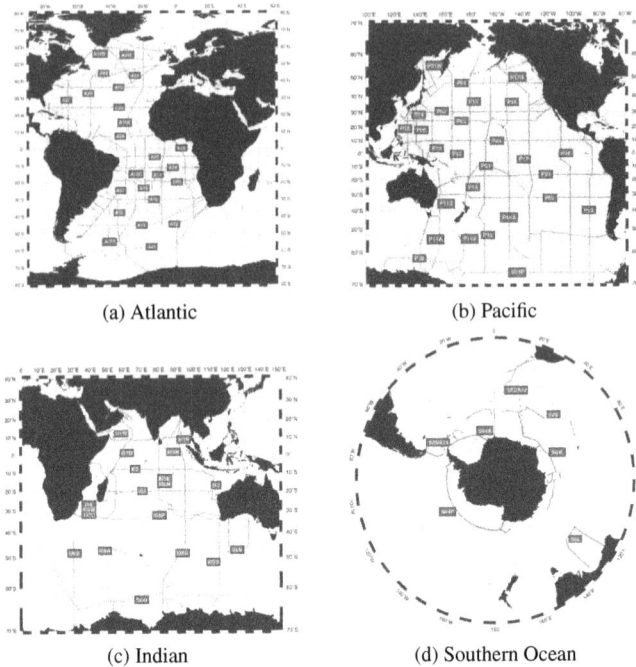

(a) Atlantic   (b) Pacific

(c) Indian   (d) Southern Ocean

Figure 13.7: WOCE sections in different oceans.

that happen near Antarctica? See on the left – surface region, top panel. The southern hemisphere surface nitrate is pretty high. So, there is less biological productivity here, although there is enough nitrate. Something else is missing which prohibits productivity and growth. Well, this may be due to other growth limiting factors such as insufficient 24-hours mean light conditions under a very strong mixing below the euphotic layer. The reader is referred to the critical depth concept by Sverdrup (1953), which says that phytoplankton blooms can only occur if the mixed layer depth is less than a critical depth (defined by a depth down to which the production of organic matter by photosynthesis equals the destruction by respiration). On the other hand, the lack of micronutrients such as iron in the Antarctic waters has been proposed as a possible cause of this limited productivity. This iron limitation does not allow the organisms to grow as much as in the mid-latitudes. Now concentrate on the mid-depth, where nitrate concentrations are high. This is due to sinking organic materials from the surface layers becoming suspended and then remineralized by bacteria. The vertical gradients of nutrients then slowly stabilize at depth and are maintained at a relatively high level throughout the rest of the water column, similar to PO_4 vertical profile in Figure 13.5.

(a) Nitrate along P15

(b) Oxygen along P15

(c) TCO2 along P15

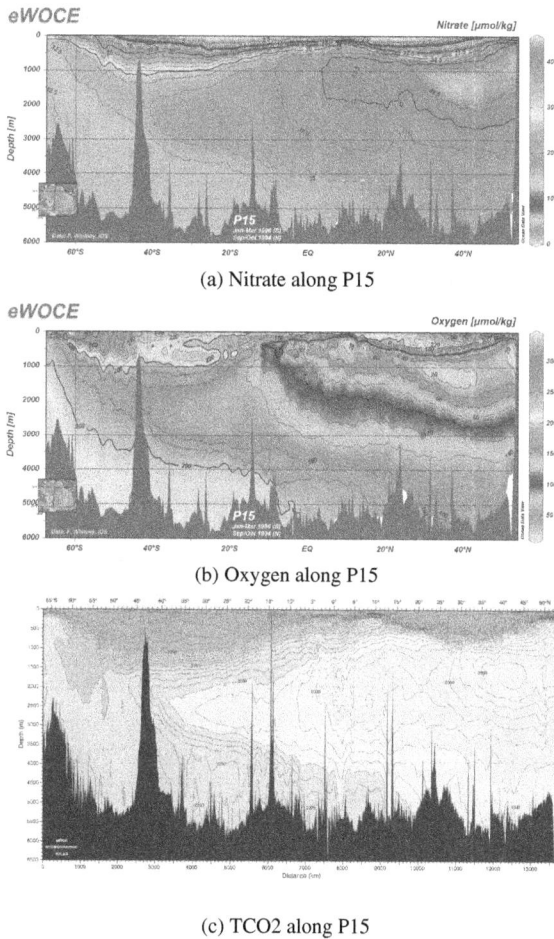

Figure 13.8: Distribution of properties along WOCE P15 Section for nitrate (top), oxygen (middle), and TCO$_2$ (bottom).

How about the horizontal gradient of nitrate? The slow decrease of nitrate from northern latitudes at mid-depth to the colder southern latitudes is driven by large-scale circulation features like the meridional overturning circulation (MOC) (see Chapter 5, Figure 5.5), which moves the tracers around the ocean interior.

Let us look at the middle panel (Figure 13.8b) for oxygen. When there is biological productivity, then nutrients are consumed and oxygen is produced. This is what happens at the surface layers (low nutrient and high oxygen). On the other hand, when there are more nutrients (more food), organisms conglomerate and respire, i.e. consume oxygen. This is what happens at mid-depth (high nutrient, low oxygen). So, the nutrient-oxygen relationship is kind of the inverse of each other (negative correlation). The subsurface *oxygen minimum zone* in the middle panel is very similar to the high-nitrate region in the top panel in Figure 13.8. However, there is another complicating factor for understanding the large-scale oxygen distribution. Near the surface, there is air-sea transfer of oxygen. Water can also hold more gases at colder temperatures. So, there is more oxygen at the surface (from the air-sea transfer) near the poles at high latitudes than at lower latitudes where the water is much warmer. Realize that the spatio-temporal distribution of dissolved oxygen is thus more complicated than a reverse relationship with nutrients, which would otherwise be valid if it were only a biological productivity-dominated system.

Now let us look at the TCO_2 distribution for the same section in the bottom panel (Figure 13.8c). At mid-depth, it looks very similar to that of the distribution of nutrients. How? Both processes are similar. The TCO_2 gets absorbed by organisms near the surface where the biological productivity is high – the organisms need carbon for the cellular structure – so the TCO_2 is consumed at the surface, sinks down to mid-depth, remineralizes there, and are made available to be lifted up by upwelling to be re-used by the surface organisms. The biological cycle works that way (see Figure 13.2). In addition, the distribution of TCO_2 has an oxygen-like pattern near the poles (relatively high values than at intermediate depths below the surface) due to air-sea transfer of CO_2. TCO_2 is higher in colder waters at higher latitudes and lower in warmer waters in the equatorial regions. Realize that different parts of the ocean show different dynamics in their biogeochemical signatures. To summarize, primary processes that govern the large-scale distribution of nutrients, oxygen, and carbon are: (i) biological productivity in the upper layers where light is available (euphotic zone) – this is where nutrients and inorganic carbon is consumed, producing organic matter and O_2; (ii) export of organic matter from euphotic zone in the form of sinking biogenic particles (detritus, $CaCO_3$ shells, etc.); (iii) remineralization of particulate and dissolved organic (=carbon) matter – this is a reverse process of productivity, it consumes oxygen; (iv) circulation processes such as horizontal advection, vertical and lateral mixing moving tracers around and (v) air-sea transfer of oxygen and CO_2 are temperature dependent. Of course, there are other regional processes such as atmospheric nutrient deposition, riverine inputs of nutrients, nitrogen fixation and denitrification, etc. Please see the books and references in the Further Reading section for some of these more advanced topics.

13.6 SIMPLE BIOGEOCHEMICAL MODELING

One of the simplest ways to model a biological process is to set up a system of
equations that describe how the nutrients and phytoplankton interact with each
other in the presence of one of their consumer, the herbivorous zooplankton.
This is known as the Nutrient-Phytoplankton-Zooplankton (NPZ) model. In
other words, the mechanism of biological productivity in the surface layers with
photosynthesis and growth can be linked with the organisms growing, decay-
ing and sinking below. A coupled set of three ordinary differential equations
describing the change of N, P, and Z with time has been formulated over the
years.

Phytoplankton

$$\frac{dP}{dt} = \mu_0 \underbrace{\left(\frac{N}{k_N + N}\right)}_{\substack{\text{Nutrient} \\ \text{limitation}}} \underbrace{(1 - e^{\alpha E/\mu_0})}_{\substack{\text{Light} \\ \text{limitation}}} P - \underbrace{g\left(\frac{P}{k_P + P}\right)Z}_{\text{grazing}} - \underbrace{m_P P}_{\text{mortality}} \qquad (13.4)$$

Zooplankton

$$\frac{dZ}{dt} = \underbrace{ag\left(\frac{P}{k_P + P}\right)Z}_{\text{grazing}} - \underbrace{m_Z Z}_{\text{mortality}} \qquad (13.5)$$

Nutrient

$$\frac{dN}{dt} = -\mu_0 \underbrace{\left(\frac{N}{k_N + N}\right)}_{\substack{\text{Nutrient} \\ \text{limitation}}} \underbrace{(1 - e^{\alpha E/\mu_0})}_{\substack{\text{Light} \\ \text{limitation}}} P - \underbrace{(1-a)g\left(\frac{P}{k_P + P}\right)Z}_{\text{grazing}} + \underbrace{m_P P}_{\text{mortality}} + \underbrace{m_Z Z}_{\text{mortality}}$$

$$(13.6)$$

Note that if add the three equations, we arrive at the conservation equation
for the biomass variables.

$$\frac{dP}{dt} + \frac{dZ}{dt} + \frac{dN}{dt} = 0 \qquad (13.7)$$

As you can see, just with three variables, N, P, Z, the system demands estimation
of all the parameters $\mu_0, \alpha, E, k_P, k_Z, m_P, m_Z, a$, and k_N. These are generally
obtained through laboratory and field experiments. One can further use the
Photosynthesis-Irradiance (P-I) curves (Aalderink & Jovin, 1997; Platt & Jassby,
1976) or nutrient uptake curves known in a region. In a regional setting one
needs to calibrate the model with field data or use previous models and/or
similar models. Remember the idea of "Validation-Calibration-Verification" for

the physical models as illustrated in Chapters 11 and 12. This methodology applies to interdisciplinary models as well.

As there are many functional types of phytoplankton and zooplankton, things can get too complicated too easily in a coupled modeling setting. One of the preferred ways to construct and understand a complicated biogeochemical model is to categorize the plankton species by their function and use representative types or groups. A general review of this approach is given by Quere et al. (2005).

13.7   BIOGEOCHEMICAL MODELS – TWO EXAMPLES

The most well-known model for ocean biogeochemistry is probably the one developed by Fasham et al. (1990). It is a seven-component model of the mixed-layer ecosystem and was the first of its kind to separate new and regenerated forms of nutrient, as well as including the microbial recycling pathways, i.e. the microbial loop (Azam et al., 1983). So, it has three different nutrients (nitrate (N_n), ammonium (N_t) and labile dissolved organic nitrogen (N_d)), phytoplankton (P, photosynthesizers), zooplankton (z, such as grazers), detritus (D), and bacteria (B, microbial loop). A flow diagram for this model is shown in Figure 13.9. For details on the different choices of model variables please see Fasham et al. (1990).

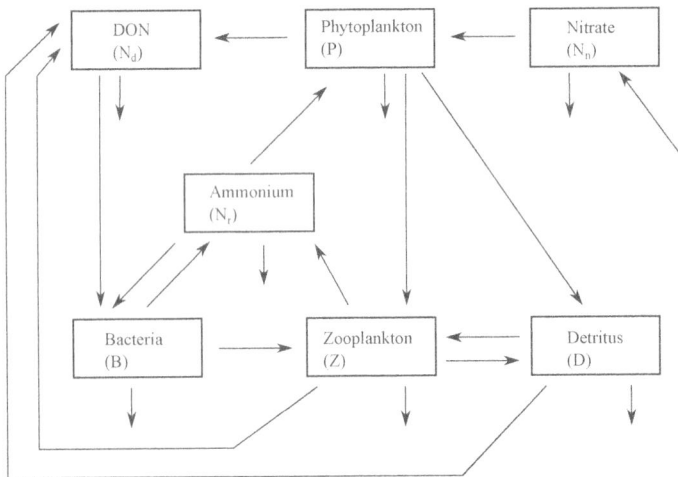

Figure 13.9: Seven-component Fasham Model. After Fasham et al. (1990).

A more recent adaptation with carbon and dilicate was developed by (Chai et al., 2002; Chai et al., 2007) and is shown in Figure 13.10. This is called the

CoSINE model (Carbon, Si(OH)$_4$, Nitrogen ecosystem). The CoSINE model consists of ten compartments, with two size classes of phytoplankton and zooplankton, detritus nitrogen and detritus silicon, silicate and TCO$_2$, and the two forms of dissolved inorganic nitrogen, nitrate, and ammonium. P1 represents small, easily grazed phytoplankton, whose biomass is regulated by micrograzers (Z1). P2 represents diatoms ($>10 \mu$m size-class) that form blooms and largely contribute to the sinking of particulate organic matter as ungrazed cells or large fecal pellets. Z2 represents larger mesozooplankton that graze on diatoms and detritus nitrogen and prey on Z1.

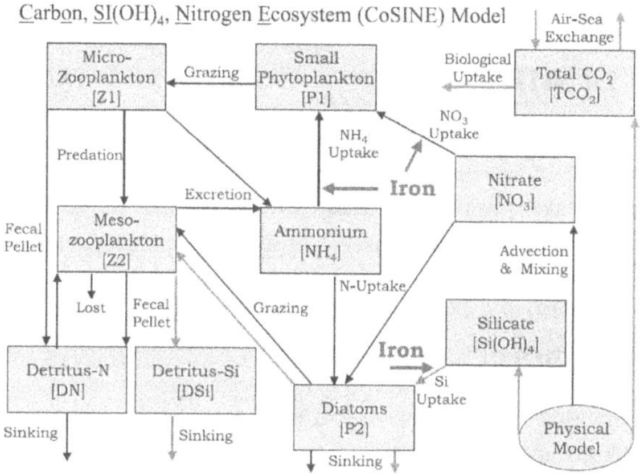

Figure 13.10: Ten-component Model. Carbon, Si(OH)$_4$, nitrogen ecosystem (CoSINE) model structure. Republished with permission of Elsevier Science and Technology Journals, from "Modeling responses of diatom productivity and biogenic silica export to iron enrichment in the equatorial Pacific Ocean," Chai et al., in Global Biogeochemical Cycles, 21 (GB3S90), 2007; permission conveyed through Copyright Clearance Center, Inc.

The detrital pool is divided between detrital nitrogen and silicon to balance supplies of nitrogen and silicon through upwelling and vertical mixing separately. The linkage from the carbon cycle to the nitrogen-based ecosystem models is done through Redfield ratios reported by L. A. Anderson and Sarmiento (1994). The role of iron is simulated by two parameters of the P-I curve: (i) the photosynthetic efficiency, α, that describes the slope of the initial light-limited phase of photosynthesis and (ii) maximum photosynthetic rate at light saturation (P^{max}). The first one depends on light-limited photosynthesis and second one is dependent on irradiance at light saturation. For more details on these model parameters please see Chai et al. (2007). Note that these models have been coupled with the physical circulation model, which provides the advection and upwelling velocities and t, s, density fields to carry out the ecosystem dynamics. The left-hand side of the ecosystem equations (terms like dN/dt, dP/dt, dZ/dt) require the velocity fields to be prognostically evolved. One can run the biological or ecosystem model offline or online depending on resources and processes to be resolved. See Figure 13.10 for the connection of the physical model to the ecosystem model and carbon cycle.

These ecosystem models are available for the community within HOPS/MSEAS, POM, ROMS, MITGCM, and other modeling frameworks. Please see individual websites for their implementation and example problems before diving into one of these coupled model exercises.

13.8   EXAMPLE BIOGEOCHEMICAL PROCESSES – ACIDIFICATION, DENITRIFICATION, HABS, AND HYPOXIA

It is important to realize that the biogeochemical models that have been discussed above are equally applicable for regional water bodies like estuaries, coastal oceans, and lakes. In fact, these are where the immediate returns of such development are useful to the society. We will discuss some of these processes here.

13.8.1   OCEAN ACIDIFICATION

Since the industrial revolution or over the last two centuries, our usage of fossil fuels and land has resulted in a steep increase of carbon dioxide into the atmosphere. About 30% of this increase is estimated to have been absorbed by the ocean. When CO_2 is absorbed by seawater, a series of reactions happen. Look at the inorganic carbon cycle equation 13.3. The carbonate chemistry of the ocean demands that the right side of that equation (in the absence of minerals, or independent of its presence, or at a much faster time-scale than the geological time scale in which that equation was cast) would simply be a reaction of CO_2 and seawater which would also generate a free hydrogen ion (a proton) first, and then another one subsequently, as follows,

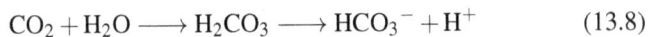

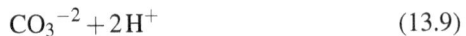

$$CO_2 + H_2O \longrightarrow H_2CO_3 \longrightarrow HCO_3^- + H^+ \qquad (13.8)$$
$$CO_3^{-2} + 2H^+ \qquad (13.9)$$

So, this simple excess of CO_2 in the ocean leads to an excess of hydrogen concentration in the water. The rising of hydrogen concentration is synonymous to the water becoming more acidic and is generally measured by a scale in partial pressure of hydrogen or "pH."

What is pH? In pure water (nothing but H_2O molecules) at 25°C and at $1atm$, a very small fraction of the H_2O molecules, about one in 10 million (or, 10^{-7}), freely breaks into a hydroxyl ion (OH^-) and a hydrogen ion (H^+). By definition, the free hydrogen ions dictate the acidity of water and is quantified by the formula,

$$pH = -\log(H^+) \tag{13.10}$$

The pH for "pure" water is simply 7. Even in pure water ions tend to form due to random processes (producing some H^+ and OH^- ions). The amount of H^+ that is made in pure water is about equal to a pH of 7. That's why 7 is neutral. The logarithmic scale reduces the fraction count by orders of magnitude as the concentrations of hydrogen ions are so much less compared to the water molecule. The negative sign brings the scale to a positive metric for pH for ease of understanding. Remember that the lower the pH the higher the concentration of hydrogen ions in the water and thus higher its acidity. The reference is pH = 7 for pure or neutral water. A table of pH of usual substances is given in Figure 13.11 associated with magnitude changes in hydrogen concentration and acidity percentage change for seawater. The range of pH for most substances fall within a scale of 1 to 14, with being the neutral (pure water), above 7 being basic (alkaline), and below 7 being acidic.

Realize that pH is the scale on which acidity is expressed, but it is not synonymous with acidity. An appropriate way to compare the acidity at two different pH values is to express the relative percentage change of the H^+ concentration at the two pH levels, as in Figure 13.12. For example, a pH decrease of 0.11 corresponds to approximately a 30% increase in acidity, which is an exact change in acidity (H^+ concentration) of 28.8% when calculated in this way. Seawater has an equilibrium pH of about 7.8 to 8.2. Scientists believe (based on data) that the pH of the ocean has changed by 0.1 pH units since the industrial revolution. This translates to a 30% increase in acidity of seawater. As the seawater becomes more acidic, carbonate ions become less abundant.

Now, carbonate ions are important building blocks for sea shells and coral skeletons. A reduction in carbonate ion supply will affect the population of calcifying organisms such as oysters, clams, sea urchins, shallow water corals, deep sea corals, and calcareous plankton such as Coccolithophorids. Furthermore, certain fish's ability to detect predators is decreased in acidic waters. When these organisms are at risk, the entire food web is at risk.

13.8.2 DENITRIFICATION

Nutrients, the essential elements necessary for autotrophic growth, are classified into macronutrients and micronutrients, necessary in high and low quantities, respectively, for primary producers. Macronutrients are primarily composed of nitrates and phosphates, while the micronutrients are metals, such as iron. Nitrification is the microbial process in which oxidation of ammonia (NH_4^+) yields nitrites (NO_2^-), which upon further oxidation yields nitrate (NO_3^-). Since this is a two-step process, two different types of microorganisms are

pH	Examples of solutions
0	Battery acid, strong hydrofluoric acid
1	Hydrochloric acid secreted by stomach lining
2	Lemon juice, gastric acid, vinegar
3	Grapefruit juice, orange juice, soda
4	Tomato juice, acid rain
5	Soft drinking water, black coffee
6	Urine, saliva
7	"Pure" water
8	Sea water
9	Baking soda
10	Great Salt Lake, milk of magnesia
11	Ammonia solution
12	Soapy water
13	Bleach, oven cleaner
14	Liquid drain cleaner

Figure 13.11: Table of pH for different liquids. From https://upload.wikimedia.org /wikipedia/commons/1/1d/2713_pH_Scale-01.jpg.

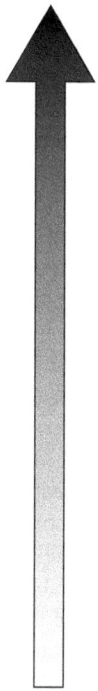

pH	H^+	Change in acidity
7.2	6.3×10^{-8}	+900%
7.3	6.3×10^{-8}	+694%
7.4	4.0×10^{-8}	+531%
7.5	3.2×10^{-8}	+401%
7.6	2.5×10^{-8}	+298%
7.7	2.0×10^{-8}	+216%
7.8	1.6×10^{-8}	+151%
7.9	1.3×10^{-8}	+100%
8.0	1.0×10^{-8}	+58%
8.1	7.9×10^{-9}	+26%
8.2	6.3×10^{-9}	

Figure 13.12: Conversion of pH to Hydrogen ion to Acidity. After PMEL/NOAA.

involved. These microorganisms utilize the energy from these redox reactions to produce organic carbon compounds. The nitrification process is controlled by the availability of oxygen and ammonia.

Denitrification is the opposite of the above reaction, in which nitrate is reduced to nitrite first under anaerobic conditions and then to molecular nitrogen gas. It plays a very important role in the nitrogen cycle, where the nitrogen is released back into the atmosphere in the forms of gases, nitrous oxide (N_2O), and nitrogen gas (N_2). It is also important for wastewater treatments. Not all forms of nitrogen are biologically available; plants can only use ammonium and nitrate, nitrogen gas is unavailable to them.

Nitrification flourishes in the pH range of 6.5 to 8 and temperature range of 20–30°C; while denitrification is favored for pH between 7 and 9 and temperature range of 26–38°C under anaerobic conditions.

13.8.3   HARMFUL ALGAL BLOOMS

Harmful Algal Blooms (HABs) occur when microalgae – which are unicellular photosynthetic organisms that live in rivers, lakes, and in the sea suddenly grow out of control and produce toxins which can accumulate on the body tissue of shellfish and fish, and ultimately, be harmful to marine mammals, birds and people that feed on them. There are different kinds of HABs, caused by different kinds of algae and toxins. HABs can occur in fresh (lake), marine (salt), and brackish (mixture of fresh and salt) water bodies around the world, where the process of eutrophication is prevalent. The gradual increase in the concentration of phosphorus, nitrogen, and other plant nutrients in an aging aquatic ecosystem such as a lake is known as eutrophication.

In freshwater systems, cyanobacteria (Remember equation 13.1), a type of photosynthetic bacteria, also known as blue-green algae, are often the cause. They grow in the presence of an excess of nutrients (called eutrophication), warm temperatures, and lots of light. They make the water blue or green. In marine and brackish waters, the dinoflagellates and diatoms, two different kinds of phytoplankton, are mostly the causes of HABs. Some species of dinoflagellates are bioluminescent like *Noctiluca scintillans*, the usual species in the famous Phosphorescent Bay in Puerto Rico. NOAA has regional models similar to those discussed before in Sections 13.4 and 13.5 (see Figures 13.9 and 13.10) for monitoring HABs. These systems provide forecasts on a regular basis for people and seagoing fishermen. Such forecast products are currently available for the Gulf of Maine, Lake Erie, and the coasts of Florida, Louisiana, and North Carolina. In addition, seasonal forecasts are made using similar models and larger-scale physical and biological outlooks.

13.8.4   HYPOXIA

Hypoxia, as the name suggests, is a condition of reduced levels of oxygen in seawater, when it can no longer sustain marine life. Typical examples of such regions are the bottom waters along the sea floor near the coast, where river runoffs bring in lighter water with massive amounts of nutrients. Marine life typically cannot survive below an oxygen level of 2 mg/liter.

Stratification in the water column, which occurs when this less dense freshwater from an estuary mixes with heavier seawater, is a typical cause of hypoxia. Limited vertical mixing between the water "layers" restricts the supply of oxygen from surface waters to more saline bottom waters, leading to hypoxic conditions in bottom habitats.

Hypoxia occurs most often, however, as a consequence of human-induced factors, especially nutrient pollution, can lead to eutrophication (increased concentration of nutrients). The causes of nutrient pollution, specifically nitrogen and phosphorus nutrients, include agricultural runoff, fossil-fuel burning, and wastewater treatment effluent.

A modest example of hypoxia is the bottom waters of the Gulf of Mexico, where about 40% of the total US industrial runoff carrying nutrients are flowing in, and during every spring, they create one of the largest "dead zones" in the northern Gulf of Mexico, where lots of marine life (fish, shellfish, marine mammals and birds) die. Another region that sees hypoxia events often is the Chesapeake Bay. NOAA has modeling and monitoring programs similar to HABs for these and other regions.

Hypoxia is also very common in the world and especially in and around the developing world's ocean. In the northern Indian Ocean, there is a large *oxygen minimum zone* and hypoxia events are often cited there. The Bay of Bengal is another region where hypoxia has been detected recently near and below the Ganges-Brahmaputra-Meghna river outflow (which carries a lot of nutrients from different sources upstream). Global distribution of naturally occurring marine hypoxia events was studied by Helly and Levin (2004), who documented some of these major events and discussed their possible causes and effects.

13.9 MODELING THE FISH POPULATION – STOCKS AND RE-CRUITMENT

One of the key challenges in interdisciplinary modeling is how to connect the physical and biogeochemical models to fish population. Fisheries is an industry, and thus there is a need to know where to go and fish every year before the season starts. The answer is tied to the knowledge of how good the adult population is this year and how the recruitment was last year. So, let us define some terms that will set the stage for discussion for modeling fish. To do this, we follow the accepted definitions in the "Guide to Fisheries Stock Assessment - From Data to Recommendation" by Professor Andrew B. Copper of the University of New Hampshire.

First, let us define what is "stock" and what is "population". A fish **population** is a group of individual fish of a single interbreeding species located in a given region, which could be as large as the Gulf of Maine, the Atlantic, or the Bay of Bengal, or as small as an estuary or a single river. A fish **stock**, on the other hand, is defined as much by regulatory or management directives depending on boundaries or harvesting locations – as by biology. The latest version of Cooper, 2006 gives an example of this as follows. "... alewives from the Taylor River are considered a separate population from those in the Lamprey River, but both are part of the Gulf of Maine Stock."

Now, to model the population size of the stock for next year (N_2), the simplest way would be to find out how the recruit is this year (R_1) and how many of them die this year (D_1). So, if we know this year's population (N_1), a simple population dynamics model would be

$$N_2 = N_1 + R_1 - D_1 \tag{13.11}$$

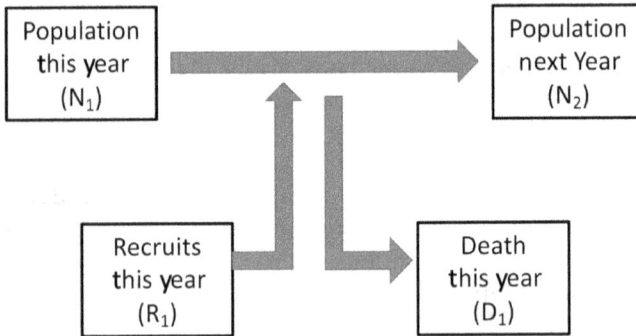

Figure 13.13: A simple Population Dynamics Model for next year's population based on this year's population, recruits, and death. After Cooper (2020).

Figure 13.13 illustrates this simple model. The idea is simple. Quoting Cooper: "Fish are born (recruitment). They grow. They reproduce. They die, either from natural causes or by fishing. That's it."

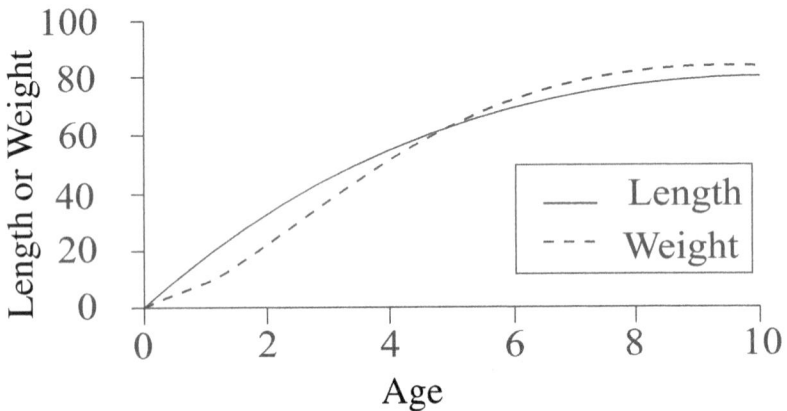

Figure 13.14: Two different depictions of the Ludwig von Bertalanffy (LVB) growth model. After J. Quinn (2014).

But life is not that simple for a fish! They encounter the environment, and there are preys and predators in the system. So, things get complicated. Scientists have thought about such processes and tried to incorporate those effects in the simple model. First of all, it is easier to assume a rate of death rather than a hard number for the death (D_1). This is the mortality rate and is given as a percentage of the population size "N." Next, the simple formula in equation 13.11 assumes that a fixed number of fish are born and survive to be counted each year. However, it is more realistic to assume that each reproductively mature adult produces, on average a fixed number of offspring that survive. This is called "net fecundity." Realize that "fecundity" is the number of offspring produced by an individual, and "net fecundity" is the number of recruits (offspring that survive to be counted or caught) produced by an individual.

Clearly, if the net fecundity rate is greater than the mortality rate, then the population would grow larger with time. This difference between the two rates is known as the "growth rate" for the stock. Since the growth rate is a percentage, it does not depend on the total size of the stock or the abundance or the density. It is also called the "intrinsic growth rate" or the density-independent growth rate. This density-independence idea allows the stock to grow or shrink at a constant rate, which works fine for small stock size but fails to follow reality for long-term modeling.

So, another idea would be to allow for an upper limit of the stock size, called the "carrying capacity." Now, you can bring in environmental conditions and constraints on the habitat and limit the growth for longer-term simulations. The density-dependent growth rates are similar to the intrinsic growth rates for small stock sizes. As the stock size grows and reaches toward the carrying capacity, the mortality rate increases or the net fecundity rate decreases, which effectively reduces the growth rate.

You can add more complexity to the simple model in its representation of mortality rate and age-structure. Modelers like to think of the mortality rate as an instantaneous mortality rate (Z) rather than an annual percentage. This helps in setting up prognostic equations with time-stepping and to model and understand the effects of events on the growth or shrinkage of a stock. Note that Z has two distinct components, a natural mortality component (M) and a fishing mortality component (F). Now, for mathematical reasons, the instantaneous mortality rate is converted to an annual survival rate. It is possible to get this annual survival rate once you know M and F. See Cooper (2020) or other advanced textbooks on this.

Let us briefly talk about the other important but much desirable complexity to add to the simple model. It is the age-structure of the life of a fish. It is a fact that most of the fish live for more than a year or two. And they grow during that time and can spawn a few times. For example, salmon spawns 2–3 times in their usual lifespan of 8–9 years. So, fish of different ages would experience different rates of mortality. But the basic simple model (equation 13.11) assumes a constant mortality rate (natural and fish mortality) for all fish. So, the models can be built in a way that the total population (N) is sub-divided into age classes. These age-structured models can then allow for separate growth rates for different age groups. Note that the number of recruits is only relevant to the first age class!

One approach to age-structure modeling is to incorporate a classical growth model developed by Ludwig von Bertalanffy growth model or the LVB growth model (Von Bertalanffy, 1938). This model assumes that the growth occurs exponentially at younger ages and gradually flattens out at a certain age. Growth is synonymous with the length of the fish and can be obtained from survey data to then determine the age. A typical LVB model is shown in Figure 13.14, where the growth is limited at length called L_∞.

Following this age-length idea, another complexity that can be added is to think about the population as total weight or biomass of the population and model that. In this approach one can incorporate the LVB model to separate the total number of the population (N) into age classes according to their lengths. This approach of modeling can incorporate uncertainties in the weights and might result in more realistic estimates of uncertainty for the weight and biomass variables through integration.

13.9.1 STOCK–RECRUITMENT RELATIONSHIP

Clearly, one of the key aspects in modeling the fish population hinges on our understanding of how the stock might be related to the recruitment at the early stages of life of a fish. This relationship, known as the stock-recruitment (S/R) relationship, is the kernel of many fish models and fundamental to the management of fish and other natural resources. You can guess that this relationship uses the idea of density-independent intrinsic growth rate (discussed above) in its simplest implementation in a functional form. More complicated functions have been developed, which allow for the density-dependent nature of growth as well.

The two most common stock-recruitment models are the Beverton–Holt (Beverton & Holt, 1957) model and the Ricker (Ricker, 1975) model. In the Beverton-Holt (B-H) model, the number of recruits increases as the mature population gets larger and then stabilizes at large stock sizes. On the other hand, the Ricker model assumes that when the spawners (S_t) or reproductively mature population reaches a certain level, the number of recruits (R_t) actually goes down. A conceptual schematic comparison between these two models is shown in Figure 13.15. Mathematically, both of these models can be represented in a general functional form using a three-parameter system as defined by Deriso, 1980 and then modified by Schnute (1985). The resulting Deriso-Schnute model can be written as (following Subbey et al., 2014),

$$R_t = \alpha S_t (1 - \gamma \beta S_t)^{\frac{1}{\gamma}} \tag{13.12}$$

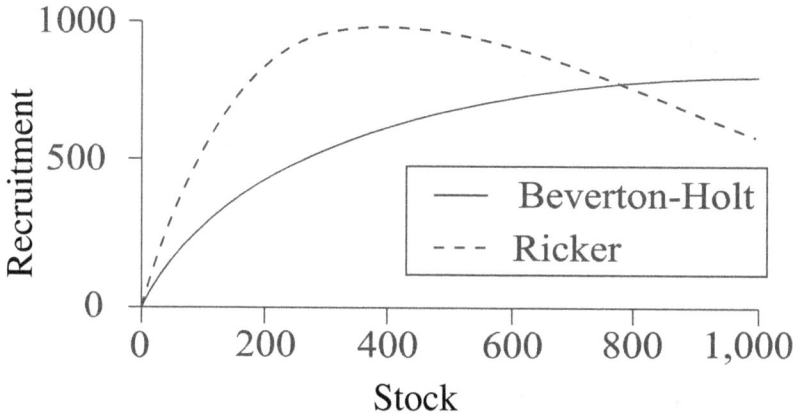

Figure 13.15: Example stock-recruitment model. After Quinn (2006)

The Ricker model is given at the limit of $\gamma- > 0$ and the B-H model is recovered when $\gamma = -1$, as follows.

$$R_t = S_t e^{\alpha - \beta S_t}$$
$$= \frac{\alpha S_t}{1 + \beta S_t} \tag{13.13}$$

Note that in this general formulation, α is the density-independent parameter, β is the density-dependent parameter, and γ is the shape parameter allowing for other variations between the two limits of the shapes presented in Figure 13.15. Note, however, that the shape parameter $\gamma = 1$ leads to other models of stock-recruitment relationship (Schaefer, 1989; Schnute, 1985) and is not limited to be within 0 and < -1.

The above general formulation allows one to expand the S-R relationship to link to the environment. For example, one can think of a number of environmental variables given by $Z_1, Z_2, Z_3...., Z_n$; and then the general formulation in equation 13.13 can be modified to:

$$R_t = S_t e^{\alpha - \beta S_t + c_1 Z - 1 + c_2 Z_2 + ... + c_n Z_n}$$
$$= \frac{\alpha S_t}{1 + \beta S_t} e^{c_1 Z - 1 + c_2 Z_2 + ... + c_n Z_n} \tag{13.14}$$

This opens up the opportunity for linking the fisheries models to the biogeochemical and physical models discussed in this book. As we understand now, many species have migrated and are in the process of migrating in response to changes in water temperature in their usual habitat. There have been studies which identified other environmental factors such as the North Atlantic Oscillation (NAO), whose interannual variability is significantly correlated with those of certain populations and their stocks and recruitment in the North Atlantic (Drinkwater, 2005). Similarly, in the Pacific, ENSO variability has been linked with fisheries in Peru through California (see Chapter 7). Furthermore, it has been well recognized that the stock-recruitment relationship is complex and transcends multiples processes acting over multiple scales in space and time (A. Robinson, 2001; Rothschild, 2000). Current numerical models try to take advantage of their high-resolution simulations and utilize the stock-recruitment relationship blended with environmental variables (similar to equation 13.14) and then project and predict fisheries variables.

One example of such a combined fish-environmental model is the study by Olsen et al., 2010 for the North Sea Cod stock population. They incorporated food availability (zooplankton, Z) and climate (sea temperature, T) in a combined Ricker-Beverton–Holt model. They constructed the following relationship between the recruitment (R) and the spawning biomass (S),

$$R = S e^{(a_0 - a_1 T)} \underbrace{\left((1 - Z)e^{-bS} + Z(1 + \gamma S)^{-1} \right)}_{\text{Density dependence}}$$

$$\underbrace{\phantom{R = S e^{(a_0 - a_1 T)} \left((1 - Z)e^{-bS} + Z(1 + \gamma S)^{-1} \right)}}_{\text{Reproductive rate}} \qquad (13.15)$$

where $\gamma = e^c / S_{max}$ and S_{max} is the maximum observed spawning biomass. Realize that the zooplankton term represents the density-dependence and the formulation is essentially the recruitment (R) being equal to a reproductive rate of the spawning biomass (S). Olsen et al. (2010) used this model to study the impact of the food and climate on the North Sea Cod by changing not only Z-terms and T-terms but also different model formulations like a traditional Ricker model or a traditional B-H model by simply omitting terms in this combined formulation. For example, taking the *log* of both sides and then eliminating terms, one gets the traditional Ricker and a Ricker model with just the zooplankton effect and no temperature effect from the general setup of equation 13.17 as follows.

$$\log(R) - \log(S) = a_0 + \log(\exp(-bS)) \qquad \text{Ricker Model} \quad (13.16)$$
$$\log(R) - \log(S) = a_0 + \log(\exp(-b(1 - Z)S)) \qquad \text{Ricker} + \text{Z Model} \quad (13.17)$$

You can do similar model derivation for the B-H set up.

Finally, it is worth noting that as the physical and biogeochemical models are getting very sophisticated and efficient in resolution they are able to resolve multiscale processes. However, the challenges these models are facing now are the impending uncertainty of the changing environment and its impact on different species. For a recent review on the challenges and recommendations for the world squid fisheries in different boundary current regions (both eastern and western) see Moustahfid et al. (2020). The physical and biogeochemical models are often being used to track and predict recruitment areas and quantify them from survey data, if available and sometimes in real-time. Such activities are possible now with advanced Lagrangian models (see Röhrs et al., 2014 for an example), where we can track the eggs and larvae and simulate their distribution weeks and months ahead if a survey is conducted. With climate change upon us, these exercises have tremendous value for both fishers and managers. They can both learn from these exercises with scientists. That is not only challenging but also interesting in that it is not often to see science and society working together for the benefit of all, including the fish population.

CONCLUSION

In this chapter, we focused on multidisciplinary multiscale modeling. The idea of the interconnectedness of biology, chemistry, and geology to the physics is provided in terms of the biosphere and its impact from and on the physical earth. We discussed simple biogeochemical modeling variables and coupled physics-biogeochemical modeling framework. Simple biogeochemical processes such as ocean acidification (OA), hypoxia, harmful algal bloom (HAB), and denitrification are discussed. The basic framework of modeling the fisheries with simple stock-recruitment relationships is presented. The connection between the S-R models and the environment is also shown at the end for further follow-up at an advanced level.

FURTHER READING

Beverton, R. J., & Holt, S. J. (1957). *On the dynamics of exploited fish populations.* Fisheries Investigation Series 2 (Volume 19), UK Ministry of Agriculture. Fisheries, and Food, London, UK.

Dolman, H. (2019). *Biogeochemical cycles and climate.* Oxford University Press, USA.

Eppley, R. W. (1972). Temperature and phytoplankton growth in the sea. *Fishery Bulletin,* 70(4), 1063–1085.

Evans, G. T., & Parslow, J. S. (1985). A model of annual plankton cycles. *Biological Oceanography, 3*(3), 327–347.

Fasham, M. J., Ducklow, H. W., & McKelvie, S. M. (1990). A nitrogen-based model of plankton dynamics in the oceanic mixed layer. *Journal of Marine Research, 48*(3), 591–639.

Hobbs, N. T., & Hooten, M. B. (2015). *Bayesian models: A statistical primer for ecologists.* Princeton University Press.

Vernadsky, V. I. (1998). *The biosphere.* Springer Science & Business Media.

Fortaleza, Brazil

14 Modeling of the Climate System

> The earth, the air, the land and the water are not an inheritance from our forefathers but on loan from our children. So we have to handover to them at least as it was handed over to us.
>
> _____
>
> Mahatma Gandhi (1869–1948)*

OVERVIEW

In the last few chapters, we have discussed a number of different modeling systems and their approaches to understand the processes in various disciplines individually and in a coupled interdisciplinary manner. In Chapters 8 and 10, we learned about physical ocean models. Applications of these models for simulation and synoptic prediction were discussed with examples in Chapters 11 and 12. Chapter 13 briefly discussed some important simple concepts of biogeochemistry and then explored some modeling frameworks for biogeochemical and fisheries problems. As mentioned in the beginning, our earth has four subsystems within the whole *Climate* system.

Thus, in this final chapter, we will learn how to model the climate system. Specifically, we will discuss the possible approaches to connect the physical, biogeochemical, and fisheries systems to the human activities. This will be done with the background of our existing knowledge of a changing climate. First, we briefly describe the idea of atmospheric modeling. This allows us to appreciate the long-term observed (in the past) and projected (for the future) climate variability in terms of temperature and carbon dioxide. The climate system models that are mentioned here are in compliance with the UCAR family of models under the IPCC guidance of *scenario modeling*. Future projections in climate change and how it might affect us are discussed in brief. The process of prediction of warming and sea-level rise and their link to the rising carbon dioxide is outlined. Nested regional and *climate downscaling* ideas using IPCC simulations are outlined as a bridge to the future.

_____

*Credit/Source: AZQuotes.com, Wind and Fly LTD, 2021. https://www.azquotes.com/quote/797882, accessed September 22, 2021.

DOI: 10.1201/9780429347221-14

14.1 ATMOSPHERIC MODELING

So far, we have talked about the hydrosphere and the biosphere in an interdisciplinary (biogeochemical) manner. Remember the four-component Climate System. Atmosphere is a critical part of the climate system that affects the human life. Let us discuss a little about atmospheric models and how they are connected with the rest of the climate system.

In fact, at about the same time when Ekman, 1905 was working on the wind-driven ocean problem, Bjerknes in 1904 (V. Bjerknes, 1904) came up with a formulation for weather forecasting as an initial-value problem using the equations of motion and proposed a "graphical calculus" method of solution. Later in 1922, Richardson attempted to create a process of numerical weather prediction (NWP) and called it a "forecast factory." Richardson divided the globe into cells and specified the dynamic variables at the center of each cell. In Chapter 11 of his book (Richardson, 1922), he describes his vision, he called it a "fantacy," an enormous building where each cell is being evaluated dynamically by a group of people calculating away the terms of the governing equations!

With the arrival of the computers in the middle of the twentieth century, Charney, Fjortoft, and von Neumann did the first successful NWP using the ENIAC system. Then in 1955, Normal Phillips (Phillips, 1956) came up with the formulation for the first atmospheric general circulation model (AGCM), which has been followed and improved through the last several decades. Figure 14.1 shows a number of processes that are resolved in an AGCM set up on the left. It also shows an example of a grid for the numerical model of an AGCM on the right.

Atmospheric models are similar in construct to the ocean models for their momentum equations, which are again the movement of air parcels on the rotating Earth. Naturally, they are governed by the Navier-Stokes equation on the rotating frame in the horizontal. In the vertical, it is usually the pressure coordinate that is chosen, and the generalized vertical coordinate transformation (see Chapter 10) is applied, which makes the continuity equation simpler in that, the continuity is independent of the density and it is time-invariant. See Jacobson and Jacobson, 2005 for many details of the art of atmospheric models and modeling with many example problems on this subject.

In contrast to ocean models, atmospheric models (air does not have salinity) need to solve for surface pressure, horizontal velocity components, temperature and water vapor in-depth layers (altitudes), and radiation (shortwave and longwave). So, explicit energy equations for thermodynamics and radiation (diagnostics) are solved in addition to momentum and continuity equations (prognostic for velocities, temperature, moisture, planetary boundary layer depth or the turbulent kinetic energy, and surface pressure). The water is generally treated as different species for different states including ice, rain, and vapor and in different concentration and types of these. Some of these mixing ratios

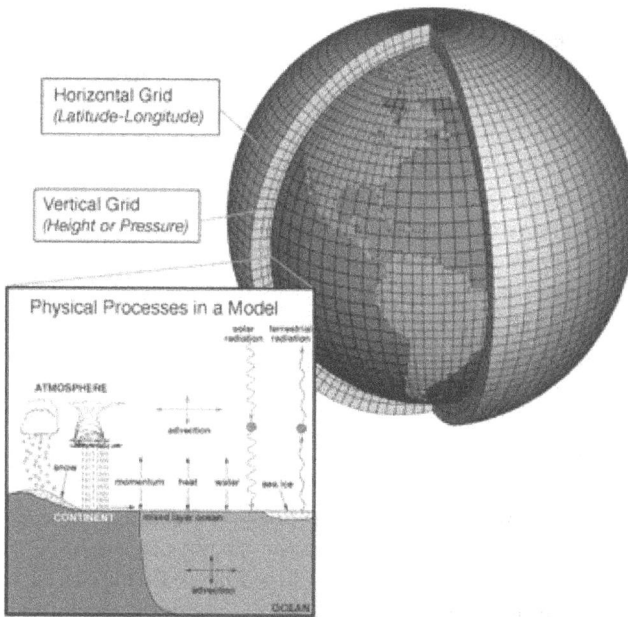

Figure 14.1: Key processes and example grid in an atmospheric model. Climate models are systems of differential equations based on the basic laws of physics, fluid motion, and chemistry. To "run" a model, scientists divide the planet into a three-dimensional grid, apply the basic equations, and evaluate the results. Atmospheric models calculate winds, heat transfer, radiation, relative humidity, and surface hydrology within each grid and evaluate interactions with neighboring points. Source: NOAA.

of water species (vapor, cloud water, cloud ice, rain, snow, etc.) are prognostic variables as well. Other atmospheric constituents such as chemical species are also treated as prognostic variables. In addition, there are diagnostic equations such as the mass continuity equation (including species continuity equations), equation of state (the ideal gas equation, in this case). Atmospheric models need to parameterize a multitude of variables including those for convection, land surface processes, albedo, cloud cover, and hydrology.

14.2   CLIMATE VARIABILITY – TEMPERATURE AND CO_2

Now that we are exposed to ocean models, interdisciplinary models and atmospheric models and their capabilities, naturally, we want to look at the whole system of climate like we discussed in Chapter 1. Our home, the Earth is in a habitable thermal equilibrium due to a delicate balance of radiation, absorption, and convection (see the models in Section 1.4). The ocean and atmosphere have their important roles in balancing this temperature through different mechanisms. The most important of these is the Greenhouse effect, which allows the atmosphere to shield part of the long-wave radiation (heat) and reflect it back to the Earth so that the Earth is warm enough to live. This is done by the so-called greenhouse gases such as water vapor, CO_2, methane (CH_4), and others. While the total composition of the Greenhouse gases is only 0.30% of the whole atmosphere, water vapor is almost 0.25%, and CO_2 is only 0.03% of the whole atmosphere (see Figure 13.1). So, why is CO_2 talked about so much with regard to climate change and emission?

The answer lies in the fact that water vapor equilibrates and adjusts within weeks in the atmosphere with temperature. So, the warmer the air, the more water vapor, and the system becomes warmer and becomes a feedback system. However, CO_2 takes hundreds to thousands of years to adjust within the atmosphere. Thus, the long-lived Greenhouse gases, such as CO_2 exert a controlling influence on climate. It was back in 1906, a Swedish Nobel Laureate, Svante Arrhenius did a simple calculation based on known physical set of equations and said, "Any doubling of the percentage of carbon dioxide in the air would raise the temperature of the earth's surface by $4°$; and if the carbon dioxide were increased fourfold, the temperature would rise by $8°$."

Scientists have put these rises in global temperature and CO_2 in the context of longer-term changes that they have been able to identify from long-term records. Figure 14.2a shows the paleo reconstruction of temperature change during the last 800,000 years. The abrupt nature of changes between the glacial and inter-glacial periods are clearly visible. Figure 14.2b shows the temperature reconstructed in more detail for the last 2000 years. It shows the well-known "hockey-stick" pattern of temperature rise in the last century from the industrial revolution compared to the last 1800 years. Finally, the bottom panel (c) of Figure 14.2 shows the dramatic rise of average temperature after the industrial revolution and even more so after 1980.

The sea level has increased in response to global warming by about 8 inches during the last century and all indications of data and verified models project a steep increase of 1–3 feet at different coastal regions by the end of the twenty-first century. See Figure 14.3a. This graph shows cumulative changes in sea level for the world's oceans since 1880, based on a combination of long-term tide gauge measurements and recent satellite measurements. This figure shows average absolute sea-level change (1880–2015), which refers to the height of the ocean surface, regardless of whether nearby land is rising or falling. Satellite data are based solely on measured sea level, while the long-term tide gauge data include a small correction factor because the size and shape of the oceans are changing slowly over time. (On average, the ocean floor has been gradually sinking since the last Ice Age peak, 20,000 years ago.) The shaded band shows

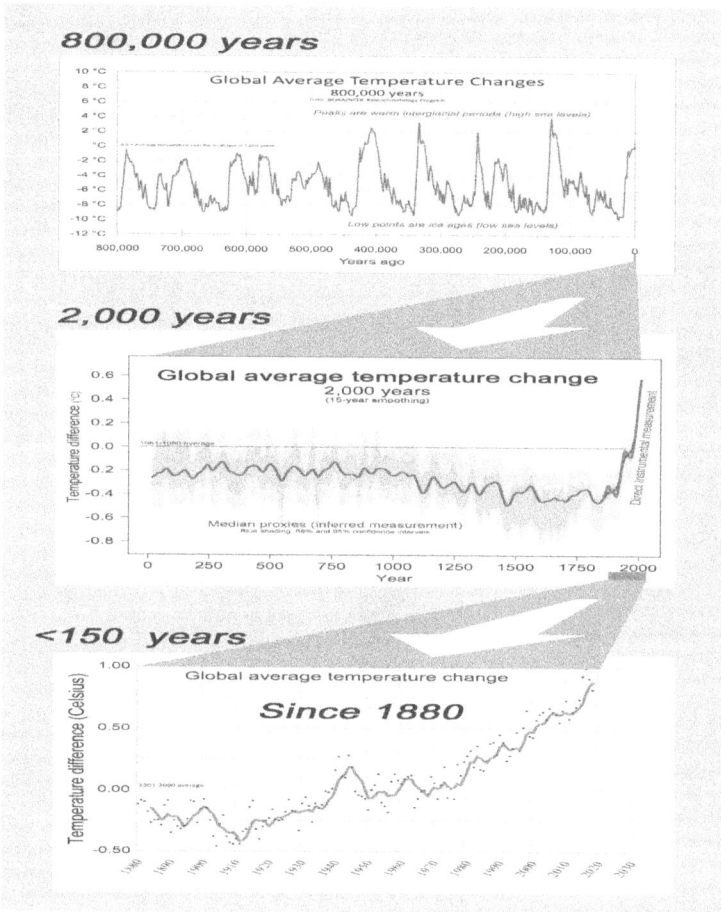

Figure 14.2: Three charts of global average temperature three respective time periods: 800,000 years, 2000 years, and 139 years, showing current global warming in the perspective of geologic time. Top chart: Earth's climate has cycled between ice ages and warm interglacial periods, with each cycle taking tens of thousands of years or more. Middle chart: Global average temperature was in a cooling trend for thousands of years before fossil fuel-based industrialization. Since then, it has increased about a full 1°C—in a time period less than 1/3000th the width of the top chart. Bottom chart: This 1°C increase, commonly called global warming, accelerated since 1980—a period less than 1/20,000th the width of the top chart. Figure Credits: Top Chart: NASA and NOAA multiple data sources. Original Source: Jouzel, J., et al. (2007). Middle Chart: Wikimedia image file. File: Temperature reconstruction last two millennia.svg by User talk: Femkemilene. Bottom chart: ncdc.noaa.gov. NOAA (National Centers for Environmental Information); Obtained from Wikimedia.

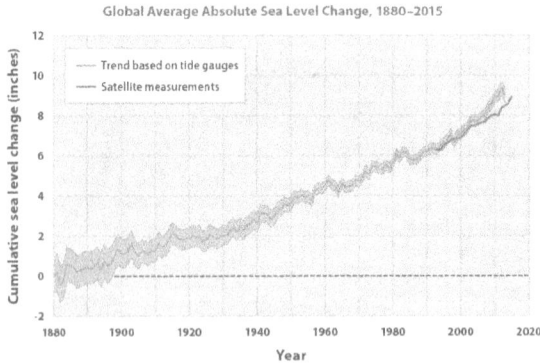

(a) Sea level rise

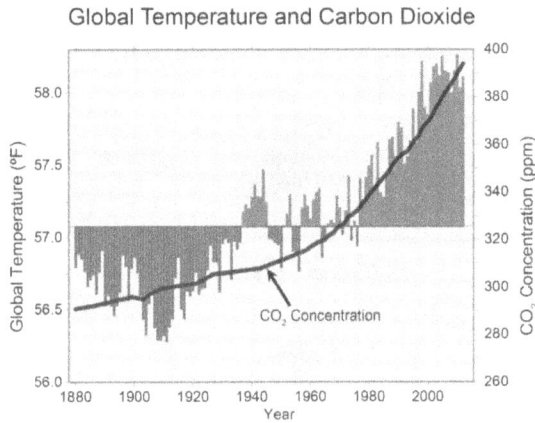

(b) Temperature and log CO_2

Figure 14.3: Sea level, temperature, and CO_2 after industrial revolution. (a) from USEPA; (b) global annual average temperature (as measured over both land and oceans) has increased by more than $1.5°F$ ($0.8°C$) since 1880 (through 2012). Red bars show temperatures above the long-term average, and blue bars indicate temperatures below the long-term average. The black line shows atmospheric carbon dioxide (CO_2) concentration in parts per million (ppm). (Figure source: updated from Karl et al. (2009); Climate Change Impacts in the United States: The Third National Climate Assessment; climatechange.gov)

the likely range of values based on the number of measurements collected and the precision of the methods used. As oceans warm, the water expands (and remember that more than 70% of the surface area of earth is water). Also, rising air temperature leads to the melting of ice sheets, polar ice caps, and resulting glaciers enter into the oceans.

We have seen a rise of almost $2°C$ in the last century in temperature and CO_2 has been accompanying it. See Figure 14.3b. The saw-tooth pattern of the CO_2 shows the seasonal cycle (driven by the biosphere). Global monthly average CO_2 concentrations have risen from around 340 ppm in 1980 to 410 ppm in 2019, which is an increase of about 20% in less than 40 years!

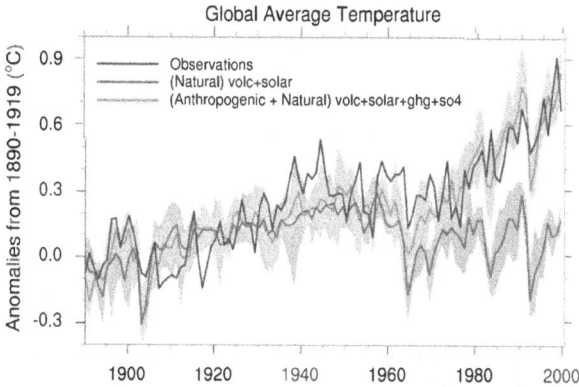

Figure 14.4: The four-member ensemble mean (red line) and ensemble member range (pink shading) for globally averaged surface air temperature anomalies (C; anomalies are formed by subtracting the 1890–1919 mean for each run from its time series of annual values) for volcanic forcing; the solid blue line is the ensemble mean and the light blue shading is the ensemble range for globally averaged temperature response to volcanic forcing calculated as a residual [(volcano + solar) – solar]; the black line is the observations after Folland et al. (2001); simulation including all forcings [(volcano + solar + GHG+ sulfate + ozone)] compared to natural forcings [(volcano + solar)]. Republished with permission of American Meteorological Society, from Meehl, G. A., Washington, W. M., Ammann, C. M., Arblaster, J. M., Wigley, T. M. L., and Tebaldi, C. (2004). Combinations of natural and anthropogenic forcings in twentieth-century climate. *Journal of Climate*, 17(19).; permission conveyed through Copyright Clearance Center, Inc.

Naturally, the question is often asked, "is the climate change natural or human-induced (anthropogenic)?" Well, there is some credible evidence to the human-induced hypothesis. And this is presented in Figure 14.4. This figure shows results from a collection of NCAR's Parallel Climate Model (PCM) ensembles for the climate for the twentieth century (Meehl et al., 2005). The observed global temperature matches quite well with the earth's natural cycle simulation (blue, with volcano and solar effects) up to the 1800s. After that, the natural cycle seems to have stayed on a flatter level than the observations. When the PCM ensemble of four different models from different organizations world-wide took the human influences into account, the simulated rise in temperature (red) followed the observation (black) reasonably well.

One can certainly debate the scientists and argue their simulations; however, given the observations of temperature rise and our understanding of the relationship between CO_2 and temperature, we should definitely think about what we can do to help heal the atmosphere in our home, the Earth, so that our future generations get to live in it comfortably.

Finally, there are many other important consequences of CO_2 increase in the ocean such as increasing of ocean acidification, coral bleaching, increasing HABs, and species migration. These are interesting research areas in the near future (see Further Reading for more on these topics).

14.3   CLIMATE SYSTEM MODELS

Climate models are based on well-documented physical processes to simulate the transfer of energy and materials through the climate system. Climate models, also known as general circulation models or GCMs, use mathematical equations to characterize how energy and matter interact in different parts of the ocean, atmosphere, and land.

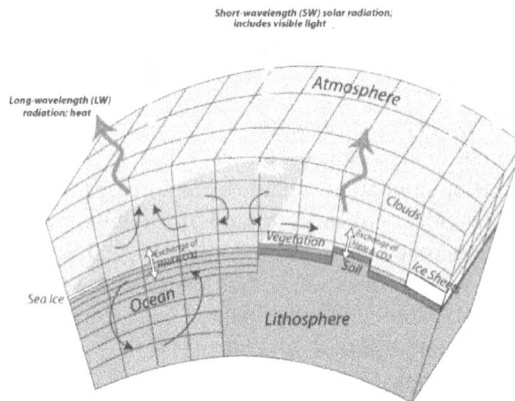

Figure 14.5: Schematic structure of a general circulation model with the major components of a climate system. Figure credit: with permission from Dr David Bice, Penn State University, licensed under CC BY-NC-SA 4.0.

Figure 14.5 shows the interconnected setup of a general circulation model with its three components: (i) Earth – lithosphere; (ii) ocean, (iii) atmosphere and how different processes can be interacting across these models in a climate modeling set up. Building and running a climate model is a complex process of identifying and quantifying earth system processes, representing them with mathematical equations, setting variables to represent initial conditions and subsequent changes in climate forcing, and repeatedly solving the equations using powerful supercomputers.

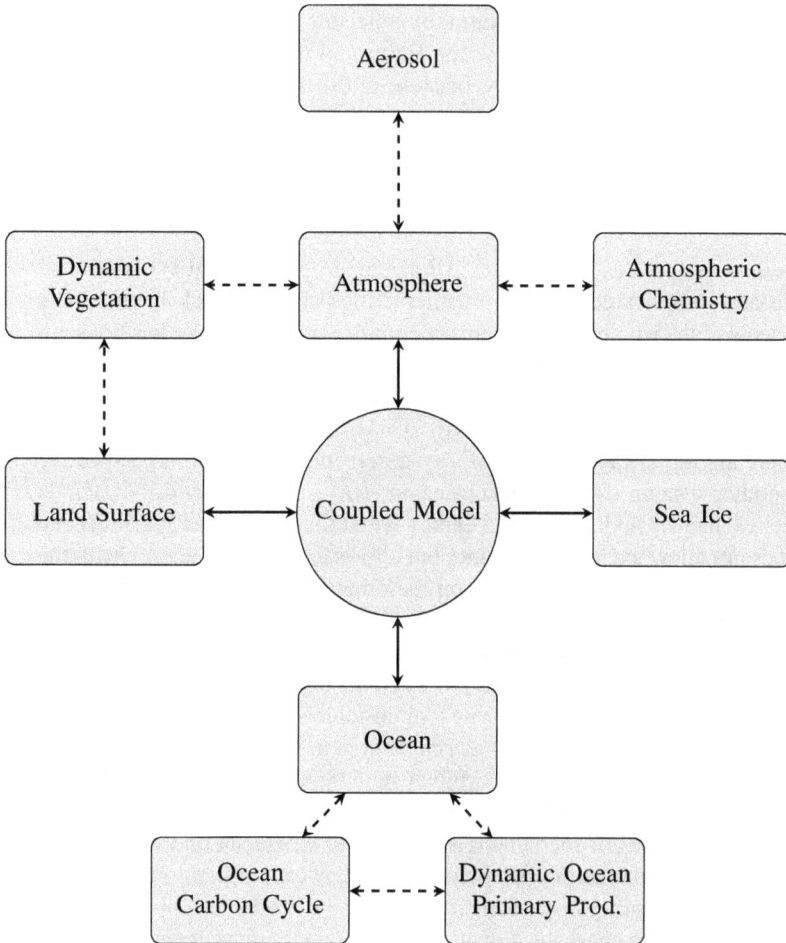

Figure 14.6: A linked system of components for a climate model.

Figure 14.6 presents a flow-chart on how different models of the sub-systems (ocean, land, atmosphere, biogeochemistry) can be linked in a "climate system model."

Modeling climate is not an easy task. There are several reasons why we need to keep on working on improving the climate models in the foreseeable future. First, the climate system model is a system of subsystems, which are inherently non-linear and thus prone to the sensitivity to initial conditions (discussed in Chapter 11) and thus could result in a "chaotic" behavior. Second, the various feedback of the important processes such as those related to clouds and water vapor are non-linear. Third, the clouds (a state of water vapor collected by different processes) are difficult to model as they both cool (by reflecting back Sun's incoming shortwaves) and warm (by reflecting back the Earth's outgoing longwave radiation) the surface of the Earth and thus play a very complex role in the climate system. Finally, because of the nature of such complexity at different levels, small differences between state variables tend to magnify rapidly during time integration.

WHAT IS FEEDBACK?

Let us discuss the idea of *feedback*. Understanding and quantifying climate change are complicated because of interactions and feedback between the subsystems of the whole Earth system are complex. Recall that the Earth system is composed of four different systems (geosphere, hydrosphere, atmosphere, and biosphere). Processes in one system influence processes in another system through exchanges of *matter* and *energy*. This exchange is called *feedback*.

There are three ways to think of a system from its interaction perspective with another system: (i) open, (ii) closed and (iii) isolated. An *open* system is one that exchanges both matter and energy with others. A *closed* system is one that does not allow exchange of matter but allows exchange of energy with the other systems. An *isolated* system, on the other hand, does not allow either matter or energy with another system.

Now, from an equilibrium perspective, the system as a whole could be stable, unstable, or metastable. A stable system would go back to its original state even if it is disturbed from its original state. An unstable system would *not* go back to its original state if disturbed. A *metastable* system would go to another stable state when it is disturbed from its original state. Scientists rationalize that our Earth is in a metastable equilibrium and moves from one stable state to another to another. Examples of such stable states would be glacial and inter-glacial periods throughout our earth's paleoclimatic journey over millions of years. The feedback between subsystems can create internal disturbances in the climate system, which would cause the climate's present state to move to another stable state through instability.

Let us think about *feedback* from such interaction and equilibrium perspectives. Interactions between systems such as ocean and atmosphere happen through exchanges of carbon dioxide, which can induce feedback on temperature, which in turn can move our planet from one stable state to another. Feedback can be of two types, positive and negative. A *positive feedback* is

Climate Forcing Increased CO_2
Concentration

Global Air
Temperature
Increase

Increased
Greenhouse
Warming from
Water Vapor

+ Net Feedback

Increased
Water Vapor
in Atmosphere

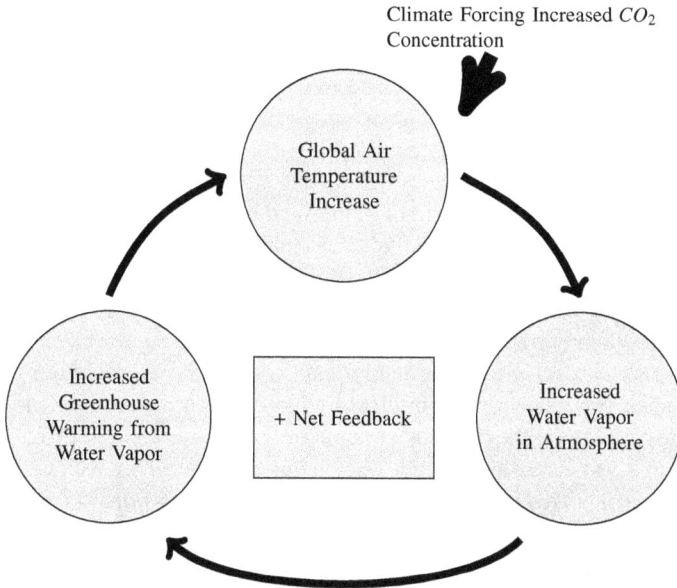

Figure 14.7: Example of a feedback system from CO_2 to temperature to water vapor which raises the temperature again.

when changes in one system cause a similar change in the other system. A positive feedback can cause runaway instability for a system. A *negative feedback*, on the other hand, means that positive change in one system causes a negative change in the other system. Let us consider the ongoing global warming scenario, where an increased level of carbon dioxide is increasing the temperature in our oceans. This increase in ocean temperature raises evaporation resulting in more water vapor in the atmosphere. As we know, water vapor is another greenhouse gas and traps additional outgoing radiation from the earth in the atmosphere. This is *positive feedback* on the temperature increase of the atmosphere. See Figure 14.7. Think about it. Increased carbon dioxide raises the temperature of the ocean, which increases evaporation, which increases the concentration of water vapor in the atmosphere, which in turn raises the temperature even more.

Similarly, there is an ice/albedo positive feedback loop that tends to contribute to global warming. In a warm climate, there is an increased amount of summer open water in the polar regions, which leads to more shortwave radiation (from the sun) being absorbed in these regions. This excess incoming radiation results in excess warming perturbation to the ice cover, which results in increased summer open water again, and in thinner ice cover, and less snow cover in winter. All three of these effects (increased summer open water, thinner ice cover, and less snow cover) are positive feedback to increase the shortwave absorption and again more heat flux and increasing temperature in summer creating more open water. The cycle goes on in a positive feedback toward more warming. Climate models separate the Earth's surface into a three-dimensional grid of cells. The results of processes modeled in each cell are passed to neighboring cells to model the exchange of matter and energy over time. Grid cell size defines the resolution of the model: The smaller the size of the grid cells, the higher the level of detail in the model. More detailed models have more grid cells, so they need more computing power.

Realize that many important processes happen at very small scales. Such processes are not resolved by the numerical model, whose resolution does not allow such processes to be simulated. One example is cumulus clouds whose scales are say on the order of a couple of miles or less. So, they remain unresolved in a 50-km resolution model. So, in this case, the effect of clouds will need to be *parameterized* in the climate model. What does this *parameterization* mean?

Parameterization is a representation of the average effect of an unresolved process on the appropriate variables (e.g. air temperature, wind speed, humidity for clouds; oceanic variables for turbulence – see Chapter 9) over the whole computational period.

Clearly, many tunable parameters exist in a very complex climate model, and these knobs will have to be tuned very carefully as the system solution is sensitive to the choices of these parameters. In a force-response system, such tuning or calibration leads to investigating the simulations in a "validate-calibrate-verify" approach discussed before (Chapter 12) and mentioned below.

Once a climate model is set up, it can be tested via a process known as "hind-casting." This process runs the model from the present time backward into the past. The model results are then compared with observed climate and weather conditions to see how well they match. This testing allows scientists to check the accuracy of the models and, if needed, revise their equations. Research teams around the world also test and compare their model outputs to observations and results from other models.

14.4 SCENARIO MODELING

Since future levels of GHG emissions are very complex to predict, and the whole climate system is yet a less-understood sensitive dynamical system driven by forces including population growth, economic development, technological progress, and global and regional geopolitics, individual accurate predictions of emissions is virtually impossible.

However, demand for such understanding in the short-term will have a profound impact on the long-term behavior of our climate and our planet.

Policy-makers need a summary of (i) what is understood about possible future GHG (ii) uncertainties in both emission models and their driving forces, so that they can make policy decisions within the bounds of such uncertainties, if possible. A solution could be a "Scenario forecasting!"

What is the definition of a scenario? A scenario is an image of future. It is neither a forecast nor a prediction for one year. Each scenario is one possible future state of a collection of anthropogenic emissions of relevant GHG species including CO_2, CH_4, N_2O, HFCs, and other similar gases.

Once a climate model can perform well in hind-casting tests, its results for simulating future climate are also assumed to be valid. To project climate into the future, the climate forcing is set to change according to a possible future scenario. Scenarios are possible stories about how quickly the human population will grow, how land will be used, how economies will evolve, and the atmospheric conditions (and therefore, climate forcing) that would result for each storyline.

In 2000, the Intergovernmental Panel on Climate Change (IPCC) issued its Special Report on Emissions Scenarios (SRES), describing four scenario families to describe a range of possible future conditions. See Figure 14.8. Referred to by letter-number combinations such as A1, A2, B1, and B2, each scenario was based on a complex relationship between the socioeconomic forces driving greenhouse gas and aerosol emissions and the levels to which those emissions would climb during the twenty-first century. The SRES scenarios have been in use for more than a decade, so many climate model results describe their inputs using the letter-number combinations.

There were seven driving forces for IPCC Fourth Assessment Report (AR4) in 2007; (i) Population prospects; (ii) economic development; (iii) energy demand and usage; (iv) resource availability; (v) technological change; (vi) prospects of energy alternatives; and (vii) land-use changes. The pathways for development were divided into four general groups: (i) A1 – for Rapid economic development with peak population reaching around mid-twenty-first century and rapid introduction of new and efficient technologies with a substantial reduction of regional difference in per-capita income; (ii) A2 – similar to A1 with regional emphasis; (iii) B1 – similar to A1 except with clean and resource-efficient technologies and global solution for sustainability; and (iv) B2 – similar to B1 with regional emphasis instead of global connectivity. The A1 scenario (most rapid development) was then further subgrouped in three different ways: (i) A1F1: Fossil fuel intensive without any attention to curb emission; (ii) A1B: Balanced emphasis on all energy sources; and (iii) A1T: Non-fossil fuel-intensive growth. So, all together, you can think of 40 different pathways within the scenario families indicated above for the future.

In 2013, climate scientists agreed upon a new set of scenarios that focused on the level of greenhouse gases in the atmosphere in 2100. Collectively, these scenarios are known as Representative Concentration Pathways or RCPs. Each

SRES Scenarios

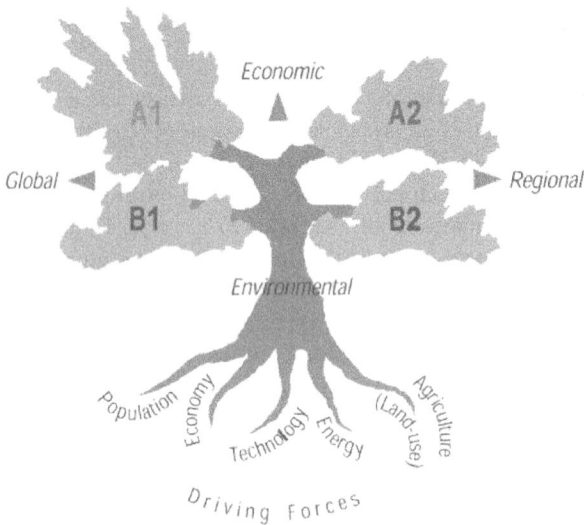

Figure 14.8: Scenarios for IPCC projections. The four SRES scenario families that share common storylines are illustrated as branches of a two-dimensional tree. The two dimensions indicate the relative orientation of the different scenario storylines toward economic or environmental concerns and global and regional scenario development patterns, respectively. There is no implication that these two are mutually exclusive or incompatible. In reality, the four scenarios share a space of a much higher dimensionality given the numerous driving forces and other assumptions needed to define any given scenario in a particular modeling approach. The A1 storyline branches out into different groups of scenarios to illustrate that alternative development paths are possible within one scenario family. Figure 4.1 from IPCC 2000: An Overview of Scenarios. In: *IPCC Special Report on Emissions Scenarios. Prepared by Working Group III of the Intergovernmental Panel on Climate Change* [Nebojsa Nakicenovic and Rob Swart (Eds.)]. Cambridge University Press, UK. pp. 570.

RCP indicates the amount of climate forcing, expressed in Watts per square meter (W/m^2), that would result from greenhouse gases in the atmosphere in 2100. The rate and trajectory of the forcing is the pathway. Like their predecessors, these values are used in setting up climate models. See Figure 14.9 for a representative flow of scenario modeling set up from emission to climate model to parameters for climate change implementation and possible adaptation.

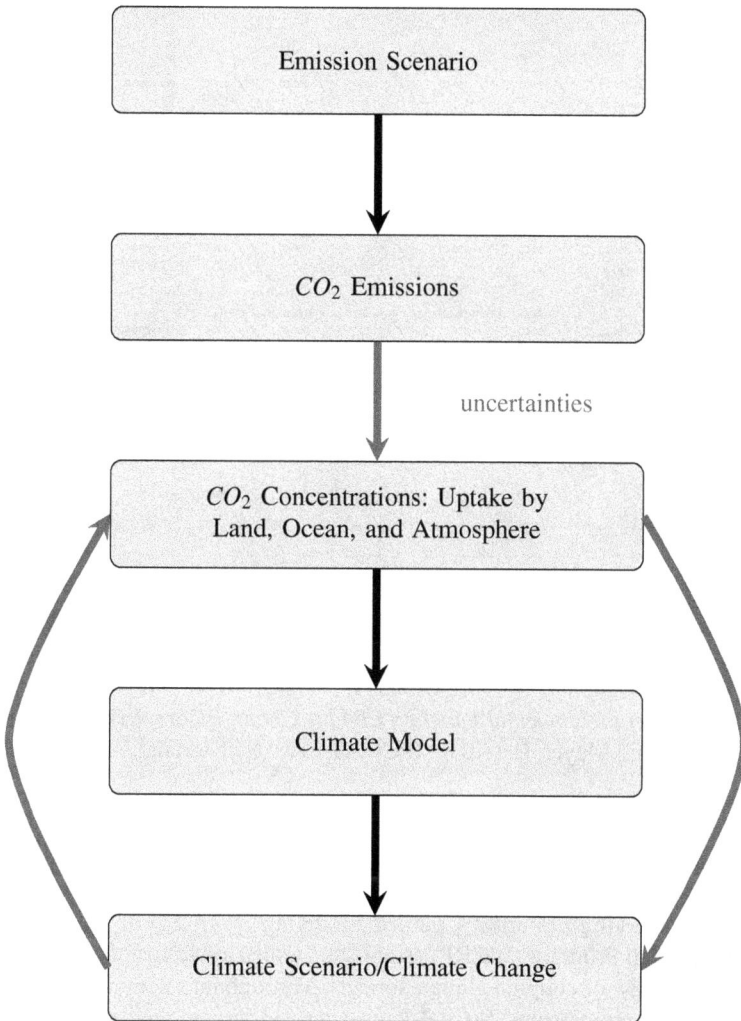

Figure 14.9: The way it works in a model – from emission to temperature.

Unlike weather forecasts, which describe a detailed picture of the expected daily sequence of conditions starting from the present, climate models are probabilistic, indicating areas with higher chances to be warmer or cooler and wetter or drier than usual. Climate models are based on global patterns in the ocean and atmosphere, and records of the types of weather that occurred under similar patterns in the past.

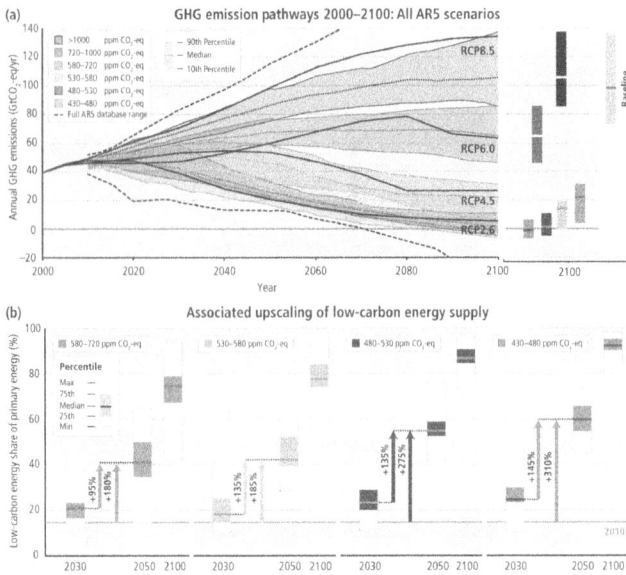

Figure 14.10: IPCC-AR5 CO_2 rise for different RCPs. Credit: Figure SPM.11 from IPCC, 2014: Summary for Policymakers. In:*Climate Change 2014: Synthesis Report. Contribution of Working Groups I, II and III to the Fifth Assessment Report of the Intergovernmental Panel on Climate Change* [Core Writing Team, Pachauri, R.K. and Meyer, L. (eds.)]. IPCC, Geneva, Switzerland.

14.5 CLIMATE PROJECTION, DOWNSCALING, AND GOVERNANCE

In 1995, the Working Group of Coupled Modeling (WGCM) of the World Climate Research Program (WCRP) established a mission to foster the development and review of coupled climate models. Through international science coordination and partnerships, WCRP contributes to advancing our understanding of the multi-scale dynamic interactions between natural and social systems that affect climate. Since there are many institutions and groups involved in running coupled ocean-atmosphere models for climate studies, the idea of intercomparison in a systematic way was put forward. Some example metrics for comparing ocean models were discussed in Chapters 11 and 12. The Coupled Model Intercomparison Project (CMIP) started in 1995 by WGCM under WCRP and have been used in the Annual Reports of the IPCC ever since. As an example of intercomparison, you can think that a number of models from different groups can be set up in the same way and use the same input. The difference between the climate projections from these different models would

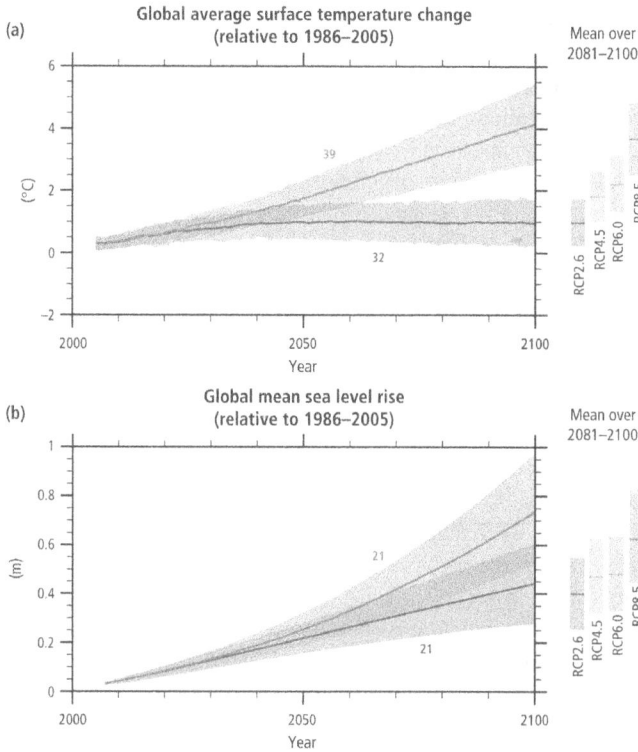

Figure 14.11: IPCC-AR5 temperature and sea level predictions up to 2100. Credit: Figures SPM.6 from IPCC, 2014: Summary for Policymakers. In:*Climate Change 2014: Synthesis Report. Contribution of Working Groups I, II and III to the Fifth Assessment Report of the Intergovernmental Panel on Climate Change* [Core Writing Team, Pachauri, R.K. and Meyer, L. (eds.)]. IPCC, Geneva, Switzerland.

then allow us to understand what is causing those differences as a first step to understand climate change projections and variability in a dynamical framework. In its early years, CMIP experiments included modeling the impact of a 1% annual increase of CO_2 in the atmosphere. In later years, more comprehensive emission scenarios, such as the RCPs discussed earlier, are adopted. A set of example projections up to 2100 from AR5 for CO_2 increase for different RCPs is as shown in Figure 14.10. Similar projections for surface temperature and sea-level rise are shown in Figure 14.11.

Multiple generations of CMIPs have evolved since 1995, and we are now looking forward to the current generation CMIP6. A CMIP6 overview paper

(Eyring et al., 2016) presents the background and rationale for the new structure of CMIP, and provides a detailed description of the "DECK" (Diagnostic, Evaluation, and Characterization of Klima). This paper also discusses the CMIP6 historical simulations, and includes a brief introduction to the 21 CMIP6-Endorsed Model intercomparison Projects (MIPs) focusing on different aspects of climate simulations addressing different scientific questions. With the Grand Science Challenges of the WCRP as its scientific backdrop, CMIP6 will address three broad questions:

– How does the Earth system respond to forcing?

– What are the origins and consequences of systematic model biases?

– How can we assess future climate changes given internal climate variability, predictability, and uncertainties in scenarios?

Around the world, different teams of scientists have built and run models to project future climate conditions under various scenarios for the next century. The model results project that global temperature will continue to increase, also show that human decisions and behavior we choose today will determine how dramatically climate will change in the future.

As climate models are becoming more and more realistic in emulating past and present climate scenarios on a large scale; our confidence in future projections is growing considerably. Naturally, it makes a case for using the projections from the climate models to infer near-term changes in ecosystem and other societal metrics. This process of connecting the regional climate model to the global climate model on the one hand (to bring information and variable down from global scale to regional scale) and then to community and local levels (to send information down to the local level) is called "dynamical downscaling." An example is shown in Figure 14.12.

In a dynamical downscaling exercise, one uses the future climate model output from a coarser Earth System Model to drive the regional ocean model at a much higher resolution (which can resolve the regional features of circulation), and thus hopefully produce high-resolution future climate projections for the region. One of the techniques often used to do this is the "*delta method.*" In this method, the difference between a future state and a current/past state is calculated from the climate model simulation. This difference field is then added to the same current/past state of the high-resolution model to obtain the future "climate-compatible" high-resolution state. That new state in the high-resolution is then forced with the output of the climate projections for different scenarios. Some nice examples of dynamical downscaling for the Gulf of Maine using CMIP5 for RCP4.5 and RCP8.5 are described by Anderson and Moore (2012), Arruda et al. (2013), and Brickman et al. (2018). The Special Report on the Ocean and Cryosphere in a Changing Climate (SROCC), together with the Special Report on Global Warming of 1.5°C (SR15) (IPCC, 2018), highlights the critical role of governance in implementing effective climate adaptation. SROCC outlines how the interlinked socio-ecological systems challenge current

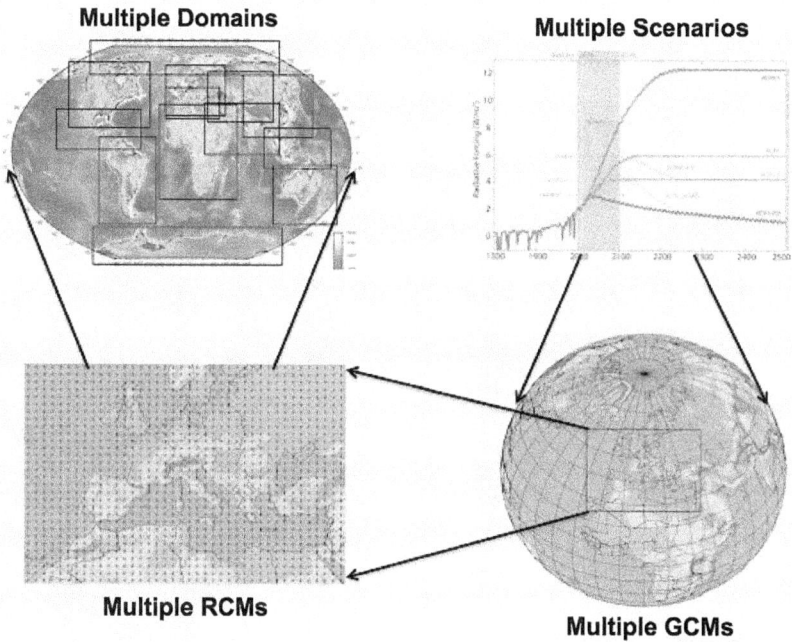

Figure 14.12: Conceptual framework for the experiments planned under the Coordinated Regional Downscaling EXperiment (CORDEX). RCM = regional climate model; GCM = global climate model. Republished with permission of John Wiley & Sons, from Filippo, G. (2009). Thirty years of regional climate modeling: Where are we and where are we going next?. Journal of Geophysical Research, Atmospheres, doi: 10.1029/2018JD030094.; permission conveyed through Copyright Clearance Center, Inc.

Cross-Chapter Box 3 (continued)

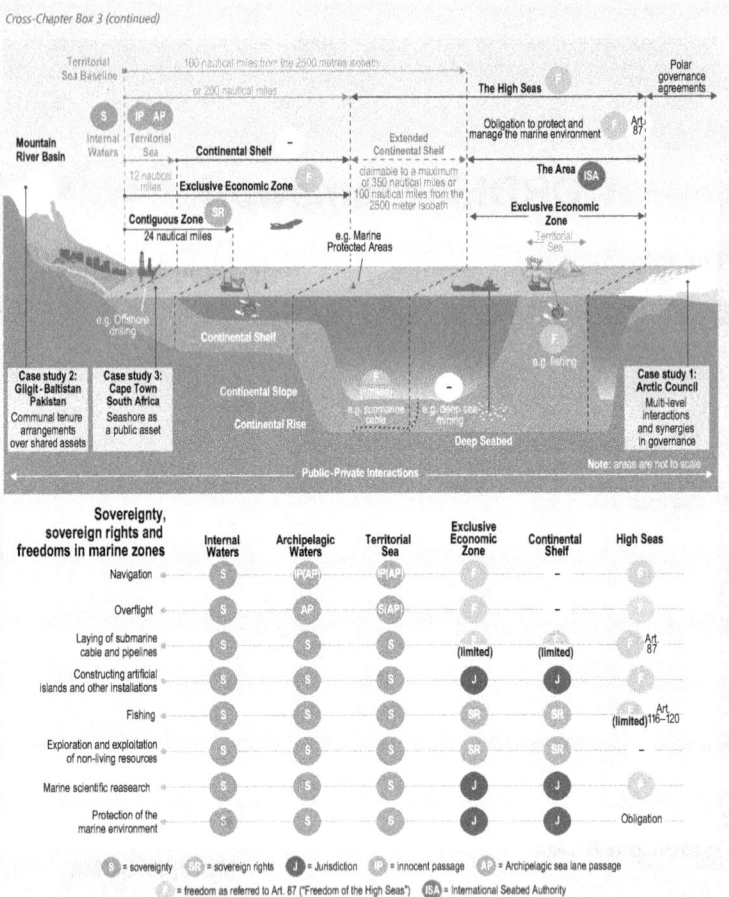

Figure CB3.1 | Spatial distribution of multi-faceted governance arrangements for the ocean, coasts and cryosphere (Panel A) sovereignty, sovereign rights, jurisdictions and freedoms defined for different ocean zones and sea by the United Nations Convention on the Law of the Sea (UNCLOS) (Panel B). *Figure CB3.1 is designed to be illustrative and is not comprehensive of all governance arrangements for the ocean, coasts and cryosphere.*

Figure 14.13: Governance of the ocean, coasts, and the cryosphere under climate change. Figure CB3.1 from Abram, N., J.-P. Gattuso, A. Prakash, L. Cheng, M.P. Chidichimo, S. Crate, H. Enomoto, M. Garschagen, N. Gruber, S. Harper, E. Holland, R.M. Kudela, J. Rice, K. Steffen, and K. von Schuckmann, 2019: Framing and Context of the Report. In: *IPCC Special Report on the Ocean and Cryosphere in a Changing Climate* [H.-O. Pörtner, D.C. Roberts, V. Masson-Delmotte, P. Zhai, M. Tignor, E. Poloczanska, K. Mintenbeck, A. Alegría, M. Nicolai, A. Okem, J. Petzold, B. Rama, N.M. Weyer (eds.)]. In press.

governmental systems in the context of climate change in the oceans. Three major aspects of these challenges are identified: (i) the scales of changes that occur in the climate and ecosystem match poorly with existing scales of governance; (ii) services provided to humans living far from mountains and coasts resulting from climate changes in the oceans and cryosphere match poorly to the existing institutions and processes of governance; and (iii) many possible governance responses to these challenges could be of limited or diminished effectiveness unless they are coordinated on scales beyond currently available governance options. Harmonizing local, regional, and global governance structures could provide an overarching policy framework for action and adaptation. The idea of polycentric governance (multiple centers of decision making with some degree of autonomy) (Carlisle & Gruby, 2019; Hamilton & Lubell, 2019; Mewhirter et al., 2018) has been put forward by this report. See Figure 14.13 for a detailed schematic of such possible governance structure and policy flow from global to the local level through different conduits which are affected by climate change.

Three important global efforts have started during 2015–2020 for mitigating the impact of climate change:

(i) The implementation of a set of sustainable development goals (SDG) from the United Nations in 2015.
(ii) The establishment of the Paris Climate Agreement in 2016.
(iii) The development of a number of vaccines and their worldwide distribution by the World Health Organization in record time to mitigate the spread and effect of the COVID-19 pandemic of 2019–2020.

First, leaders from 193 countries came together and created a plan called the SDG in 2015. The set of 17 goals imagines a future in 2030 that would be rid of poverty and hunger, provide basic education to all, and be safe from the worst effects of climate change. Then, the Paris Climate Agreement was signed by 197 countries to work together under the UN framework for climate change mitigation, adaptation, and finance. Third, the lessons learned during the pandemic of 2019–2020 should guide us to face new challenges of climate change as a global society. These three events have identified the challenges and needs for our global society today and the promise of our willingness to work together to meet those global challenges. Scientific research, technological innovation, effective communication, and active global participation can take us toward a goal of zero emissions by 2050, as proposed by some including Bill Gates (Gates, 2021). The goal of achieving a sustainable Earth system may be challenging, but the human race has endured such challenges before and will do so again as our climate – the final frontier – keeps changing.

CONCLUSION

In this final chapter, we learned some basic ideas on how to model the climate system. Specifically, we discussed some of the possible approaches to connect the physical, biogeochemical and fisheries system to the human activities. First, we briefly described the idea of atmospheric modeling. This allowed us to appreciate the long-term observed (in the past) and projected (for the future) climate variability in terms of temperature and carbon dioxide. The climate system models that are mentioned here are in compliance with the UCAR family of models under the IPCC guidance of *scenario modeling*. The process of prediction of warming and sea-level rise and their link to the rising carbon dioxide is briefly discussed. Nested regional and *climate downscaling* ideas using IPCC simulations are outlined. In closing, a governance roadmap was presented from IPCC in a changing climate for a sustainable earth.

FURTHER READING

Abram, N., Gattuso, J.-P., Prakash, A., Cheng, L., Chidichimo, M. P., Crate, S., Enomoto, H., Garschagen, M., Gruber, N., Harper, S., et al. (2019). Framing and context of the report. *IPCC special report on the ocean and cryosphere in a changing climate,* 766 pp, 73–129.

Emanuel, K. (2018). *What we know about climate change.* MIT Press.

Nakicenovic, N., Alcamo, J., Davis, G., Vries, B. d., Fenhann, J., Gaffin, S., Gregory, K., Grubler, A., Jung, T. Y., Kram, T., et al. (2000). Special report on emissions scenarios. Cambridge University Press, Cambridge, 608 pp.

Pachauri, R. K., Allen, M. R., Barros, V. R., Broome, J., Cramer, W., Christ, R., Church, J. A., Clarke, L., Dahe, Q., Dasgupta, P., et al. (2014). *Climate change 2014: Synthesis report. Contribution of Working Groups I, II and III to the fifth Assessment report of the Intergovernmental Panel on Climate Change.* IPCC.

Peixoto, J. P., & Oort, A. H. (1992). *Physics of climate*, first edition (p.520). American Institute of Physics, New York.

Pörtner, H.-O., Roberts, D. C., Masson-Delmotte, V., Zhai, P., Tignor, M., Poloczanska, E., Mintenbeck, K., Nicolai, M., Okem, A., Petzold, J., et al. (2019). *IPCC special report on the ocean and cryosphere in a changing climate*, 765 pp. IPCC Intergovernmental Panel on Climate Change: Geneva, Switzerland, 1(3).

Bibliography

Aalderink, R., & Jovin, R. (1997). Estimation of the photosynthesis/irradiance (p/i) curve parameters from light and dark bottle experiments. *Journal of Plankton Research, 19*(11), 1713–1742.

Abram, N., Gattuso, J.-P., Prakash, A., Cheng, L., Chidichimo, M. P., Crate, S., Enomoto, H., Garschagen, M., Gruber, N., Harper, S., et al. (2019). Framing and context of the report. *IPCC special report on the ocean and cryosphere in a changing climate,* 766 pp, 73–129.

Adcroft, A., Anderson, W., Balaji, V., Blanton, C., Bushuk, M., Dufour, C. O., Dunne, J. P., Griffies, S. M., Hallberg, R., Harrison, M. J., et al. (2019). The GFDL global ocean and sea ice model om4. 0: Model description and simulation features. *Journal of Advances in Modeling Earth Systems, 11*(10), 3167–3211.

Adcroft, A., & Hallberg, R. (2006). On methods for solving the oceanic equations of motion in generalized vertical coordinates. *Ocean Modelling, 11*(1–2), 224–233.

Adcroft, A., Hallberg, R., & Harrison, M. (2008). A finite volume discretization of the pressure gradient force using analytic integration. *Ocean Modelling, 22*(3–4), 106–113.

Agarwal, A., & Lermusiaux, P. F. (2011). Statistical field estimation for complex coastal regions and archipelagos. *Ocean Modelling, 40*(2), 164–189.

Allen, J., Beardsley, R., Blanton, J., Boicourt, W., Butman, B., Coachman, L., Huyer, A., Kinder, T., Royer, T., Schumacher, J., Smith, R., Sturges, W., & Winant, C. (1983). Physical oceanography of continental shelves. *Reviews of Geophysics and Space Physics, 21,* 1149–1181.

Anderson, B. D., & Moore, J. B. (2012). *Optimal filtering.* Courier Corporation.

Anderson, L. A., & Sarmiento, J. L. (1994). Redfield ratios of remineralization determined by nutrient data analysis. *Global Biogeochemical Cycles, 8*(1), 65–80.

Antonov, J. I. (1998). *World ocean atlas, 1998. volume 2, temperature of the Pacific Ocean* (tech. rep.). NOAA.

Arruda, W. Z., Campos, E. J. D., Zharkov, W., Soutelino, R. G., & Silveira, I. C. A. (2013). Events of equatorward translation of the Vitoria Eddy. *Continental Shelf Research, 70,* 61–73.

Azam, F., Fenchel, T., Field, J. G., Gray, J., Meyer-Reil, L., & Thingstad, F. (1983). The ecological role of water-column microbes in the sea. *Marine Ecology Progress Series, 10,* 257–263.

Bakun, A. (1973). Coastal upwelling indices, west coast of North America, 1946–71. *NOAA Tech. Report.*

Baracchini, T., Chu, P. Y., Šukys, J., Lieberherr, G., Wunderle, S., Wüest, A., & Bouffard, D. (2020). Data assimilation of in situ and satellite remote sensing data to 3d hydrodynamic lake models: A case study using delft3d-flow v4. 03 and openda v2. 4. *Geoscientific Model Development, 13*(3), 1267–1284.

Barnes, S. L. (1964). A technique for maximizing details in numerical weather map analysis. *Journal of Applied Meteorology and Climatology*, *3*(4), 396–409.

Batchelor, C. K., & Batchelor, G. (2000). *An introduction to fluid dynamics*. Cambridge University Press.

Beardsley, R. C. (1997). *Physical oceanography of the Gulf of Maine: An update*. In *Proceedings of the Gulf of Maine Ecosystem Dynamics Scientific Symposium and Workshop, 1997, 1997*, 39–52. Regional Association for Research in the Gulf of Maine.

Belkin, I. M., Cornillon, P. C., & Sherman, K. (2009). Fronts in large marine ecosystems. *Progress in Oceanography*, *81*(1–4), 223–236.

Bendat, J. S., & Piersol, A. G. (2011). *Random data: Analysis and measurement procedures* (Vol. 729). John Wiley & Sons.

Bennett, A. F. (1992). *Inverse methods in physical oceanography*. Cambridge University Press.

Beverton, R. J., & Holt, S. J. (1957). *On the dynamics of exploited fish populations*. Fisheries Investigation Series 2 (Volume 19), UK Ministry of Agriculture. Fisheries, and Food, London, UK.

Bhaskaran, P. K., Dube, S. K., Murty, T. S., Gangopadhyay, A., Chaudhury, A., & Rao, A. (2005). Tsunami travel time atlas for the Indian Ocean. *Indian Institute of Technology, Kharagpur, India*, 273–292.

Biastoch, A., Böning, C. W., Getzlaff, J., Molines, J.-M., & Madec, G. (2008). Causes of interannual–decadal variability in the meridional overturning circulation of the midlatitude North Atlantic Ocean. *Journal of Climate*, *21*(24), 6599–6615.

Bisagni, J., Beardsley, R., Ruhsam, C., Manning, J., & Williams, W. (1996). Historical and recent evidence concerning the presence of Scotian Shelf water on southern Georges Bank. *Deep-Sea Research*, *II 43*, 1439–1471.

Bisagni, J., & Cornillon, P. (1984). The synoptic sound-speed field of a warmcore GS ring. *Journal of the Acoustical Society of America*, *76*, 532–539.

Bisagni, J., & Smith, P. (1998). Eddy-induced flow of Scotian Shelf water across Northeast Channel, Gulf of Maine. *Continental Shelf Research*, *18*, 515–539.

Bjerknes, J. (1969). Atmospheric teleconnections from the equatorial pacific. *Monthly Weather Review*, *97*(3), 163–172.

Bjerknes, V. (1904). The problem of weather prediction, as seen from the standpoints of mechanics and physics. *Meteorologische Zeitschrif, 21, 1*, 7.

Bleck, R. (2002). An oceanic general circulation model framed in hybrid isopycnic-cartesian coordinates. *Ocean Modelling*, *37*, 55–88.

Blumberg, A. F., & Mellor, G. L. (1987). A description of a three-dimensional coastal ocean circulation model. *Three-Dimensional Coastal Ocean Models*, *4*, 1–16.

Bolton, T., & Zanna, L. (2019). Applications of deep learning to ocean data inference and subgrid parameterization. *Journal of Advances in Modeling Earth Systems*, *11*(1), 376–399.

Booij, N., Ris, R. C., & Holthuijsen, L. H. (1999). A third-generation wave model for coastal regions: 1. Model description and validation. *Journal of Geophysical Research: Oceans*, *104*(C4), 7649–7666.

Boyer, T., Levitus, S., Garcia, H., Locarnini, R. A., Stephens, C., & Antonov, J. (2005). Objective analyses of annual, seasonal, and monthly temperature and salinity for the world ocean on a 0.25 grid. *International Journal of Climatology: A Journal of the Royal Meteorological Society, 25*(7), 931–945.

Bretherton, F. P., Davis, R. E., & Fandry, C. B. (1976, July). A technique for objective analysis and design of oceanographic experiments applied to MODE-73. In *Deep Sea Research and Oceanographic Abstracts, 23*(7), 559–582, Elsevier.

Brickman, D., Hebert, D., & Wang, Z. (2018). Mechanism for the recent ocean warming events on the Scotian shelf of Eastern Canada. *Continental Shelf Research, 156*, 11–22.

Brink, K. (1982). A comparison of long coastal trapped wave theory with observations off Peru. *Journal of Physical Oceanography, 12*(8), 897–913.

Brink, K. H., & Robinson, A. R. (2005). *The global coastal ocean-regional studies and syntheses* (Vol. 11). Harvard University Press.

Broecker, W. S. (1987). Natural history. *The Biggest Chill, 96*, 74–82.

Broecker, W. S. (1991). The great ocean conveyor. *Oceanography, 4*(2), 79–89.

Broecker, W. S. (1997). Thermohaline circulation, the Achilles heel of our climate system: Will man-made CO_2 upset the current balance? *Science, 278*(5343), 1582–1588.

Brooks, D. A. (1985). Vernal circulation in the Gulf of Maine. *Journal of Geophysical Research: Oceans, 90*(C3), 4687–4706.

Brooks, D. A. (1987). The influence of warm-core rings on slope water entering the Gulf of Maine. *Journal of Geophysical Research: Oceans, 92*(C8), 8183–8196.

Brooks, D. A. (1990). Currents at lindenkohl sill in the southern Gulf of Maine. *Journal of Geophysical Research: Oceans, 95*(C12), 22173–22192.

Brooks, D. A., & Townsend, D. W. (1989). Variability of the coastal current and nutrient pathways in the eastern Gulf of Maine. *Journal of Marine Research, 47*(2), 303–321.

Brown, O., Cornillon, P., Emmerson, S., & Carle, H. (1986). Gulf Stream warm rings: A statistical study of their behavior. *Deep Sea Research, 33*, 1459–1473.

Brown, W. (1998). Wind-forced pressure response of the Gulf of Maine. *Journal of Geophysical Research, 103*, 30661–30678.

Brown, W. S., Gangopadhyay, A., Bub, F. L., Yu, Z., Strout, G., & Robinson, A. R. (2007). An operational circulation modeling system for the Gulf of Maine/Georges Bank Region—Part I: Basic elements. *IEEE Journal of Oceanic Engineering, 32*(4), 807–822.

Brown, W. S., Gangopadhyay, A., & Yu, Z. (2007). An operational circulation modeling system for the Gulf of Maine/Georges Bank Region—Part II: Applications. *IEEE Journal of Oceanic Engineering, 32*(4), 823–838.

Brown, W., & Beardsley, R. (1978). Winter circulation in the western Gulf of Maine, Part 1. Cooling and water mass formation. *Journal of Physical Oceanography, 8*, 265–277.

Brown, W., & Irish, J. (1992). The annual evolution of geostrophic flow in the Gulf of Maine, 1986–1987. *Journal of Physical Oceanography, 22*, 445–473.

Brown, W., & Irish, J. (1993). The annual variation of water mass structure in the Gulf of Maine, 1986–1987. *Journal of Marine Research, 51*, 53–107.

Bryan, K. (1969a). Climate and the ocean circulation. *Monthly Weather Review, 97*(11), 806–827.

Bryan, K. (1969b). A numerical model for the study of the circulation of the world ocean. *Journal of Computational Physics*, *4*, 347–376.

Bryan, K. (1997). A numerical method for the study of the circulation of the world ocean. *Journal of Computational Physics*, *135*(2), 154–169.

Bryan, K., & Cox, M. D. (1972). The circulation of the world ocean: A numerical study. Part I, a homogeneous model. *Journal of Physical Oceanography*, *2*(4), 319–335.

Butman, B., & Beardsley, R. (1987). Long-term observations on the southern flank of Georges Bank, Part I, A description of the seasonal cycle of currents, temperature, stratification and wind stress. *Journal of Physical Oceanography*, *17*, 367–384.

Butman, B., Loder, J., & Beardsley, R. (1987). The seasonal mean circulation, observation and theory (R. Backus, Ed.) [MIT Press, Cambridge, MA, USA]. *Georges Bank*, 125–138.

Caesar, L., McCarthy, G. D., Thornalley, D. J. R., Cahill, N., & Rahmstorf, S. (2021). Current Atlantic meridional overturning circulation weakest in last millennium. *Nature Geoscience*, *14*(3), 118–120.

Calado, L., Da Silveira, I., Gangopadhyay, A., & De Castro, B. (2010). Eddy-induced upwelling off cape são tomé (22 s, brazil). *Continental Shelf Research*, *30*(10–11), 1181–1188.

Calado, L., Gangopadhyay, A., & da Silveira, I. C. A. (2006). A parametric model for the Brazil Current meanders and eddies off southeastern Brazil. *Geophysical Research Letters*, *33*(12), L12602. https://doi.org/10.1029/2006GL026092

Calado, L., Gangopadhyay, A., & Da Silveira, I. (2008). Feature-oriented regional modeling and simulations (forms) for the western South Atlantic: Southeastern Brazil Region. *Ocean Modelling*, *25*(1–2), 48–64.

Cane, M. A., Zebiak, S. E., & Dolan, S. C. (1986). Experimental forecasts of El Nino. *Nature*, *321*(6073), 827–832.

Carlisle, K., & Gruby, R. L. (2019). Polycentric systems of governance: A theoretical model for the commons. *Policy Studies Journal*, *47*(4), 927–952.

Carpenter Jr, R. L., Droegemeier, K. K., Woodward, P. R., & Hane, C. E. (1990). Application of the piecewise parabolic method (ppm) to meteorological modeling. *Monthly Weather Review*, *118*(3), 586–612.

Carrier, G., & Robinson, A. (1962). On the theory of the wind-driven ocean circulation. *Journal of Fluid Mechanics*, *12*(1), 49–80.

Carter, E. F., & Robinson, A. R. (1987a). Analysis models for the estimation of oceanic fields. *Journal of Atmospheric and Oceanic Technology*, *4*(1), 49–74.

Carter, E. F., & Robinson, A. R. (1987b). Analysis models for the estimation of oceanic fields. *Journal of Atmospheric and Oceanic Technology*, *4*(1), 49–74.

Chai, F., Dugdale, R., Peng, T.-H., Wilkerson, F., & Barber, R. (2002). One-dimensional ecosystem model of the equatorial Pacific upwelling system. Part I: Model development and silicon and nitrogen cycle. *Deep Sea Research Part II: Topical Studies in Oceanography*, *49*(13–14), 2713–2745.

Chai, F., Jiang, M.-S., Chao, Y., Dugdale, R., Chavez, F., & Barber, R. (2007). Modeling responses of diatom productivity and biogenic silica export to iron enrichment in the equatorial pacific ocean. *Global Biogeochemical Cycles*, *21*(3), 1–16.

Chao, Y., Gangopadhyay, A., Bryan, F. O., & Holland, W. R. (1996). Modeling the gulf stream system: How far from reality? *Geophysical Research Letters*, *23*(22), 3155–3158.

Chapman, D., & Beardsley, R. (1989). On the origin of shelf water in the Middle Atlantic Bight. *Journal of Physical Oceanography*, *19*, 384–391.

Charney, J. G. (1990). The dynamics of long waves in a baroclinic westerly current. In *Lindzen R.S., Lorenz E.N., Platzman G.W. (eds) The Atmosphere—A Challenge.* American Meteorological Society, Boston, MA. https://doi.org/10.1007/978-1-944970-35-2_13.

Charney, J. G., Fjörtoft, R., & Neumann, J. v. (1950). Numerical integration of the barotropic vorticity equation. *Tellus*, *2*(4), 237–254.

Charney, J. G., & Flierl, G. R. (1981). Oceanic analogues of large-scale atmospheric motions. MIT Press, Cambridge, MA.

Charney, J. G., & Stern, M. (1962). On the stability of internal baroclinic jets in a rotating atmosphere. *Journal of the Atmospheric Sciences*, *19*(2), 159–172.

Chassignet, E. P., Hurlburt, H. E., Smedstad, O. M., Halliwell, G. R., Hogan, P. J., Wallcraft, A. J., Baraille, R., & Bleck, R. (2007). The HYCOM (hybrid coordinate ocean model) data assimilative system. *Journal of Marine Systems*, *65*(1–4), 60–83.

Chaudhuri, A. H., Gangopadhyay, A., & Bisagni, J. J. (2011a). Contrasting response of the eastern and western North Atlantic circulation to an episodic climate event. *Journal of Physical Oceanography*, *41*(9), 1630–1638.

Chaudhuri, A. H., Gangopadhyay, A., & Bisagni, J. J. (2011b). Response of the gulf stream transport to characteristic high and low phases of the North Atlantic oscillation. *Ocean Modelling*, *39*(3–4), 220–232.

Chave, A. D., Weidelt, P., & Jones, A. (2012). The theoretical basis for electromagnetic induction. *AD Chave, & AG Jones, The Magnetotelluric Method*, 19–44. Cambridge University Press.

Chen, C., Beardsley, R. C., & Cowles, G. (2006). Finite volume coastal ocean. *Oceanography*, *19*(1), 78.

Chi, L., Wolfe, C. L., & Hameed, S. (2019). The distinction between the gulf stream and its north wall. *Geophysical Research Letters*, *46*(15), 8943–8951.

Churchill, J. H., & Cornillon, P. C. (1991). Gulf stream water on the shelf and upper slope north of Cape Hatteras. *Continental Shelf Research*, *11*(5), 409–431.

Cirano, M., Mata, M. M., Campos, E. J., & Deiró, N. F. (2006). A circulacão oceânica de larga-escala na região oeste do atlântico sul com base no modelo de circulacão global occam. *Revista Brasileira de Geofísica*, *24*(2), 209–230.

Clancy, R. (1983). The effect of observational error correlations on objective analysis of ocean thermal structure. *Deep Sea Research Part A. Oceanographic Research Papers*, *30*(9), 985–1002.

Colella, P., & Woodward, P. R. (1984). The piecewise parabolic method (ppm) for gas-dynamical simulations. *Journal of Computational Physics*, *54*(1), 174–201.

Conkright, M. E., Locarnini, R. A., Garcia, H. E., O'Brien, T. D., Boyer, T. P., Stephens, C., & Antonov, J. I. (2002). *World ocean atlas 2001: Objective analyses, data statistics, and figures: CD-ROM documentation* (tech. rep.). NOAA.

Cooper, A. B. (2006). *A guide to fisheries stock assessment: From data to recommendations*. University of New Hampshire, Sea Grant College Program.

Cornillon, P., Weyer, R., & Flierl, G. (1989). Translational velocity of warm core rings relative to the slope water. *Journal of Physical Oceanography, 19*, 1317–1332.

Courtier, P., Thépaut, J.-N., & Hollingsworth, A. (1994). A strategy for operational implementation of 4d-var, using an incremental approach. *Quarterly Journal of the Royal Meteorological Society, 120*(519), 1367–1387.

Cox, M. D. (1970). A mathematical model of the Indian Ocean. In *Deep Sea Research and Oceanographic Abstracts, 17*(1), 47–75. Elsevier.

Cox, M. D. (1984). A primitive equation, 3-dimensional model of the ocean. *GFDL Ocean Group Technical Report No 1, GFDL, Princeton University.*

Cox, M. D., & Bryan, K. (1984). A numerical model of the ventilated thermocline. *Journal of Physical Oceanography, 14*(4), 674–687.

Cressie, N. (1990). The origins of kriging. *Mathematical Geology, 22*(3), 239–252.

Cressman, G. P. (1959). An operational objective analysis system. *Monthly Weather Review, 87*(10), 367–374.

Crutzen, P. J. (2002). The "anthropocene". In *Journal de physique iv (proceedings), 12*(10), 1–5. EDP sciences.

Cummings, J. A., & Smedstad, O. M. (2013). Variational data assimilation for the global ocean. *Data assimilation for atmospheric, oceanic and hydrologic applications (*Vol. ii). Springer.

Cushman-Roisin, B., & Beckers, J.-M. (2011). *Introduction to geophysical fluid dynamics: Physical and Numerical Aspects.* Academic Press.

Cushman-Roisin, B. (1994). *Introduction to geophysical fluid dynamics.* Englewood Cliffs, NJ, Prentice-Hall.

da Silva, A., Young, C. C., & Levitus, S. (1994). Atlas of Surface Marine Data 1994. vol. 1: Algorithms and procedures. *NOAA Atlas NESDIS, 6*(83), U.S. Department of Commerce, Washington, DC.

Daley, R. (1991). Atmospheric data analysis, Cambridge Atmospheric and *Space Science Series*, 457 pp. Cambridge University Press, Cambridge.

Davis, R. E. (1991). Observing the general circulation with floats. *Deep Sea Research Part A. Oceanographic Research Papers, 38,* S531–S571.

Davis, R. E. (2005). Intermediate-depth circulation of the Indian and South Pacific oceans measured by autonomous floats. *Journal of Physical Oceanography, 35*(5), 683–707.

Davis, R., Regier, L., Dufour, J., & Webb, D. (1992). The autonomous Lagrangian circulation explorer (ALACE). *Journal of Atmospheric and Oceanic Technology, 9*(3), 264–285.

Davis, R., Sherman, J., & Dufour, J. (2001). Profiling ALACEs and other advances in autonomous subsurface floats. *Journal of Atmospheric and Oceanic Technology, 18*(6), 982–993.

Deriso, R. B. (1980). Harvesting strategies and parameter estimation for an age-structured model. *Canadian Journal of Fisheries and Aquatic Sciences, 37*(2), 268–282.

Di Lorenzo, E., Moore, A. M., Arango, H. G., Cornuelle, B. D., Miller, A. J., Powell, B., Chua, B. S., & Bennett, A. F. (2007). Weak and strong constraint data assimilation in the inverse regional ocean modeling system (ROMs): Development and application for a baroclinic coastal upwelling system. *Ocean Modelling, 16*(3–4), 160–187.

Dickey, T. D., & Bidigare, R. R. (2005). Interdisciplinary oceanographic observations: The wave of the future. *Scientia Marina, 69*(S1), 23–42.

Dolman, H. (2019). *Biogeochemical cycles and climate*. Oxford University Press, USA.

Drinkwater, K. F. (2005). The response of Atlantic cod (Gadus morhua) to future climate change. *ICES Journal of Marine Science, 62*(7), 1327–1337.

Duda, T. F., Lin, Y.-T., Newhall, A. E., Helfrich, K. R., Lynch, J. F., Zhang, W. G., Lermusiaux, P. F., & Wilkin, J. (2019). Multiscale multiphysics data-informed modeling for three-dimensional ocean acoustic simulation and prediction. *The Journal of the Acoustical Society of America, 146*(3), 1996–2015.

Dukowicz, J. K., & Smith, R. D. (1994). Implicit free-surface method for the Bryan-Cox-Semtner ocean model. *Journal of Geophysical Research: Oceans, 99*(C4), 7991–8014.

Dukowicz, J. K., Smith, R. D., & Malone, R. C. (1993). A reformulation and implementation of the Bryan-Cox-Semtner ocean model on the connection machine. *Journal of Atmospheric and Oceanic Technology, 10*(2), 195–208.

Edwards, M. (1989). *Global gridded elevation and bathymetry on 5-minute geographic grid (etopo5)*. NOAA, National Geophysical Data Center, Boulder, Colorado, USA.

Ekman, V. W. (1905). *On the influence of the earth's rotation on ocean-currents*. Almqvist & Wiksells Boktryckeri, A.-B.

Eliassen, A. (1983). The Charney—Stern Theorem on Barotropic—baroclinic Instability in: Knopoff L., Keilis-Borok V.I., Puppi G. (eds). *Instabilities in Continuous Media*. Contributions to Current Research in Geophysics. Birkhäuser, Basel. https://doi.org/10.1007/978-3-0348-6608-8_11.

Emanuel, K. (2005). *Divine wind: the history and science of hurricanes*. Oxford University Press,

Emanuel, K. (2018). *What we know about climate change*. MIT Press.

Emery, W., Lee, W., & Magaard, L. (1984). Geographic and seasonal distributions of Brunt-Väisälä frequency and Rossby radii in the North Pacific and North Atlantic. *Journal of Physical Oceanography, 14*(2), 294–317.

Eppley, R. W. (1972). Temperature and phytoplankton growth in the sea. *Fishery Bulletin, 70*(4), 1063–1085.

Evans, G. T., & Parslow, J. S. (1985). A model of annual plankton cycles. *Biological Oceanography, 3*(3), 327–347.

Eyring, V., Bony, S., Meehl, G. A., Senior, C. A., Stevens, B., Stouffer, R. J., & Taylor, K. E. (2016). Overview of the Coupled Model Intercomparison Project Phase 6 (CMIP6) experimental design and organization. *Geoscientific Model Development, 9*(5), 1937–1958.

Fasham, M. J., Ducklow, H. W., & McKelvie, S. M. (1990). A nitrogen-based model of plankton dynamics in the oceanic mixed layer. *Journal of Marine Research, 48*(3), 591–639.

Flagg, C. (1987). Hydrographic structure and variability. In R. Backus & D. Bourne (Eds.), *Georges Bank*. MIT Press.

Frajka-Williams, E., Ansorge, I. J., Baehr, J., Bryden, H. L., Chidichimo, M. P., Cunningham, S. A., Danabasoglu, G., Dong, S., Donohue, K. A., Elipot, S., et al. (2019). Atlantic meridional overturning circulation: Observed transport and variability. *Frontiers in Marine Science*, *6*, 260.

French, A. P. (2001). *Vibrations and waves*, 316 pp. American Association of Physics Teachers.

Frisch, U., & Kolmogorov, A. N. (1995). *Turbulence: The legacy of A.N. Kolmogorov*, 296 pp. Cambridge University Press.

Galerkin, B. (1915). On electrical circuits for the approximate solution of the Laplace equation. *Vestnik Inzh*, *19*, 897–908.

Gandin, L. S. (1965). The objective analysis of meteorological field, Israel program for scientific translations. *Quarterly Journal of the Royal Meteorological Society: Jerusalem, Israel*, *93*(395).

Gangopadhyay, A., & Robinson, A. (1997). Circulation and dynamics of the western North Atlantic. Part III: Forecasting the meanders and rings. *Journal of Atmospheric and Oceanic Technology*, *14*, 1352–1365.

Gangopadhyay, A., & Robinson, A. (2002). Feature oriented regional modeling of Oceanic Fronts. *Dynamics of Atmospheres and Oceans*, *36*(1–3), 201–232.

Gangopadhyay, A., Robinson, A., & Arango, H. (1997). Circulation and dynamics of the western North Atlantic, I, Multiscale feature models. *Journal of Atmospheric and Oceanic Technology*, *146*, 1314–1332.

Gangopadhyay, A., Bharat Raj, G., Chaudhuri, A. H., Babu, M., & Sengupta, D. (2013). On the nature of meandering of the springtime western boundary current in the Bay of Bengal. *Geophysical Research Letters*, *40*(10), 2188–2193.

Gangopadhyay, A., Cornillon, P., & Watts, D. R. (1992). A test of the Parsons–Veronis hypothesis on the separation of the Gulf Stream. *Journal of Physical Oceanography*, *22*(11), 1286–1301.

Gangopadhyay, A., Gawarkiewicz, G., Silva, E. N. S., Monim, M., & Clark, J. (2019). An observed regime shift in the formation of warm core rings from the gulf stream. *Scientific Reports*, *9*(1), 1–9.

Gangopadhyay, A., Gawarkiewicz, G., Silva, E. N. S., Silver, A. M., Monim, M., & Clark, J. (2020). A census of the warm-core rings of the gulf stream: 1980–2017. *Journal of Geophysical Research: Oceans*, *125*(8), e2019JC016033.

Gangopadhyay, A., Lermusiaux, P. F., Rosenfeld, L., Robinson, A. R., Calado, L., Kim, H. S., Leslie, W. G., & Haley Jr, P. J. (2011). The California Current System: A multiscale overview and the development of a feature-oriented regional modeling system (forms). *Dynamics of Atmospheres and Oceans*, *52*(1–2), 131–169.

Gangopadhyay, A., & Robinson, A. R. (2002). Feature-oriented regional modeling of oceanic fronts. *Dynamics of Atmospheres and Oceans*, *36*(1–3), 201–232.

Gangopadhyay, A., Robinson, A. R., Haley, P. J., Leslie, W. G., Lozano, C. J., Bisagni, J. J., & Yu, Z. (2003). Feature-oriented regional modeling and simulations in the Gulf of Maine and Georges Bank. *Continental Shelf Research*, *23*(3–4), 317–353.

Gangopadhyay, A., Schmidt, A., Agel, L., Schofield, O., & Clark, J. (2013). Multiscale forecasting in the western North Atlantic: Sensitivity of model forecast skill to glider data assimilation. *Continental Shelf Research*, *63*, S159–S176.

Gangopadhyay, A., Shen, C. Y., Marmorino, G. O., Mied, R. P., & Lindemann, G. J. (2005). An extended velocity projection method for estimating the subsurface current and density structure for coastal plume regions: An application to the Chesapeake Bay outflow plume. *Continental Shelf Research*, *25*(11), 1303–1319.

Gangopadhyay, A., & Chao, Y. (2000). Sensitivity of the gulf stream path on the cyclonic wind stress curl. *The Global Atmosphere and Ocean System*, *7*(1), 151–178.

Garvine, R. W. (1995). A dynamical system for classifying buoyant coastal discharges. *Continental Shelf Research*, *15*(13), 1585–1596.

Garvine, R. W., & Monk, J. D. (1974). Frontal structure of a river plume. *Journal of Geophysical Research*, *79*(15), 2251–2259.

Gates, B. (2021). *How to avoid a climate disaster: The solutions we have and the breakthroughs we need*. Knopf.

Gawarkiewicz, G., Bahr, F., Beardsley, R., & Brink, K. (2001). Interaction of a slope eddy with the shelfbreak front in the Middle Atlantic Bight. *Journal of Physical Oceanography*, *31*, 2783–2796.

Gawarkiewicz, G., Chen, K., Forsyth, J., Bahr, F., Mercer, A. M., Ellertson, A., Fratantoni, P., Seim, H., Haines, S., & Han, L. (2019). Characteristics of an advective marine heatwave in the middle Atlantic bight in early 2017. *Frontiers in Marine Science*, *6*, 712.

Gawarkiewicz, G., & Plueddemann, A. J. (2020). Scientific rationale and conceptual design of a process-oriented shelfbreak observatory: The OOI Pioneer array. *Journal of Operational Oceanography*, *13*(1), 19–36.

Gelb, A. (1974). *Applied optimal estimation,* 304 pp. MIT Press.

Gill, A. E. (1982). *Atmosphere-ocean dynamics (international geophysics series)*. Academic Press.

Glenn, S., & Robinson, A. (1995). *Validation of an operational Gulf Stream forecasting model. Qualitative Skill Assessment for Coastal Models* (Vol. 47). American Geophysical Union.

Gordon, A. L. (1986). Interocean exchange of thermocline water. *Journal of Geophysical Research: Oceans*, *91*(C4), 5037–5046.

Griffies, S. (2018). *Fundamentals of ocean climate models,* 528 pp. Princeton University Press.

Haidvogel, D. B., Arango, H., Budgell, W. P., Cornuelle, B. D., Curchitser, E., Di Lorenzo, E., Fennel, K., Geyer, W. R., Hermann, A. J., Lanerolle, L., et al. (2008). Ocean forecasting in terrain-following coordinates: Formulation and skill assessment of the regional ocean modeling system. *Journal of Computational Physics*, *227*(7), 3595–3624.

Haidvogel, D. B., & Beckmann, A. (1999). *Numerical ocean circulation modeling*. World Scientific.

Haley, P. J., & Lermusiaux, P. F. (2010). Multiscale two-way embedding schemes for free-surface primitive equations in the "multidisciplinary simulation, estimation and assimilation system". *Ocean dynamics, 60*(6), 1497–1537.

Haley Jr, P. J., Agarwal, A., & Lermusiaux, P. F. (2015). Optimizing velocities and transports for complex coastal regions and archipelagos. *Ocean Modelling, 89,* 1–28.

Halkin, D., & Rossby, T. (1985). The structure and transport of the Gulf Stream at 73°W. *Journal of Physical Oceanography, 15,* 1439–1452.

Hall, M., & Bryden, H. (1985). Profiling the GS with a current meter mooring. *Geophysical Research Letters, 12,* 203–206.

Hamilton, M. L., & Lubell, M. (2019). Climate change adaptation, social capital, and the performance of polycentric governance institutions. *Climatic Change, 152*(3), 307–326.

Hecht, M. W., & Hasumi, H. (2013). *Ocean modeling in an eddying regime* (Vol. 177). John Wiley & Sons.

Hellerman, S., & Rosenstein, M. (1983). Normal monthly wind stress over the world ocean with error estimates. *Journal of Physical Oceanography, 13*(7), 1093–1104.

Helly, J. J., & Levin, L. A. (2004). Global distribution of naturally occurring marine hypoxia on continental margins. *Deep Sea Research Part I: Oceanographic Research Papers, 51*(9), 1159–1168.

Hobbs, N. T., & Hooten, M. B. (2015). *Bayesian models: A statistical primer for ecologists.* Princeton University Press.

Hogg, N. G. (1983). A note on the deep circulation of the western North Atlantic: Its nature and causes. *Deep Sea Research Part A. Oceanographic Research Papers, 30*(9), 945–961.

Hogg, N. G. (1992). On the transport of the gulf stream between Cape Hatteras and the grand banks. *Deep Sea Research Part A. Oceanographic Research Papers, 39*(7–8), 1231–1246.

Hogg, N. G., & Huang, R. X. (1995). *Collected works of Henry M. Stommel* [in 3 volumes]. American Meteorological Society.

Hogg, N. G., Pickart, R. S., Hendry, R. M., & Smethie Jr, W. J. (1986). The northern recirculation gyre of the gulf stream. *Deep Sea Research Part A. Oceanographic Research Papers, 33*(9), 1139–1165.

Hogg, N. G., & Stommel, H. (1985). On the relation between the deep circulation and the gulf stream. *Deep Sea Research Part A. Oceanographic Research Papers, 32*(10), 1181–1193.

Holboke, M. J., & Lynch, D. R. (1995). Simulations of the Maine Coastal Current. *Estuarine and Coastal Modeling,* 156–167. ASCE.

Holton, J. R. (1973). An introduction to dynamic meteorology. *American Journal of Physics, 41*(5), 752–754.

Holton, J. (2016). *The dynamic meteorology of the stratosphere and mesosphere* (Vol. 15). Springer.

Hood, R., Beal, L., Benway, H., Chandler, C., Coles, V., Cutter, G., Dick, H., Gangopadhyay, A., Goes, J., Humphris, S., et al. (2018). United States contributions to the Second International Indian Ocean Expedition (USIIOE-2). *WHOI Tech Report,* Wood Hole, MA, USA.

Hood, R. R., Urban, E. R., McPhaden, M. J., Su, D., & Raes, E. (2016). The 2nd International Indian Ocean Expedition (IIOE-2): Motivating new exploration in a poorly understood basin. *Limnology and Oceanography Bulletin, 25*(4), 117–124.

Houghton, R., Schlitz, R., Beardsley, R., Butman, R., & Lockwood Chamberlin, J. (1982). The Middle Atlantic Bight cold pool, Evolution of the temperature structure during summer 1979. *Journal of Physical Oceanography, 12*, 1019–1029.

Huang, X.-Y., Xiao, Q., Barker, D. M., Zhang, X., Michalakes, J., Huang, W., Henderson, T., Bray, J., Chen, Y., Ma, Z., et al. (2009). Four-dimensional variational data assimilation for WRF: Formulation and preliminary results. *Monthly Weather Review, 137*(1), 299–314.

Huntford, R. (2012). *Nansen: The explorer as hero.* Abacus.

Illig, S., Cadier, E., Bachèlery, M.-L., & Kersalé, M. (2018). Subseasonal coastal-trapped wave propagations in the Southeastern Pacific and Atlantic Oceans: 1. a new approach to estimate wave amplitude. *Journal of Geophysical Research: Oceans, 123*(6), 3915–3941.

J. Quinn, T. (2014). *Population dynamics.* Wiley StatsRef: Statistics Reference Online.

Jackson, L. B. (2013). *Digital filters and signal processing: With MATLAB exercises.* Springer Science & Business Media.

Jacobson, M. Z., & Jacobson, M. Z. (2005). *Fundamentals of atmospheric modeling.* Cambridge University Press.

Jacox, M., & Edwards, C. (2012). Upwelling source depth in the presence of nearshore wind stress curl. *Journal of Geophysical Research: Oceans, 117*(C5), 1–8.

Jacox, M. G., Edwards, C. A., Hazen, E. L., & Bograd, S. J. (2018). Coastal upwelling revisited: Ekman, Bakun, and improved upwelling indices for the US west coast. *Journal of Geophysical Research: Oceans, 123*(10), 7332–7350.

Jana, S., Gangopadhyay, A., & Chakraborty, A. (2015). Impact of seasonal river input on the Bay of Bengal simulation. *Continental Shelf Research, 104*, 45–62.

Jana, S., Gangopadhyay, A., Lermusiaux, P. F., Chakraborty, A., Sil, S., & Haley Jr, P. J. (2018). Sensitivity of the Bay of Bengal upper ocean to different winds and river input conditions. *Journal of Marine Systems, 187*, 206–222.

Kalman, R. E. (1963). Mathematical description of linear dynamical systems. *Journal of the Society for Industrial and Applied Mathematics, Series A: Control, 1*(2), 152–192.

Kantha, L. H., & Clayson, C. A. (2000). *Numerical models of oceans and oceanic processes.* Elsevier.

Katz, R. W. (2002). Sir Gilbert Walker and a connection between El Nino and statistics. *Statistical Science, 17*(1), 97–112.

Kelly, S. M., & Lermusiaux, P. F. (2016). Internal-tide interactions with the Gulf Stream and middle Atlantic bight shelfbreak front. *Journal of Geophysical Research: Oceans, 121*(8), 6271–6294.

Kelly, S. M., Lermusiaux, P. F., Duda, T. F., & Haley Jr, P. J. (2016). A coupled-mode shallow-water model for tidal analysis: Internal tide reflection and refraction by the gulf stream. *Journal of Physical Oceanography, 46*(12), 3661–3679.

Knauss, J. A. (1969). A note on the transport of the GS. *Deep Sea Research, 16*(Suppl.), 117–123.

Knauss, J. A., & Garfield, N. (2016). *Introduction to physical oceanography*. Waveland Press.

Kolmogorov, A. N. (1991). The local structure of turbulence in incompressible viscous fluid for very large Reynolds numbers. *Proceedings of the Royal Society of London. Series A: Mathematical and Physical Sciences, 434*(1890), 9–13.

Krasnopolsky, V. M. (2007). Neural network emulations for complex multidimensional geophysical mappings: Applications of neural network techniques to atmospheric and oceanic satellite retrievals and numerical modeling. *Reviews of Geophysics, 45*(3), 1–34.

Krelling, A. P. M., Gangopadhyay, A., da Silveira, I. C. A., & Vilela-Silva, F. (2020). Development of a feature-oriented regional modelling system for the North Brazil Undercurrent region (1–11S) and its application to a process study on the genesis of the Potiguar Eddy. *Journal of Operational Oceanography*. https://doi.org/10.1080/1755876X.2020.1743049

Krige, D. G. (1951). A statistical approach to some basic mine valuation problems on the Witwatersrand. *Journal of the Southern African Institute of Mining and Metallurgy, 52*(6), 119–139.

Kundu, P., Cohen, I., & Dowling, D. (2012). *Fluid mechanics*, 920 pp. Academic Press, New York, NY.

Lai, D., & Richardson, P. L. (1977). Distribution and movement of GS rings. *Journal of Physical Oceanography, 7*, 670–683.

Lam, F.-P. A., Haley Jr, P. J., Janmaat, J., Lermusiaux, P. F., Leslie, W. G., Schouten, M. W., te Raa, L. A., & Rixen, M. (2009). At-sea real-time coupled four-dimensional oceanographic and acoustic forecasts during battlespace preparation 2007. *Journal of Marine Systems, 78*, S306–S320.

Langmuir, C. H., & Broecker, W. (2012). *How to build a habitable planet: The story of earth from the big bang to humankind-revised and expanded edition*. Princeton University Press.

Large, W. G., McWilliams, J. C., & Doney, S. C. (1994). Oceanic vertical mixing: A review and a model with a nonlocal boundary layer parameterization. *Reviews of Geophysics, 32*(4), 363–403.

Large, W., & Yeager, S. (2009). The global climatology of an interannually varying air–sea flux data set. *Climate Dynamics, 33*(2–3), 341–364.

LeBlond, P. H., & Mysak, L. A. (1981). *Waves in the ocean*. Elsevier.

Lermusiaux, P. (1999). Data assimilation via error sub- space statistical estimation, Part II, Mid-Atlantic Bight Shelfbreak front simulations and ESSE validation. *Monthly Weather Review, 1278*, 1408–1432.

Lermusiaux, P. F. (2007). Adaptive modeling, adaptive data assimilation and adaptive sampling. *Physica D: Nonlinear Phenomena, 230*(1–2), 172–196.

Lermusiaux, P., Chiu, C.-S., & Robinson, A. (2001). Modeling uncertainties in the prediction of acoustic wave-field in a shelfbreak environment. *Proceedings of the Fifth International Conference on Theoretical and Computational Acoustics*, Beijing, China.

Lermusiaux, P. F., Haley Jr, P. J., Jana, S., Gupta, A., Kulkarni, C. S., Mirabito, C., Ali, W. H., Subramani, D. N., Dutt, A., Lin, J., et al. (2017). Optimal planning and sampling predictions for autonomous and Lagrangian platforms and sensors in the Northern Arabian Sea. *Oceanography, 30*(2), 172–185.

Lermusiaux, P. F., Haley Jr, P. J., Leslie, W. G., Agarwal, A., Logutov, O. G., & Burton, L. J. (2011). Multiscale physical and biological dynamics in the Philippine archipelago: Predictions and processes. *Oceanography, 24*(1), 70–89.

Lermusiaux, P. F., & Robinson, A. (1999). Data assimilation via error subspace statistical estimation. Part I: Theory and schemes. *Monthly Weather Review, 127*(7), 1385–1407.

Leslie, W., Robinson, A., Haley Jr, P., Logutov, O., Moreno, P., Lermusiaux, P., & Coelho, E. (2008). Verification and training of real-time forecasting of multi-scale ocean dynamics for maritime rapid environmental assessment. *Journal of Marine Systems, 69*(1–2), 3–16.

Levitus, S. (1982). Climatological atlas of the world ocean. *NOAA Professional Paper 13, U.S. Govt* (p. 173). Printing Office.

Levitus, S., & Boyer, T. P. (1994). *World ocean atlas 1994. Volume 4. Temperature* (tech. rep.). National Environmental Satellite, Data, and Information Service, Washington.

Levitus, S., Antonov, J. I., Baranova, O. K., Boyer, T. P., Coleman, C. L., Garcia, H. E., Grodsky, A. I., Locarnini, R. A., Mishonov, A. V., Johnson, D. R., et al. (2013). The world ocean database. *Data Science Journal, 12*, WDS229–WDS234.

Levitus, S., Locarnini, R. A., Boyer, T. P., Mishonov, A. V., Antonov, J. I., Garcia, H. E., Baranova, O. K., Zweng, M. M., Johnson, D. R., & Seidov, D. (2010). *World ocean atlas 2009* (tech. rep.). NOAA.

Lighthill, M. J., & Lighthill, J. (2001). *Waves in fluids*. Cambridge University Press.

Linder, C., & Gawarkiewicz, G. (1998). A climatology of the shelfbreak front in the Middle Atlantic Bight. *Journal of Geophysical Research, 103*(C9), 18405–18424.

Lindzen, R. S., Lorenz, E. N., & Platzman, G. W. (1990). *The atmosphere-a challenge: The science of Jule Gregory Charney*. Springer.

Locarnini, M., Mishonov, A., Baranova, O., Boyer, T., Zweng, M., Garcia, H., Seidov, D., Weathers, K., Paver, C., Smolyar, I., et al. (2018). *World ocean atlas 2018. Volume 1: Temperature*. NOAA.

Loder, J., Brickman, D., & Horne, E. (1992). Detailed structure of currents and hydrography on the northern side of Georges Bank. *Journal of Geophysical Research, 97*, 14331–14351.

Longshore, D. (1998). *Encyclopedia of hurricanes, typhoons, and cyclones*, 372. Facts on File. Inc., New York.

Lorenc, A., Ballard, S., Bell, R., Ingleby, N., Andrews, P., Barker, D., Bray, J., Clayton, A., Dalby, T., Li, D., et al. (2000). The met. Office global three-dimensional variational data assimilation scheme. *Quarterly Journal of the Royal Meteorological Society, 126*(570), 2991–3012.

Lorenz, E. N. (1963). Deterministic nonperiodic flow. *Journal of Atmospheric Sciences, 20*(2), 130–141.

Lozier, M. S. (2010). Deconstructing the conveyor belt. *Science, 328*(5985), 1507–1511.

Lozier, M. S. (2012). Overturning in the North Atlantic. *Annual Review of Marine Science, 4*, 291–315.

Luettich, R. A., Westerink, J. J., Scheffner, N. W., et al. (1992). ADCIRC: An advanced three-dimensional circulation model for shelves, coasts, and estuaries. Report 1, theory and methodology of ADCIRC-2DD1 and ADCIRC-3DL.

Lynch, D. (1999). Coupled physical/biological models for the coastal ocean. *Naval Research Reviews, 52*(2), 2–15.

Lynch, D., Ip, J., Naimie, C., & Werner, F. (1996). Comprehensive coastal circulation model with application to the Gulf of Maine. *Continental Shelf Research, 16*, 875–906.

Lynch, D. R., & Werner, F. E. (1987). Three-dimensional hydrodynamics on finite elements. Part I: Linearized harmonic model. *International Journal for Numerical Methods in Fluids, 7*(9), 871–909.

Lynch, D. R., & Werner, F. E. (1991). Three-dimensional hydrodynamics on finite elements. Part II: Non-linear time-stepping model. *International Journal for Numerical Methods in Fluids, 12*(6), 507–533.

Lynch, D., Werner, F., Greenberg, D., & Loder, J. (1992). Diagnostic model for baroclinic, wind-driven and tidal circulation in shallow seas. *Continental Shelf Research, 12*, 37–64.

Majumder, S., Schmid, C., & Halliwell, G. (2016). An observations and model-based analysis of meridional transports in the South Atlantic. *Journal of Geophysical Research: Oceans, 121*(8), 5622–5638.

Manabe, S., & Bryan, K. (1969). Climate calculations with a combined ocean-atmosphere model. *Journal of the Atmospheric Sciences, 26*(4), 786–789.

Manabe, S., Bryan, K., & Spelman, M. J. (1975). A global ocean-atmosphere climate model. Part I. The atmospheric circulation. *Journal of Physical Oceanography, 5*(1), 3–29.

Mandal, S., Sil, S., Gangopadhyay, A., Jena, B. K., & Venkatesan, R. (2020). On the nature of tidal asymmetry in the Gulf of Khambhat, Arabian Sea using HF radar surface currents. *Estuarine, Coastal and Shelf Science, 232*, 106481.

Mandal, S., Sil, S., Gangopadhyay, A., Jena, B. K., Venkatesan, R., & Gawarkiewicz, G. (2021). Seasonal and tidal variability of surface currents in the Western Andaman sea using HF radars and Buoy observations during 2016–2017. *IEEE Transactions on Geoscience and Remote Sensing, 59*(9), 7235–7244.

Mandal, S., Sil, S., Gangopadhyay, A., Murty, T., & Swain, D. (2018). On extracting high-frequency tidal variability from HF radar data in the Northwestern Bay of Bengal. *Journal of Operational Oceanography, 11*(2), 65–81.

Mariano, A. J., & Brown, O. B. (1992). Efficient objective analysis of dynamically heterogeneous and nonstationary fields via the parameter matrix. *Deep Sea Research Part A. Oceanographic Research Papers, 39*(7–8), 1255–1271.

Marks, K., & Smith, W. (2006). An evaluation of publicly available global bathymetry grids. *Marine Geophysical Researches, 27*(1), 19–34.

Marshall, J., Adcroft, A., Hill, C., Perelman, L., & Heisey, C. (1997). A finite-volume, incompressible Navier Stokes model for studies of the ocean on parallel computers. *Journal of Geophysical Research: Oceans, 102*(C3), 5753–5766.

Marshall, J., C., Perelman, L., & Adcroft, A. (1997). Hydrostatic, quasi-hydrostatic, and nonhydrostatic ocean modeling. *Journal of Geophysical Research: Oceans, 102*(C3), 5733–5752.

Marshall, J., & Plumb, R. A. (2016). *Atmosphere, ocean and climate dynamics: An introductory text.* Academic Press.

Marshall, D. P., & Tansley, C. E. (2001). An implicit formula for boundary current separation. *Journal of Physical Oceanography, 31*(6), 1633–1638.

Maury, M. F. (1856). *The physical geography of the sea: With illustrative charts and diagrams*. Sampson Low, Son & Company.

Mavor, T. P., & Huq, P. (1996). Propagation velocities and instability development of a coastal current. *Buoyancy Effects on Coastal and Estuarine Dynamics, 53*, 59–69.

Mazumder, S. (2015). *Numerical methods for partial differential equations: Finite difference and finite volume methods*. Academic Press.

McDougall, T. J., & Barker, P. M. (2011). Getting started with TOES-10 and the Gibbs SeaWater (GSW) oceanographic toolbox. *SCOR/IAPSO WG, 127*, 1–28.

McPhaden, M. J., Timmermann, A., Widlansky, M. J., Balmaseda, M. A., & Stockdale, T. N. (2015). The curious case of the el niño that never happened: A perspective from 40 years of progress in climate research and forecasting. *Bulletin of the American Meteorological Society, 96*(10), 1647–1665.

McWilliams, J., & Flierl, G. (1979). On the evolution of isolated, nonlinear vortices. *Journal of Physical Oceanography, 9*, 1155–1182.

Meehl, G. A., Washington, W. M., Collins, W. D., Arblaster, J. M., Hu, A., Buja, L. E., Strand, W. G., & Teng, H. (2005). How much more global warming and sea level rise? *Science, 307*(5716), 1769–1772.

Mellor, G. L., & Yamada, T. (1982). Development of a turbulence closure model for geophysical fluid problems. *Reviews of Geophysics, 20*(4), 851–875.

Mewhirter, J., Lubell, M., & Berardo, R. (2018). Institutional externalities and actor performance in polycentric governance systems. *Environmental Policy and Governance, 28*(4), 295–307.

Mielke, C., Frajka-Williams, E., & Baehr, J. (2013). Observed and simulated variability of the AMOC at 26 N and 41 N. *Geophysical Research Letters, 40*(6), 1159–1164.

Miles, J. W. (1961). On the stability of heterogeneous shear flows. *Journal of Fluid Mechanics, 10*(4), 496–508.

Millero, F. J. (2010). History of the equation of state of seawater. *Oceanography, 23*(3), 18–33.

Millero, F. J., & Poisson, A. (1981). International one-atmosphere equation of state of seawater. *Deep Sea Research Part A. Oceanographic Research Papers, 28*(6), 625–629.

Moore, A. M., Arango, H. G., Di Lorenzo, E., Cornuelle, B. D., Miller, A. J., & Neilson, D. J. (2004). A comprehensive ocean prediction and analysis system based on the tangent linear and adjoint of a regional ocean model. *Ocean Modelling, 7*(1–2), 227–258.

Moore, D. W., & Philander, S. (1977). Modeling of the tropical oceanic circulation. *The Sea, 6*, 319–361.

Mountain, D., & Jessen, P. (1987). Bottom waters of the Gulf of Maine, 1978–1983. *Journal of Marine Research, 45*, 319–345.

Mountain, D., & Manning, J. (1994). Seasonal and interannual variability in the properties of the surface waters of the Gulf of Maine. *Continental Shelf Research, 14*, 1555–1581.

Moustahfid, H., Hendrickson, L. C., Arkhipkin, A., Pierce, G. J., Gangopadhyay, A., Kidokoro, H., Markaida, U., Nigmatullin, C., Sauer, W. H., Jereb, P., et al. (2020). Ecological-fishery forecasting of squid stock dynamics under climate variability and change: Review, challenges, and recommendations. *Reviews in Fisheries Science & Aquaculture, 28*(4), 1–36.

Munk, W. H. (1950). On the wind-driven ocean circulation. *Journal of Meteorology, 7*(2), 80–93.

Munk, W. H., & Carrier, G. F. (1950). The wind-driven circulation in ocean basins of various shapes. *Tellus, 2*(3), 158–167.

Naimie, C. (1995). *On the modeling of the seasonal variation in the three-dimensional circulation near Georges Bank* (Ph.D. Thesis) (Volume: 03755). Dartmouth College. Hanover, NH.

Naimie, C. (1996). Georges Bank residual circulation during weak and strong stratification periods, prognostic numerical model results. *Journal of Geophysical Research, 101,* 6469–6486.

Naimie, C., Loder, J., & Lynch, D. (1994). Seasonal variation of the three-dimensional residual circulation on Georges Bank. *Journal of Geophysical Research, 99,* 15967–15989.

Nakicenovic, N., Alcamo, J., Davis, G., Vries, B. d., Fenhann, J., Gaffin, S., Gregory, K., Grubler, A., Jung, T. Y., Kram, T., et al. (2000). Special report on emissions scenarios. Cambridge University Press, Cambridge, 608 pp.

Neill, S. P., & Hashemi, M. R. (2018). *Fundamentals of ocean renewable energy: Generating electricity from the sea.* Academic Press.

Newland, D. E. (2012). *An introduction to random vibrations, spectral & wavelet analysis.* Courier Corporation.

Niiler, P., & Robinson, A. (1967). The theory of free inertial jets: II. A numerical experiment for the path of the Gulf Stream. *Tellus, 19*(4), 601–619.

O'Brien, J. (1971). A two-dimensional model of the wind-driven North Pacific. *Inv. Presq., 35,* 331–349.

Olson, D. (1980). The physical oceanography of two rings observed by the Cyclonic Ring Experiment. Part II: Dynamics. *Journal of Physical Oceanography, 10,* 514–528.

Olsen, E., Aanes, S., Mehl, S., Holst, J. C., Aglen, A., & Gjøsæter, H. (2010). Cod, haddock, saithe, herring, and capelin in the Barents Sea and adjacent waters: A review of the biological value of the area. *ICES Journal of Marine Science, 67*(1), 87–101.

Onken, R., Álvarez, A., Fernández, V., Vizoso, G., Basterretxea, G., Tintoré, J., Haley Jr, P., & Nacini, E. (2008). A forecast experiment in the Balearic Sea. *Journal of Marine Systems, 71*(1–2), 79–98.

Pachauri, R. K., Allen, M. R., Barros, V. R., Broome, J., Cramer, W., Christ, R., Church, J. A., Clarke, L., Dahe, Q., Dasgupta, P., et al. (2014). *Climate change 2014: Synthesis report. Contribution of Working Groups I, II and III to the fifth Assessment report of the Intergovernmental Panel on Climate Change.* IPCC.

Pandey, P., Khare, N., & Sudhakar, M. (2006). Oceanographic research: Indian efforts and preliminary results from the southern ocean. *Current Science, 90*(7), 978–984.

Parsons, A. (1969). A two-layer model of Gulf Stream separation. *Journal of Fluid Mechanics, 39*(3), 511–528.

Pedlosky, J. (2013a). *Geophysical fluid dynamics.* Springer Science & Business Media.

Pedlosky, J. (2013b). *Ocean circulation theory.* Springer Science & Business Media.

Peixoto, J. P., & Oort, A. H. (1992). *Physics of climate*, first edition (p.520). American Institute of Physics, New York.

Pettigrew, N., Townsend, D., Xue, H., Wallinga, J., Brickley, P., & Hetland, R. (1998). Observations of the Eastern Maine Coastal Current and its offshore extensions in 1994. *Journal of Geophysical Research*, *103*, 30623–30639.

Philander, S. G. (1989). El Niño, La Niña, and the southern oscillation. *International geophysics series*, 308 pp. Academic Press.

Philander, S., Yamagata, T., & Pacanowski, R. (1984). Unstable air-sea interactions in the tropics. *Journal of the Atmospheric Sciences*, *41*(4), 604–613.

Phillips, N. A. (1954). Energy transformations and meridional circulations associated with simple baroclinic waves in a two-level, quasi-geostrophic model. *Tellus*, *6*(3), 274–286.

Phillips, N. A. (1956). The general circulation of the atmosphere: A numerical experiment. *Quarterly Journal of the Royal Meteorological Society*, *82*(352), 123–164.

Pickart, R. S. (1992a). Space—time variability of the deep western boundary current oxygen core. *Journal of Physical Oceanography*, *22*(9), 1047–1061.

Pickart, R. S. (1992b). Water mass components of the North Atlantic deep western boundary current. *Deep Sea Research*, *39*, 1553–1572.

Pickart, R. S., & Hogg, N. G. (1989). A tracer study of the deep gulf stream cyclonic recirculation. *Deep Sea Research Part A. Oceanographic Research Papers*, *36*(6), 935–956.

Pickart, R. S., & Watts, D. R. (1990). Deep western boundary current variability at Cape Hatteras. *Journal of Marine Research*, *48*(4), 765–791.

Platt, T., & Jassby, A. D. (1976). The relationship between photosynthesis and light for natural assemblages of coastal marine phytoplankton 1. *Journal of Phycology*, *12*(4), 421–430.

Pond, S., & Pickard, G. L. (1983). *Introductory dynamical oceanography*. Gulf Professional Publishing.

Pope, S. B. (2001). Turbulent flows.

Pörtner, H.-O., Roberts, D. C., Masson-Delmotte, V., Zhai, P., Tignor, M., Poloczanska, E., Mintenbeck, K., Nicolai, M., Okem, A., Petzold, J., et al. (2019). *IPCC special report on the ocean and cryosphere in a changing climate*, 765 pp. IPCC Intergovernmental Panel on Climate Change: Geneva, Switzerland, *1*(3).

Price, J. F., Mooers, C. N., & Van Leer, J. C. (1978). Observation and simulation of storm-induced mixed-layer deepening. *Journal of Physical Oceanography*, *8*(4), 582–599.

Price, J. F., Weller, R. A., & Pinkel, R. (1986). Diurnal cycling: Observations and models of the upper ocean response to diurnal heating, cooling, and wind mixing. *Journal of Geophysical Research: Oceans*, *91*(C7), 8411–8427.

Quere, C. L., Harrison, S. P., Colin Prentice, I., Buitenhuis, E. T., Aumont, O., Bopp, L., Claustre, H., Cotrim Da Cunha, L., Geider, R., Giraud, X., et al. (2005). Ecosystem dynamics based on plankton functional types for global ocean biogeochemistry models. *Global Change Biology*, *11*(11), 2016–2040.

Rabier, F., Järvinen, H., Klinker, E., Mahfouf, J.-F., & Simmons, A. (2000). The ECMWF operational implementation of four-dimensional variational assimilation. I:

Experimental results with simplified physics. *Quarterly Journal of the Royal Meteorological Society, 126*(564), 1143–1170.

Ramp, S., Schlitz, R., & Wright, W. (1985). The deep flow through the Northeast Channel, Gulf of Maine. *Journal of Physical Oceanography, 15,* 1790–1808.

Rao, R., Kumar, M. G., Ravichandran, M., Rao, A., Gopalakrishna, V., & Thadathil, P. (2010). Interannual variability of Kelvin Wave propagation in the wave guides of the equatorial Indian Ocean, the coastal Bay of Bengal and the Southeastern Arabian Sea during 1993–2006. *Deep Sea Research Part I: Oceanographic Research Papers, 57*(1), 1–13.

Rapp, B. E. (2016). *Microfluidics: Modeling, mechanics and mathematics.* William Andrew.

Rennell, J. (1788). *Memoir of a map of Hindoostan; or, the mogul empire: With an introduction, illustrative of the geography and present division of that country: And a map of the countries situated between the head of the Indus, and the Caspian Sea.* M. Brown.

Rennell, J. (1832). *An investigation of the currents of the Atlantic Ocean: And of those which prevail between the Indian Ocean and the Atlantic.* Lady Rodd.

Rennell, J. et al. (2014). *A chart of the bank of Lagullus and southern coast of Africa [material cartografico]: Inscribed to Sir George Wombwell, Bart.* [London] Published according to Act of Parliament by J. Rennell.

Richardson, L. (1922). Weather prediction by numerical process (Cambridge University Press). Dover Publications edition (1965), New York.

Richardson, P. L. (1980). Benjamin Franklin and Timothy Folger's first printed chart of the Gulf Stream. *Science, 207*(4431), 643–645.

Richardson, P. L. (1985). Average velocity and transport of the Gulf Stream near 55W. *Journal of Marine Research, 43,* 83–111.

Richardson, P. L. (2008). On the history of meridional overturning circulation schematic diagrams. *Progress in Oceanography, 76*(4), 466–486.

Ricker, W. E. (1975). Computation and interpretation of biological statistics of fish populations. *Bulletin - Fisheries Research Board of Canada, 191,* 1–382.

Robinson, A. (2001). Forecasting and simulating coastal ocean processes and variabilities with the Harvard Ocean Prediction System. In C. Mooers (Ed.), *coastal ocean prediction. AGU Coastal and Estuarine Studies Series* (pp. 77–100). AGU.

Robinson, A. R. (1963). *Wind-driven ocean circulation: A collection of theoretical studies.* Blaisdell Publishing Company.

Robinson, A. R. (2012). *Eddies in marine science.* Springer Science & Business Media.

Robinson, A., Glenn, S., Spall, M., Walstad, L., Gardner, G., & Leslie, W. (1989). Forecasting meanders and Rings. *EOS, Oceanography Report, 7045,* 1464–1473.

Robinson, A. R., McCarthy, J. J., & Rothschild, B. J. (2005). *Biological-physical interactions in the sea* (Vol. 12). Harvard University Press.

Robinson, A. R., & Brink, K. H. (2005). *The global coastal ocean: Multiscale interdisciplinary processes* (Vol. 13). Harvard University Press.

Robinson, A., & Gangopadhyay, A. (1997). Circulation and dynamics of the western north Atlantic, II, Dynamics of meanders and rings. *Journal of Atmospheric and Oceanic Technology, 146,* 1333–1351.

Robinson, A., Spall, M., & Pinardi, N. (1988). Gulf Stream simulations and the dynamics of ring and meander processes. *Journal of Physical Oceanography, 18*, 1811–1853.

Robinson, A., & Leslie, W. (1985). Estimation and prediction of oceanic fields. *Progress in Oceanography, 14*, 485–510.

Röhrs, J., Christensen, K. H., Vikebø, F., Sundby, S., Saetra, Ø., & Broström, G. (2014). Wave-induced transport and vertical mixing of pelagic eggs and larvae. *Limnology and Oceanography, 59*(4), 1213–1227.

Rossby, C.-G. (1936). *Dynamics of steady ocean currents in the light of experimental fluid mechanics.* Massachusetts Institute of Technology; Woods Hole Oceanographic Institution.

Rossby, C.-G., & Montgomery, R. B. (1936). *On the momentum transfer at the sea surface. I. On the frictional force between air and water and on the occurrence of a laminar boundary layer next to the surface of the sea. II. Measurements of vertical gradient of wind over water. III. Transport of surface water due to the wind system over the North Atlantic.* Massachusetts Institute of Technology; Woods Hole Oceanographic Institution.

Rothschild, B. J. (2000). "Fish stocks and recruitment": The past thirty years. *ICES Journal of Marine Science, 57*(2), 191–201.

Sadhuram, Y., Sarma, V., Murthy, T. R., & Rao, B. P. (2005). Seasonal variability of physico-chemical characteristics of the Haldia channel of Hooghly Estuary, India. *Journal of Earth System Science, 114*(1), 37–49.

Saji, N., Ambrizzi, T., & Ferraz, S. (2005). Indian Ocean dipole mode events and austral surface air temperature anomalies. *Dynamics of Atmospheres and Oceans, 39*(1–2), 87–101.

Sanchez-Franks, A., Hameed, S., & Wilson, R. E. (2016). The Icelandic low as a predictor of the Gulf Stream north wall position. *Journal of Physical Oceanography, 46*(3), 817–826.

Schaefer, W. (1989). Hooking mortality of walleyes in a Northwestern Ontario Lake. *North American Journal of Fisheries Management, 9*(2), 193–194.

Schmidt, A., Gangopadhyay, A., & Brickley, P. (2012). An operational modeling implementation for the Trinidad-Venezuela region using feature models. In *2012 Oceans*, pp. 1–7. IEEE.

Schmidt, A., & Gangopadhyay, A. (2013). An operational ocean circulation prediction system for the Western North Atlantic: Hindcasting during July–September of 2006. *Continental Shelf Research, 63*, S177–S192.

Schmitz, W., Jr. (1976). Eddy kinetic energy in the deep western North Atlantic. *Journal of Geophysical Research, 81*, 4981–4982.

Schmitz, W. Jr. (1995). On the interbasin-scale thermohaline circulation. *Reviews of Geophysics, 33*(2), 151–173.

Schnute, J. (1985). A general theory for analysis of catch and effort data. *Canadian Journal of Fisheries and Aquatic Sciences, 42*(3), 414–429.

Schofield, O., Glenn, S., Orcutt, J., Arrott, M., Meisinger, M., Gangopadhyay, A., Brown, W., Signell, R., Moline, M., Chao, Y., et al. (2010). Automated sensor network to advance ocean science. *Eos, Transactions American Geophysical Union, 91*(39), 345–346.

Semtner, A. J. (1986). Finite-difference formulation of a world ocean model. In *Advanced Physical Oceanographic Numerical Modelling*, 187–202, Springer, Dordrecht.

Shaji, C., & Gangopadhyay, A. (2007). Synoptic modeling in the Eastern Arabian Sea during the southwest monsoon using upwelling feature models. *Journal of Atmospheric and Oceanic Technology*, *24*(5), 877–893.

Shee, A., Sil, S., Gangopadhyay, A., Gawarkiewicz, G., & Ravichandran, M. (2019). Seasonal evolution of oceanic upper layer processes in the northern Bay of Bengal following a single Argo float. *Geophysical Research Letters*, *46*(10), 5369–5377.

Siedler, G., & Peters, H. (1986). *Physical properties (general) of sea water*. Springer.

Silva, E. N. S., Gangopadhyay, A., Fay, G., Welandawe, M. K., Gawarkiewicz, G., Silver, A. M., Monim, M., & Clark, J. (2020). A survival analysis of the Gulf Stream warm core rings. *Journal of Geophysical Research: Oceans*, *125*(10), e2020JC016507.

Silveira, I. C. A. d., Lima, J. A. M., Schmidt, A. C. K., Ceccopieri, W., Sartori, A., Franscisco, C. P. F., & Fontes, R. F. C. (2008). Is the meander growth in the Brazil Current system off Southeast Brazil due to baroclinic instability? *Dynamics of Atmospheres and Oceans*, *45*(3-4), 187–207. https://doi.org/10.1016/j.dynatmoce.2008.01.002

Silver, A., Gangopadhyay, A., Gawarkiewicz, G., Silva, E. N. S., & Clark, J. (2021a). Interannual and seasonal asymmetries in Gulf Stream ring formations from 1980 to 2019. *Scientific Reports*, *11*(1), 1–7.

Silver, A., Gangopadhyay, A., Gawarkiewicz, G., Taylor, A., & Sanchez-Franks, A. (2021b). Forecasting the Gulf Stream Path using buoyancy and wind forcing over the North Atlantic. *Journal of Geophysical Research: Oceans*, *126*(8), https://doi.org/10.1029/2021JC017614

Sloan, N., III. (1996). *Dynamics of a shelf slope front, Process Studies and data-driven simulations in the Middle Atlantic Bight* (Ph.D. Thesis). Harvard University. Cambridge, MA, USA.

Smagorinsky, J. (1963). General circulation experiments with the primitive equations: I. the basic experiment. *Monthly Weather Review*, *91*(3), 99–164.

Smith, P. C. (1989). Seasonal and interannual variability of current, temperature and salinity off southwest nova scotia. *Canadian Journal of Fisheries and Aquatic Sciences*, *46*(S1), s4–s20.

Smith, W. H., & Sandwell, D. T. (1997). Global sea floor topography from satellite altimetry and ship depth soundings. *Science*, *277*(5334), 1956–1962.

Song, Y., & Haidvogel, D. (1994). A semi-implicit ocean circulation model using a generalized topography-following coordinate system. *Journal of Computational Physics*, *115*(1), 228–244.

Soutelino, R., Da Silveira, I., Gangopadhyay, A., & Miranda, J. (2011). Is the Brazil Current eddy-dominated to the north of 20 S? *Geophysical Research Letters*, *38*(3), 1–5.

Soutelino, R. G., Gangopadhyay, A., & Silveira, I. C. A. (2013). The roles of vertical shear and topography on the eddy formation near the site of origin of the Brazil Current. *Continental Shelf Research*, *70*, 46–60.

Spall, M. A., & Robinson, A. R. (1989). A new open ocean, hybrid coordinate primitive equation model. *Mathematics and Computers in Simulation*, *31*(3), 241–269.

Spall, M., & Robinson, A. (1990). Regional primitive equation studies of the GS meander and ring formation region. *Journal of Physical Oceanography*, *20*, 985–1016.

Sprintall, J., & Tomczak, M. (1992). Evidence of the barrier layer in the surface layer of the tropics. *Journal of Geophysical Research: Oceans*, *97*(C5), 7305–7316.

Stephenson, D. B., Wanner, H., Bronnimann, S., & Luterbacher, J. (2003). The history of scientific research on the North Atlantic oscillation. *Geophysical Monograph-American Geophysical Union*, *134*, 37–50.

Stern, M. E. (1975). Ocean circulation physics, 246 pp. Academic Press, London.

Stommel, H. (1948). The westward intensification of wind-driven ocean currents. *Eos, Transactions American Geophysical Union*, *29*(2), 202–206.

Stommel, H. (1957). A survey of ocean current theory. *Deep Sea Research*, *4*, 149–184.

Stommel, H. M. (1965). *The Gulf Stream: A physical and dynamical description*. University of California Press.

Stommel, H. M. (1987). *A view of the sea: A discussion between a chief engineer and an oceanographer about the machinery of the ocean circulation*. Princeton University Press.

Stommel, H., & Arons, A. B. (1960). On the abyssal circulation of the world ocean – I: Stationary planetary flow patterns on a sphere. *Deep Sea Research*, *6*, 140–154.

Strang, G., & Freund, L. (1986). Introduction to applied mathematics. *Journal of Applied Mechanics*, *53*(2), 480.

Subbey, S., Devine, J. A., Schaarschmidt, U., & Nash, R. D. (2014). Modelling and forecasting stock–recruitment: Current and future perspectives. *ICES Journal of Marine Science*, *71*(8), 2307–2322.

Subramani, D. N., Haley Jr, P. J., & Lermusiaux, P. F. (2017). Energy-optimal path planning in the coastal ocean. *Journal of Geophysical Research: Oceans*, *122*(5), 3981–4003.

Sverdrup, H. U., Johnson, M. W., Fleming, R. H., et al. (1942). *The oceans: Their physics, chemistry, and general biology* (Vol. 7). Prentice-Hall, New York.

Swallow, J. C. (1955). A neutral-buoyancy float for measuring deep currents. *Deep Sea Research (1953)*, *3*(1), 74–81.

Swallow, J., & Worthington, L. (1961). An observation of a deep countercurrent in the western North Atlantic. *Deep Sea Research*, *8*, 1–19.

Syvitski, J. P., Vörösmarty, C. J., Kettner, A. J., & Green, P. (2005). Impact of humans on the flux of terrestrial sediment to the global coastal ocean. *Science*, *308*(5720), 376–380.

Talley, L. D. (2011). *Descriptive physical oceanography: An introduction*. Academic Press.

Talley, L. D. (2013). Closure of the global overturning circulation through the Indian, Pacific, and Southern Oceans: Schematics and transports. *Oceanography*, *26*(1), 80–97.

Taylor, K. E. (2001). Summarizing multiple aspects of model performance in a single diagram. *Journal of Geophysical Research: Atmospheres*, *106*(D7), 7183–7192.

Taylor, A. H. (2011). *The dance of air and sea: How oceans, weather, and life link together*. Oxford University Press.

Taylor, A. H., & Stephens, J. A. (1998). The North Atlantic oscillation and the latitude of the Gulf Stream. *Tellus A: Dynamic Meteorology and Oceanography, 50*(1), 134–142.

Tennekes, H., & Lumley, J. L. (2018). *A first course in turbulence.* MIT Press.

Thorpe, S. A. (2007). *An introduction to ocean turbulence* (Vol. 10). Cambridge University Press, Cambridge.

Tomczak, M., & Godfrey, J. S. (2013). *Regional Oceanography: An introduction.* Elsevier.

Ueckermann, M. P., & Lermusiaux, P. F. (2012). 2.29 finite volume MATLAB framework documentation. *Reports in Ocean Science and Engineering, 14*(42).

Ulaby, F. T., Moore, R. K., & Fung, A. K. (1986). *Microwave remote sensing: Active and passive. Volume 3. From theory to applications.* Addison-Wesley Publishing Company, Advanced Book Program/World Science Division.

Umlauf, L., & Burchard, H. (2005). Second-order turbulence closure models for geophysical boundary layers. A review of recent work. *Continental Shelf Research, 25*(7–8), 795–827.

UNESCO, I., & SCOR, I. (1981). Background papers and supporting data on the international equation of state of seawater 1980. UNESCO technical papers in Marine Science, 192 pp. Paris, France.

Vallis, G. K. (2017). *Atmospheric and oceanic fluid dynamics.* Cambridge University Press.

Van Aken, H. M. (2007). *The oceanic thermohaline circulation: An introduction* (Vol. 39). Springer Science & Business Media.

Veleda, D., Araujo, M., Zantopp, R., & Montagne, R. (2012). Intraseasonal variability of the North Brazil Undercurrent forced by remote winds. *Journal of Geophysical Research, 117*(C11), C11024. https://doi.org/10.1029/2012JC008392

Venkatesan, R., Tandon, A., D'Asaro, E., & Atmanand, M. (2018). *Observing the oceans in real time.* Springer.

Vernadsky, V. I. (1998). *The biosphere.* Springer Science & Business Media.

Veronis, G. et al. (1973). Large scale ocean circulation. *Advances in Applied Mechanics, 13*(1), 92.

Von Bertalanffy, L. (1938). A quantitative theory of organic growth (inquiries on growth laws. II). *Human Biology, 10*(2), 181–213.

Vörösmarty, C. J., Fekete, B. M., & Tucker, B. A. (1998). *Global River Discharge, 1807–1991,* Version 1.1 (RIVDIS), https://daac.ornl.gov/RIVDIS/guides/rivdis_guide.html, Oak Ridge Natl. Lab. Distributed Arch. Cent., Oak Ridge, Tenn.

Walker, G. T. (1910). Correlations in seasonal variations of weather. *Memoirs of the Indian Meteorological Department, 21*, 22–45.

Walker, G. T. (1924). Correlations in seasonal variations of weather. i. a further study of world weather. *Memoirs of the Indian Meteorological Department., 24*, 275–332.

Walker, S. G. T., & Bliss, E. (1928). *World weather, III.* Edward Stanford.

Wallace, J. M., & Gutzler, D. S. (1981). Teleconnections in the geopotential height field during the northern hemisphere winter. *Monthly Weather Review, 109*(4), 784–812.

Wallcraft, A. J., & Moore, D. R. (1997). The NRL layered ocean model. *Parallel Computing, 23*(14), 2227–2242.

Wang, Z. A., Moustahfid, H., Mueller, A. V., Michel, A. P., Mowlem, M., Glazer, B. T., Mooney, T. A., Michaels, W., McQuillan, J. S., Robidart, J. C., et al. (2019). Advancing observation of ocean biogeochemistry, biology, and ecosystems with cost-effective in situ sensing technologies. *Frontiers in Marine Science*, *6*, 519.

Watts, D. R. (1985). *Gulf stream variability*. Eddies in Marine Science.

Watts, D., Tracy, K., & Friedlander, A. (1989). Producing accurate maps of the Gulf Stream thermal front using objective analysis. *Journal of Geophysical Research*, *94*, 8040–8052.

Webb, D. J., Coward, A., de Cuevas, B., & Gwilliam, C. (1998). *The first main run of the OCCAM global ocean model*. Southampton Oceanography Centre, Southampton, UK.

White, L., Adcroft, A., & Hallberg, R. (2009). High-order regridding–remapping schemes for continuous isopycnal and generalized coordinates in ocean models. *Journal of Computational Physics*, *228*(23), 8665–8692.

Wolter, K., & Timlin, M. S. (2011). El Niño/southern oscillation behaviour since 1871 as diagnosed in an extended multivariate ENSO index (mei. ext). *International Journal of Climatology*, *31*(7), 1074–1087.

Worthington, L. (1976). On the North Atlantic circulation [Publisher: The Johns Hopkins University Press]. *The Johns Hopkins Oceanographic Studies Monogr*, *6*, 100.

Wright, W. (1976). The limits of shelf water south of Cape Cod, 1941 to 1972. *Journal of Marine Research*, *34*, 1–14.

Wright, D., Greenberg, D., Loder, J., & Smith, P. (1986). The steady-state barotropic response of the Gulf of Maine and adjacent regions to surface wind stress. *Journal of Physical Oceanography*, *16*, 947–966.

Xu, J., Lermusiaux, P., Haley, P., Leslie, W., & Logutov, O. (2008). Spatial and temporal variations in acoustic propagation in Dabob Bay during Plusnet'07 exercise. *Journal of the Acoustical Society of America*, *123*(5), 3894.

Xue, H., Chai, F., & Pettigrew, N. (2000). A model study of the seasonal circulation in the Gulf of Maine. *Journal of Physical Oceanography*, *30*, 1111–1135.

Yu, L., Jin, X., & Weller, R. A. (2008). 2008: Multidecade global flux datasets from the objectively analyzed air-sea fluxes (OAFlux) project: Latent and sensible heat fluxes, ocean evaporation, and related surface meteorological variables. Woods Hole Oceanographic Institution OAFlux project Technical Report OA-2008-01.

Yu, L., & Weller, R. A. (2007). Objectively analyzed air–sea heat fluxes for the global ice-free oceans (1981–2005). *Bulletin of the American Meteorological Society*, *88*(4), 527–540.

Zhao, J., & Johns, W. (2014a). Wind-driven seasonal cycle of the Atlantic meridional overturning circulation. *Journal of Physical Oceanography*, *44*(6), 1541–1562.

Zhao, J., & Johns, W. (2014b). Wind-forced interannual variability of the Atlantic meridional overturning circulation at 26.5 N. *Journal of Geophysical Research: Oceans*, *119*(4), 2403–2419.

Zienkiewicz, O. C., & Taylor, R. L. (2005). *The finite element method for solid and structural mechanics*, 476 pp. Elsevier.

Alphabetical Index

For Product Safety Concerns and Information please contact our EU
representative GPSR@taylorandfrancis.com
Taylor & Francis Verlag GmbH, Kaufingerstraße 24, 80331 München, Germany

www.ingramcontent.com/pod-product-compliance
Lightning Source LLC
Chambersburg PA
CBHW060744220326
41598CB00022B/2322